Progress in Brain Research
Volume 225

New Horizons in Neurovascular Coupling: A Bridge Between Brain Circulation and Neural Plasticity

Serial Editor

Progress in Brain Research
Volume 225

New Horizons in Neurovascular Coupling: A Bridge Between Brain Circulation and Neural Plasticity

Edited by

Kazuto Masamoto
Brain Science Inspired Life Support Research Center, University of Electro-Communications, Tokyo, Japan

Hajime Hirase
RIKEN Brain Science Institute, Wako, Saitama, Japan

Katsuya Yamada
Hirosaki University Graduate School of Medicine, Hirosaki, Aomori, Japan

ELSEVIER
AMSTERDAM • BOSTON • HEIDELBERG • LONDON • NEW YORK • OXFORD
PARIS • SAN DIEGO • SAN FRANCISCO • SINGAPORE • SYDNEY • TOKYO

Elsevier
Radarweg 29, PO Box 211, 1000 AE Amsterdam, Netherlands
The Boulevard, Langford Lane, Kidlington, Oxford OX5 1GB, UK
50 Hampshire Street, 5th Floor, Cambridge, MA 02139, USA

First edition 2016

Notices

Knowledge and best practice in this field are constantly changing. As new research and experience broaden our understanding, changes in research methods, professional practices, or medical treatment may become necessary.

Practitioners and researchers must always rely on their own experience and knowledge in evaluating and using any information, methods, compounds, or experiments described herein. In using such information or methods they should be mindful of their own safety and the safety of others, including parties for whom they have a professional responsibility.

To the fullest extent of the law, neither the Publisher nor the authors, contributors, or editors, assume any liability for any injury and/or damage to persons or property as a matter of products liability, negligence or otherwise, or from any use or operation of any methods, products, instructions, or ideas contained in the material herein.

ISBN: 978-0-444-63704-8
ISSN: 0079-6123

For information on all Elsevier publications visit our website at https://www.elsevier.com/

Publisher: Zoe Kruze
Acquisition Editor: Kirsten Shankland
Editorial Project Manager: Hannah Colford
Production Project Manager: Magesh Kumar Mahalingam
Designer: Greg Harris

Contributors

K. Arai
Neuroprotection Research Laboratory, Massachusetts General Hospital and Harvard Medical School, Charlestown, MA, United States

D. Coman
Magnetic Resonance Research Center (MRRC), Yale University, New Haven, CT, United States

N. Egawa
Neuroprotection Research Laboratory, Massachusetts General Hospital and Harvard Medical School, Charlestown, MA, United States

M. Fukuda
University of Pittsburgh, Pittsburgh, PA, United States

P. Herman
Magnetic Resonance Research Center (MRRC), Yale University, New Haven, CT, United States

E. Hillman
Kavli Institute for Brain Science; Mortimer B. Zuckerman Institute for Mind Brain and Behavior, Columbia University, New York, NY, United States

H. Hirase
RIKEN Brain Science Institute, Wako, Saitama, Japan

Y. Hoshi
Institute for Medical Photonics Research, Preeminent Medical Photonics Education & Research Center, Hamamatsu University School of Medicine, Hamamatsu, Japan

H. Hotta
Tokyo Metropolitan Institute of Gerontology, Tokyo, Japan

F. Hyder
Magnetic Resonance Research Center (MRRC), Yale University, New Haven, CT, United States

I. Kanno
Molecular Imaging Center, National Institute of Radiological Sciences, Chiba, Japan

S.-G. Kim
Center for Neuroscience Imaging Research, Institute for Basic Science, Sungkyunkwan University, Suwon, South Korea

M. Kozberg
Columbia University, New York, NY, United States

T. Kurihara
Keio University School of Medicine, Tokyo, Japan

J. Lok
Neuroprotection Research Laboratory; Massachusetts General Hospital and Harvard Medical School, Charlestown, MA, United States

K. Masamoto
Brain Science Inspired Life Support Research Center, University of Electro-Communications, Tokyo, Japan

T. Nishijima
Tokyo Metropolitan University, Tokyo, Japan

M. Nuriya
Keio University, Shinjuku, Tokyo, Japan

A.J. Poplawsky
University of Pittsburgh, Pittsburgh, PA, United States

B.G. Sanganahalli
Magnetic Resonance Research Center (MRRC), Yale University, New Haven, CT, United States

C.Y. Shu
Yale University, New Haven, CT, United States

H. Soya
University of Tsukuba, Tsukuba, Japan

I. Torres-Aleman
Cajal Institute, Madrid, Spain

K. Yamada
Hirosaki University Graduate School of Medicine, Hirosaki, Aomori, Japan

Contents

Preface

The ability of assessing neural activity by measuring brain circulation has revolutionized the way we study the brain. Since cerebral hemodynamics can be measured noninvasively, ie, without physical damages to the brain, neurovascular coupling has become the principal means for understanding brain function as shown by modern imaging techniques such as positron emission tomography (PET), functional magnetic resonance imaging (fMRI), and near-infrared spectroscopy (fNIRS). Nevertheless, the mechanisms underlying the neurovascular coupling have been still wrapped in a fascinating mystery. Recent evidences have suggested that neurovascular coupling participates in the maintenance of not only brain metabolism but also central nervous system plasticity. In this volume, we feature 11 review articles on our latest understandings of neurovascular coupling mechanisms as well as physiology from multiple aspects. The first three chapters provide "A physiological basis of neurovascular coupling," namely Hotta (Chapter 1), Nuriya (Chapter 2), and Yamada (Chapter 3) put perspectives on the latest findings in neurogenic, gliogenic, and vasculogenic mechanisms of neurovascular coupling, respectively. The second topics titled "Methodology for measurements of brain circulation" are covered by Kanno (Chapter 4), Hyder (Chapter 5), Fukuda (Chapter 6), and Hoshi (Chapter 7) who argue technological aspects of neurovascular and neurometabolic imaging tools specifically on the signal source issues in macroscopic and microscopic blood flow imaging modalities, calibrated and submillimeter-resolution, and fNIRS, respectively. Finally, the last four chapters provide the latest views on the rationale of neurovascular coupling actively participating in cell-to-cell communication to support neural plasticity in development, exercise, and aging processes, titled "Plastic changes in neurovascular coupling." A new conceptual frame of trophic coupling among divergent brain cells is reviewed from the viewpoints of neurovascular development by Arai (Chapter 8) and Hillman (Chapter 10) and their colleagues. Kurihara (Chapter 9) illustrates how the neurovascular coupling develops along with hypoxic signaling in the retina, which is considered one of the most accessible areas in the central nervous system. Moreover, plasticity on neurovascular coupling triggered by physical exercises is reviewed in depth by Nishijima (Chapter 11). Finally, given the current progress in the field of neurovascular coupling, we provide a future perspective: what further progress might lead to breakthroughs.

Kazuto Masamoto
Hajime Hirase
Katsuya Yamada

PART

I

A Physiological Basis of Neurovascular Coupling

CHAPTER

Neurogenic control of parenchymal arterioles in the cerebral cortex

1

H. Hotta[1]

Tokyo Metropolitan Institute of Gerontology, Tokyo, Japan

[1]Corresponding author: Tel.: +81-3-39643241x4343; Fax: +81-3-35794776, e-mail address: hhotta@tmig.or.jp

Abstract

Central neural vasomotor mechanisms act on the parenchymal vasculature of the brain to regulate regional cerebral blood flow (rCBF). Among the diverse components of the local neural circuits of the cerebral cortex, many may contribute to the regulation of rCBF. For example, the cholinergic vasodilative system that originates in the basal forebrain acts on the neocortex and hippocampus. Notably, rCBF is reduced in the elderly and patients with dementia. The vasodilatory response, independent of changes in blood pressure and glucose metabolism in the brain, occurs in the parenchymal arterioles to produce a significant increase in cortical rCBF. Recent studies illuminate the physiological role of the cholinergic vasodilator system related to neurovascular coupling, neuroprotection, and promotion of the secretion of nerve growth factor. In this review, cellular mechanisms and species differences in the neurogenic control of vascular systems, as well as benefits of the cholinergic vasodilatory systems against cerebral ischemia- and age-dependent impairment of neurovascular plasticity, are further discussed.

Keywords

Cerebral cortex, Basal forebrain, Cholinergic, Aging, Neuroprotection

1 INTRODUCTION

Cerebral blood flow (CBF) is an important factor that maintains brain function, and a prolonged insufficiency causes degeneration and irreversible impairment of brain function. In the brain parenchyma, there is a wealth of blood vessels. Approximately 15% of cardiac output flows through the brain that accounts for only 2% of body weight. Various mechanisms maintain CBF to support brain activity, and one important mechanism is neural regulation of the cardiovascular system. As with any body organ, brain blood flow is determined by perfusion pressure and vascular resistance.

Progress in Brain Research, Volume 225, ISSN 0079-6123, http://dx.doi.org/10.1016/bs.pbr.2016.03.001

The baroreceptor reflex, mediated by the autonomic nervous system connecting the heart and peripheral vasculature, prevents excessive decreases in blood pressure to ensure a sufficient blood supply to the brain. The brain vasculature can also react to local conditions to adjust blood flow. A major third source of vascular control in the brain is the neurogenic control of cerebral blood vessels governed by the surrounding vasoactive nerves (Fig. 1).

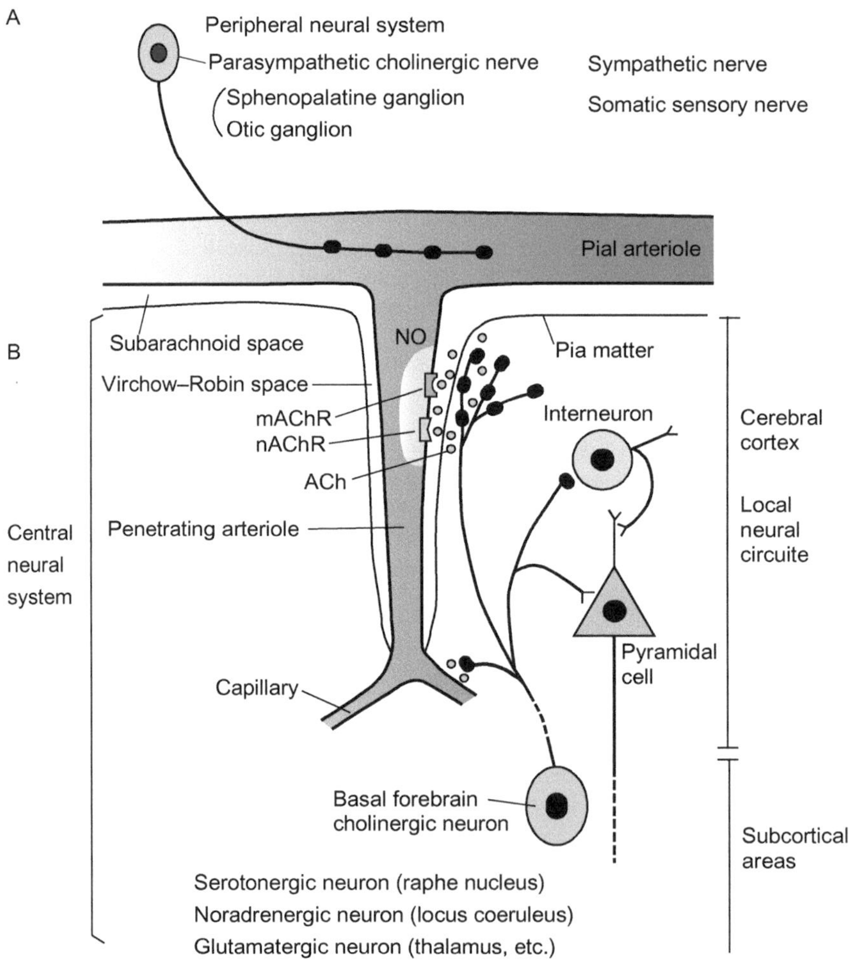

FIG. 1

Neurogenic control of cerebral blood vessels. (A) The peripheral neural system innervates large intracranial and pial vessels on the surface of the brain. (B) The central neural system comprises nerves originating in the brain that pass through the brain, reaching the parenchymal vessels (penetrating arterioles and capillaries).

The neural system controlling cerebral blood vessels is divided into peripheral and central neural systems. The peripheral neural system comprises nerves originating in the peripheral ganglia outside the skull, ie, sympathetic and parasympathetic autonomic and somatic sensory nerves (Goadsby and Edvinsson, 2002). The peripheral neural system innervates large intracranial and pial vessels on the surface of the brain (Fig. 1A) and is sufficient for regulating overall blood flow to the brain, which occurs in the autonomic vascular regulation of peripheral nerves (eg, sciatic nerve; Sato et al., 1994).

In contrast, the central neural system comprises nerves originating in the brain that pass through the brain, reaching the parenchymal vessels (Fig. 1B). Because brain functions are compartmentalized, regional (r)CBF must be appropriately allocated. The rCBF can be regulated by changes in the diameter of the penetrating arteriole that connects the pial arteriole on the surface of the brain to the intraparenchymal capillary. The activities of parenchymal neurons of local neural circuits (see Section 2.1) contribute to the regulation of rCBF in association with those of other cells, such as astrocytes (see Nuriya and Hirase, 2016, in this volume), vascular cells, or both (see Yamada, 2016, in this volume). Cellular organization differs among each area of the brain parenchyma, and the mechanisms of local regulation of parenchymal blood vessels vary accordingly. For example, one component of the central neural system is the cholinergic vasodilative system that originates in the basal forebrain and acts specifically on the cortex and hippocampus that is vulnerable to transient ischemia, aging, and neurodegenerative diseases (Sato and Sato, 1992). The vasodilative response, independent of changes in blood pressure and glucose metabolism in the brain, occurs at the parenchymal penetrating arterioles (Hotta et al., 2013) to markedly increase cortical rCBF. Importantly, the physiological role of the cholinergic vasodilative system related to neurovascular coupling (Piché et al., 2010) and neuroprotection (Hotta et al., 2002) is also associated with increased secretion of the nerve growth factor (NGF; Hotta et al., 2007a, 2009a).

This review is principally focused on the cholinergic vasodilative system that originates in the basal forebrain and recent studies related to neural regulation of the cerebral cortical (partly hippocampal) parenchymal arterioles.

2 NEUROGENIC CONTROL OF INTRACORTICAL rCBF

2.1 LOCAL NEURAL CIRCUITS OF THE CEREBRAL CORTEX

Local neural circuits of the cerebral cortex comprise pyramidal cells, nonpyramidal cells, excitatory fibers from other cortical areas and thalamus, and other afferent fibers such as cholinergic fibers from the basal forebrain (nucleus basalis of Meynert [NBM]), serotonergic fibers from the raphe nucleus of the midbrain, noradrenergic fibers from the locus ceruleus, and dopaminergic fibers from the ventral tegmental area (Nieuwenhuys et al., 2008). Many of these neural components may contribute to the regulation of rCBF (see reviews of Sato and Sato, 1992; Hillman, 2014).

Pyramidal cells are glutamatergic excitatory output cells located in layers II/III, V, and VI. Excitatory cells in layer IV are mainly spiny stellate and star pyramidal cells. The activities of these excitatory output cells are regulated by inhibitory non-pyramidal cells through their inhibitory neurotransmitter gamma aminobutyric acid (GABA). These inhibitory interneurons, which are distributed through all six layers, represent approximately 10–30% of the neuronal population (the percentages vary among cortical layers, areas, and species) and are classified into different subtypes based on morphology (eg, basket, chandelier, and Martinotti cells), firing characteristics (eg, fast or irregular spiking), and expression of specific molecular markers (eg, vasoactive intestinal peptide [VIP], parvalbumin, and somatostatin [SOM]; Fig. 2) (DeFelipe et al., 2013; Kubota et al., 2011).

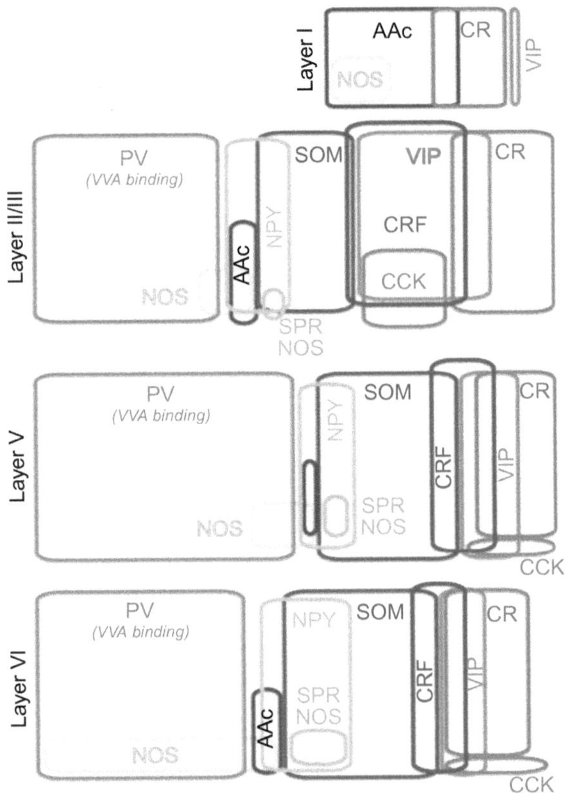

FIG. 2

Signaling molecules expressed by GABAergic cells in the frontal cortex. The relative number of cells that express each molecule is proportional to the size of the box in each layer. *Deep* (*gray* in the print version) and *light blue* (*light gray* in the print version) indicate strong and weak NOS expression, respectively (Kubota et al., 2011).

2.2 CHANGES IN rCBF INDUCED BY THE ACTIVITY OF CORTICAL NEURONS

The electrical activity of the brain correlates strongly with changes in rCBF, and subthreshold synaptic processes correlate more closely to rCBF than the spike rates of principal neurons (Lauritzen et al., 2012; see Fukuda et al., 2016, in this volume). When pyramidal cells are selectively activated by optogenetic stimulation, synaptic activity (local field potential) and action potentials (multiunit activity) are tightly related to hemodynamic signals (Ji et al., 2012). An increase in cortical rCBF in mice, induced by optogenetic stimulation of pyramidal cells, is reduced by a cyclooxygenase-2 (COX2) inhibitor, suggesting that COX2-generated prostaglandin E2 produced by pyramidal neurons contributes to neurovascular coupling in the cortex (Lacroix et al., 2015).

Among various subtypes of cortical GABA interneurons (Fig. 2), specific subsets control parenchymal vessel diameter (Cauli et al., 2004). In slices of brain harvested from neonatal rats, blood vessels in the plane from layers I–III with diameters ranging from 5 to 30 μm were selected, and single interneurons (layers I–III) within 40 μm of the selected vessel were recorded in whole-cell configuration. The firing of single interneurons (≥8 Hz induced by current for 30 or 120 s) either dilates or constricts neighboring microvessels in 13/149 neurons tested. The 13 interneurons were subjected to single-cell reverse transcriptase-multiplex polymerase chain reaction analysis, and the data show that interneurons that induced dilatation express VIP or nitric oxide synthase (NOS), whereas SOM is expressed by those that induce contraction. Further, the results of in vivo experiments show that direct optogenetic activation of cortical inhibitory neurons increases local rCBF (Anenberg et al., 2015). In mice that express channelrhodopsin-2 in GABAergic neurons, optogenetic cortical stimulation greatly attenuates spontaneous cortical spikes, whereas laser speckle contrast imaging revealed that blood flow is increased. The optogenetically evoked rCBF responses are not affected by application to the cortex of glutamatergic (NBQX and MK-801) and GABA-A receptor (picrotoxin) antagonists. These results suggest that activation of cortical inhibitory interneurons mediates large changes in blood flow independent of ionotropic glutamatergic or GABAergic synaptic transmission, likely by releasing coexpressed vasoactive transmitters.

2.3 CHOLINERGIC VASODILATION INDUCED BY AFFERENT FIBERS FROM THE BASAL FOREBRAIN

Stimulation of basal forebrain cholinergic nuclei produces an increase in rCBF in the cortical parenchyma through the activation of muscarinic (mAChR) and nicotinic (nAChR) cholinergic receptors within the blood–brain barrier (BBB; Biesold et al., 1989). Further, synthesis of nitric oxide (NO) is essential for this response (Adachi et al., 1992b; Raszkiewicz et al., 1992). The significant increase in cortical rCBF during basal forebrain stimulation, independent of changes in systemic blood pressure, is uncoupled from cortical glucose metabolism in anesthetized (Hallström et al., 1990; Kimura et al., 1990) and unanesthetized (Vaucher et al., 1997) rats. Therefore, cholinergic projection from the basal forebrain was proposed as an important system for vascular control in the cerebral cortex (see reviews of Sato and Sato, 1992, 1995).

The responses of cortical parenchymal rCBF induced by focal electrical or chemical stimulation of the cholinergic areas of the basal forebrain were measured using a variety of techniques, including laser Doppler (Biesold et al., 1989) and laser speckle (Fig. 3) (Hotta et al., 2011) flowmetry as well as the [^{14}C]iodoantipyrine method (Adachi et al., 1990b; Vaucher et al., 1997) and the clearance of helium (Lacombe et al., 1989). When rCBF is increased during focal electrical stimulation of NBM in the basal forebrain, the diameters of the pial arterioles on the cortical surface are unchanged (Adachi et al., 1992a). Therefore, NBM activity produces an effective increase in rCBF through vasodilation in the brain parenchyma. Studies conducted in vivo using two-photon microscopy identified the type of parenchymal blood vessel that dilates to increase rCBF during stimulation of the basal forebrain (Hotta et al., 2013). Electrical stimulation of NBM (0.5 ms, 30–50 μA, 50 Hz) dilates the cortical penetrating arterioles in the frontal cortex of mice anesthetized with urethane (Fig. 4). Moreover, electron microscopic studies of rats show that the magnitude of the changes in penetrating arterioles (approximately 11% of the basal diameter) determined by two-photon microscopy is consistent with that of the increase in diameter of parenchymal microvessels located 60 μm below the cortical surface (mean inner diameters = 4.9 and 5.5 μm in unstimulated and NBM-stimulated rats, respectively) (Hotta et al., 2004).

Dilation of the penetrating arteriole would cause a similar enlargement of the microvessels connected to the arteriole. Although the possibility of active dilation of smaller branched arterioles and capillaries after NBM stimulation cannot be excluded, NBM stimulation, which has less effect on surface arterioles, induces dilation of the penetrating arteriole that drives an increase of the cortical rCBF. Further, a faster response of the penetrating arterioles in the upper layers compared with the lower layers during NBM stimulation (see later) supports a regulatory mechanism of rCBF that initiates in the penetrating arterioles that irrigate a larger tissue volume. This mechanism is in contrast to regulation at the level of the capillaries that propagate upstream to microarteries (Itoh and Suzuki, 2012).

The diameters of penetrating arterioles increase throughout different layers of the cortex (examined up to a depth of 800 μm, layers I–V), except at the cortical surface and upper surface of layer V where the diameter of penetrating arterioles increases only slightly during NBM stimulation (Fig. 4). Hypercapnia causes significant dilation of the penetrating arterioles in all cortical layers, including the surface pial arterioles. The diameters of penetrating arterioles begin to increase within 1 s after the onset of NBM stimulation in the upper cortical layers and later in lower layers. Histological studies of rats show that major projections from NBM with a relatively high density of terminal boutons are present in layers I, II, and VI of the frontal cortex (Luiten et al., 1985, 1987). These results indicate that activation of NBM dilates cortical penetrating arterioles in a layer-specific manner in magnitude and latency that is

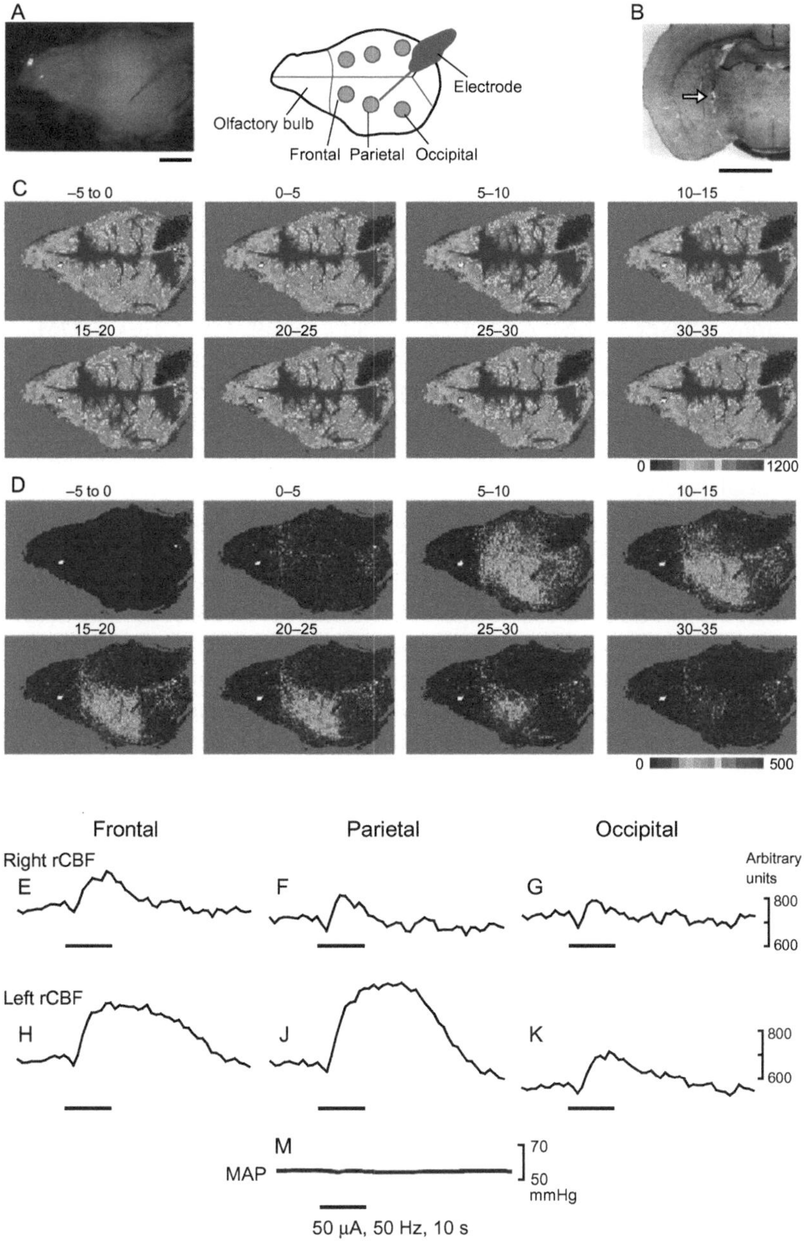

FIG. 3

Laser speckle flowmetry analysis of the spatiotemporal changes in regional cerebral blood flow (rCBF) evoked by focal electrical stimulation of the unilateral nucleus basalis of Meynert (NBM) in an anesthetized mouse. Electrical stimulation of the basal forebrain (as indicated in B) was performed at 50 μA, 0.5 ms, and 50 Hz for 10 s. (A) Image (*left*) and diagram (*right*) of the

(Continued)

likely to be associated with the density of cholinergic nerve terminals from NBM. The effect of changes in the luminal diameter on rCBF, according to Poiseuille's law, indicates that the resistance to flow decreases as a function of the fourth power of changes in lumen diameter, and therefore an 11% increase in lumen diameter can explain >50% increases in rCBF.

2.3.1 Effect of NBM stimulation on the activity of cortical neurons

In rats, FOS serves as a marker for increased neuronal activity, and approximately 30% of excitatory COX2-positive pyramidal neurons are activated by stimulation of the basal forebrain throughout the layers of the ipsilateral cortex, except for pyramidal neurons in layer V (Lecrux et al., 2012). In contrast, layers II–VI SOM- and/or NPY-containing and layer I GABA interneurons are selectively activated, NOS-containing interneurons are weakly and bilaterally activated, whereas VIP- or ACh-containing GABA interneurons are not activated (Kocharyan et al., 2008). However, studies of mice using two-photon calcium imaging show that stimulation of the basal forebrain (50 impulses, 75–150 μA, 0.1 ms, 100 Hz) bidirectionally modulates the activity of a small population of excitatory neurons and several subtypes of inhibitory interneurons in layers I and II/III of the visual cortex (Alitto and Dan, 2013) (Table 1). Five percent of excitatory neurons and 25% of parvalbumin-positive neurons are activated through mAChRs at low levels of cortical desynchronization and suppressed through nAChRs when cortical desynchronization is strong. In contrast, VIP-positive and layer I interneurons are preferentially activated through nAChRs during strong cortical desynchronization. The values of the responses of 41% and 25.5% sulforhodamine 101-labeled astrocytes in layers I and II/III, respectively, were negative, in contrast to the 15% and 6.6% positive responses of those in layers I and II/III, respectively. These cortical neurons that are activated or inhibited by NBM stimulation appear to be involved in basal forebrain-mediated changes in electroencephalogram (EEG) activity (Lee and Dan, 2012; Metherate et al., 1992) and promote NGF secretion from cortical neurons (see Section 5.3). The contribution to the arteriolar response, in part, may be possible, but not essential, because unilateral removal of local cortical neurons (using the excitotoxin, ibotenic acid) does not affect the rCBF responses in lesioned cortices (Linville et al., 1993).

FIG. 3—CONT'D

viewing field, which represents the entire dorsal surface of the brain with the olfactory bulb and occipital cortex on the *left* and *right sides* of each image, respectively. (B) Coronal section of the brain on the *left side* 0.9-mm posterior to the bregma showing the position of the tip of the stimulating electrode (*arrow*). Scale bar = 2 mm. (C) Averaged flow images over selected intervals of 5 s, as indicated above each image (stimulus onset = 0 s). (D) Differential signal change after subtracting the baseline control signal (−5 to 0 s) from subsequent images. (E–K) rCBF trace of the frontal (E and H), parietal (F and J), and occipital (G and K) cortices contralateral (E–G) and ipsilateral (H–K) to the site of stimulation, which were extracted from the region of interests indicated by the *gray circles* (A). (M) Mean arterial pressure (MAP) was simultaneously recorded (Hotta et al., 2011).

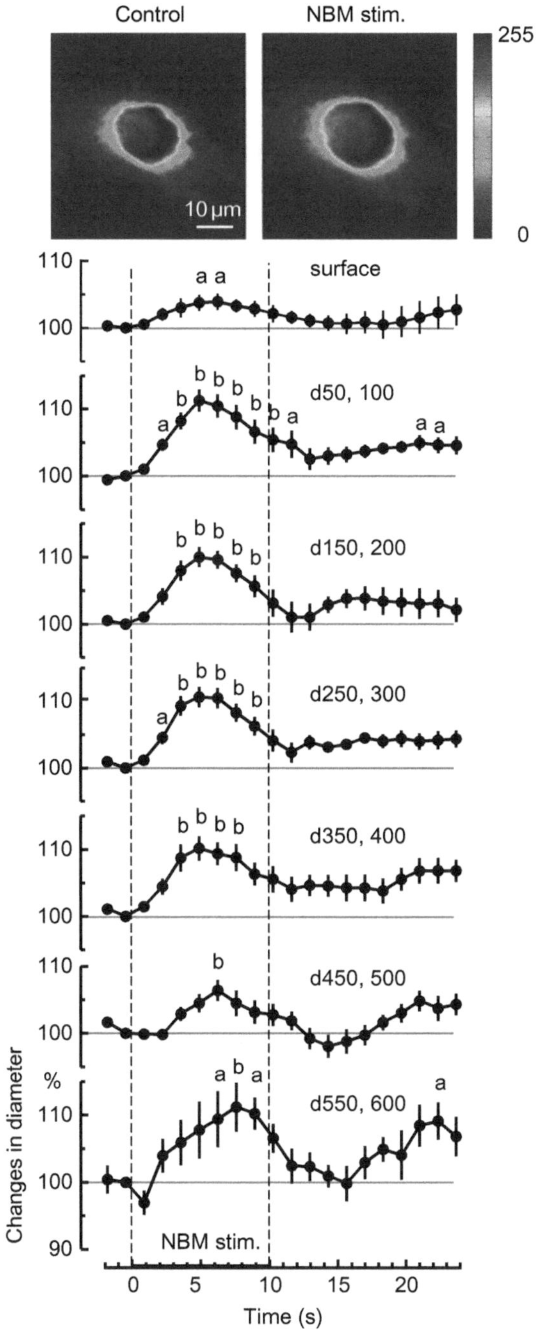

FIG. 4

Two-photon microscopic analysis of the changes in the diameter of penetrating arterioles in response to stimulation of the nucleus basalis of Meynert (NBM). *Upper inset*:

(Continued)

Table 1 Classification and Characterization of Neurons in the Cortex

Type[a]		Pyramidal Neuron	Nonpyramidal Neuron			
Transmitter[a]		Glutamatergic (excitatory)	GABAergic (inhibitory)			
% Cortical neurons		60–85%[b]	10–30%[c]			
Subtype[d]			Layers II–VI			Layer I
			VIP	PV	SOM	
			15%	39.3%	23.4%	4%
Response to BF stimulation[e]	(+)	5.4%[M]	84.2%[MN]	25.2%[M]	8.5%	88.6%[N]
	(−)	2.6%	0%	29.8%[MN]	6.4%	ca. 5%

Response to basal forebrain (BF) stimulation: (+) positive, (−) negative, [M]muscarinic, and [N]nicotinic.
Coexpressed peptides: PV, *parvalbumin;* SOM, *somatostatin;* VIP, *vasoactive intestinal polypeptide.*
[a]*Peters and Jones (1984).*
[b]*Nieuwenhuys et al. (2008).*
[c]*DeFelipe et al. (2013).*
[d]*Uematsu et al. (2008).*
[e]*Alitto and Dan (2013).*

2.3.2 Properties of cholinergic terminals in the cortex and a possible mechanism of dilation of the penetrating arteriole induced by NBM

Cholinergic innervation in the cortex (determined using immunohistochemical analysis of choline acetyltransferase [ChAT]) is predominantly nonjunctional. Specifically, in each layer a relatively low proportion of ChAT-immunostained varicosities exhibits synaptic membrane differentiation as follows: 10%, layer I; 14%, layers II–III; 11%, layer IV; 21%, layer V; and 14%, layer VI in the cortex (Umbriaco et al., 1994). These findings suggest that this system as well as the peripheral autonomic nervous system depends predominantly on volume transmission to exert its modulatory effects on cortical cells (De Lima and Singer, 1986;

FIG. 4—CONT'D

Representative images of a penetrating arteriole at a depth of 250 μm before (*left*) and during (*right*) stimulation of NBM. Signal intensity levels are shown as an eight-bit color (*different shades of gray* in the print version) scale (indicated *right*). *Graphs*: The change in diameter at various depths, including the surface pial arteriole, is expressed as the percentage of the prestimulus value (ordinate). The *dashed lines* and *heavy bar* on the abscissa indicate the time during which NBM was stimulated (0.5 ms, 30–50 μA, 50 Hz, for 10 s). The onset of electrical stimulation of NBM is expressed as zero (abscissa). Each point represents the mean ± standard error of the mean. Significant differences from the diameter of prestimulus control are indicated by a ($p<0.05$) and b ($p<0.01$).

Modified from Hotta, H., Masamoto, K., Uchida, S., Sekiguchi, Y., Takuwa, H., Kawaguchi, H., Shigemoto, K., Sudo, R., Tanishita, K., Ito, H., Kanno, I., 2013. Layer-specific dilation of penetrating arteries induced by stimulation of the nucleus basalis of Meynert in the mouse frontal cortex. J. Cereb. Blood Flow Metab. 33, 1440–1447.

Umbriaco et al., 1994). The majority of cholinergic fibers in the cerebral cortex originate in basal forebrain nuclei with an ipsilateral projection (Lehmann et al., 1980; Luiten et al., 1987).

The three-dimensional axonal morphologies of individual forebrain cholinergic neurons of mice that were imaged using genetically directed (CreER/loxP) sparse labeling reveal that the mean volume of axon arbor territories was approximately 1 mm^3 and that mammalian forebrain cholinergic neurons are among the largest and most complex neurons, according to the axon length and number of branch points (Wu et al., 2014). Considering such widespread distribution of cholinergic varicosities, most cellular elements in the cortex could be exposed to a certain basal concentration of ACh. The presence of acetylcholinesterase in the intercellular space would control this basal level rather than rapidly eliminating ACh from the immediate vicinity of ACh varicosities (Descarries et al., 1997; Umbriaco et al., 1994).

Fiber terminals from basal forebrain cholinergic areas (detected using an anterograde tracer) make intimate contacts not only with dendrites or cell bodies of neurons in the cerebral cortex but also with cortical parenchymal blood vessels, including the penetrating arteriole and capillary (Luiten et al., 1987; Vaucher and Hamel, 1995). Electron microscopic studies found that >60% of the perivascular terminals of rats are located within 1 μm from the vessel in the frontoparietal cortex and that some of the perivascular terminals engage in junctional contacts with adjacent neuronal elements (Vaucher and Hamel, 1995). In humans, nerve fibers that express ChAT form dense plexuses at the boundary between the pia mater and the cortex and in the tunica adventitia of the penetrating arterioles, and networks of ChAT-positive nerve fibers are present within the tunica muscularis of the radially directed arterioles that cross the intermediate and deep cortical laminae as well as their transverse and recurrent branches (Benagiano et al., 2000). Considering the predominant vasodilative response of the penetrating arteriole (Hotta et al., 2013), these morphological cholinergic fibers may be derived from the basal forebrain, although there are no published data that precisely localize the terminals from the basal forebrain around the penetrating arteriole.

We hypothesized that NBM-induced dilation of the penetrating arteriole (Hotta et al., 2013) occurs mainly through volume transmission of ACh (Kurosawa et al., 1989) that is released from perivascular cholinergic varicosities surrounding the penetrating arterioles to mAChRs (Elhusseiny and Hamel, 2000) and nAChRs (Clifford et al., 2008; Kalaria et al., 1994). Thus, ACh level is increased in the wall of penetrating arteriole, which in turn activates NOS (Adachi et al., 1992b; Raszkiewicz et al., 1992) and likely endothelial NOS (Zhang et al., 1995) (Fig. 1). Robust colocalization of mAChRs and glial fibrillary acidic protein on the astrocyte process (glia limitans) surrounding the penetrating arteriole, particularly in layers I and II (Moro et al., 1995; Van Der Zee et al., 1993), suggests a possible link between the astrocytes and arteriolar response via a Ca^{2+}-dependent release of vasoactive gliotransmitters (see Nuriya and Hirase, 2016, in this volume). However, Takata et al. (2013) found that changes in rCBF induced by NBM stimulation are unchanged in mice lacking astrocytic inositol triphosphate type-2 receptors (required for generating increases in intracellular Ca^{2+}), indicating that increases of intracellular Ca^{2+} in astrocytes are not critical for the arteriolar response.

3 CONTRIBUTION OF THE NEUROGENIC VASODILATIVE SYSTEM TO NEUROVASCULAR COUPLING

Brain function is closely related to rCBF. During neuronal activation in the brain, glucose and oxygen consumption and rCBF increase. Because the relative increase in oxygen consumption is less than that of rCBF, the ratio of deoxygenated hemoglobin to total hemoglobin decreases, which is called the blood oxygenation level-dependent (BOLD) effect that is detected using functional magnetic resonance imaging (fMRI), which is widely used to detect the functional localization of brain activity (Ogawa et al., 1990; see Fukuda et al., 2016, in this volume). However, the mechanism that regulates local neural activity that is associated with an increase in local rCBF (neurovascular coupling) is unknown (Hillman, 2014). Vascular changes are tightly coupled to neuronal activity through neuronal glucose consumption or local release of vasoactive agents. However, the nonlinear relationship between synaptic activity and hemodynamic responses suggests that significant activity may occur outside the hotspot (Lauritzen et al., 2012). Because cortical rCBF is stringently regulated by NBM, changes in rCBF associated with neuronal activity may be partly controlled by NBM.

An interesting example of using restrictions of brain blood flow to impact cortical activity comes from studies on rats exhibiting kainic acid-induced seizures. Severe bradyarrhythmia induced by vagal stimulation or chemical vasodilators decrease brain blood flow and stop seizure activity (Hotta et al., 2009b). Transient common carotid artery occlusion could ipsilaterally suppress seizure activity (Saito et al., 2006).

3.1 POSSIBLE INVOLVEMENT OF THE CHOLINERGIC RECEPTOR IN rCBF

Cholinergic mechanisms may play an important role in mediating rCBF in the somatosensory cortex, which is induced by nonnoxious vibrotactile stimulation of the contralateral forepaw of anesthetized cats. For example, the rCBF response (but not glucose metabolism) is abolished by intravenous injection of the mAChR blocker scopolamine (Ogawa et al., 1994) or intracortical injection of 3-bromopyruvate, an inhibitor of acetyl-CoA synthesis (Fukuyama et al., 1996). Further, conscious monkeys respond similarly (Tsukada et al., 1997). In contrast, scopolamine does not alter the rCBF response evoked by whisker stimulation of unanesthetized (Nakao et al., 1999) or urethane-anesthetized rats (Lecrux et al., 2011). Although whisker stimulation induces FOS expression in GABA interneurons that specifically express VIP and ACh, blockade of vasodilative mAChR or VIP receptors using antagonists [scopolamine or VIP(6–28), intracisternal injection] does not affect this rCBF response

(Lecrux et al., 2011). These results do not support the involvement of muscarinic receptors in the mechanism of enhancement of rCBF by functional neuronal activation, at least in the whisker-barrel cortex sensory pathway of unanesthetized (Nakao et al., 1999) and anesthetized (Lecrux et al., 2011) rats. The discrepancy between these results may be explained by differences in species (cats and monkeys vs rats) or differences in the stimulated areas of the skin (limb vs face).

3.2 POSSIBLE INVOLVEMENT OF AFFERENT INPUT FROM NBM

The possibility that changes in rCBF associated with neuronal activity may be partly controlled by NBM was evaluated using laser speckle contrast imaging of urethane-anesthetized rats (Piché et al., 2010). The contribution of NBM to the changes in rCBF was examined by injecting the GABAergic agonist muscimol into the right part of NBM to compare somatosensory-evoked cortical rCBF responses before and after NBM inactivation. Brushing of a hindlimb induces a robust increase in rCBF in the contralateral parietal cortex, over the representation of the hindlimb, without affecting blood pressure (Fig. 5A–E). Inactivation of NBM using muscimol reduces the rCBF response approximately 40% in the hemisphere ipsilateral to muscimol-inactivated NBM compared with vehicle (Fig. 5F). In the left part of the parietal cortex (contralateral to the inactivated NBM), rCBF changes are not significantly affected by muscimol, indicating that GABAergic inhibition of stimulus-evoked rCBF alterations induced by muscimol specifically affects the cortex ipsilateral to the injection site. Further, basal rCBF is unaffected by muscimol injected into either hemisphere. These findings can be explained by the inactivation of the basalocortical vasodilative system comprising ipsilateral cholinergic projections (Adachi et al., 1990a; Biesold et al., 1989; Lacombe et al., 1989; Sato and Sato, 1992; Vaucher et al., 1997). The data indicate that a relative contribution of NBM to the somatosensory-evoked rCBF changes is of $\geq$40%.

The possibility that NBM is activated during brushing is supported by findings of increased extracellular levels of cortical ACh (Kurosawa et al., 1992) and NGF (via nAChRs; Hotta et al., 2014) during and after brushing, respectively. Further, electrical or chemical stimulation of NBM induces these changes in a similar manner (Kurosawa et al., 1989; Hotta et al., 2007a; see Section 5.3). Moreover, an fMRI study of urethane-anesthetized rats found that the BOLD signal in the right part of NBM is significantly higher during brushing of a left hindlimb compared with baseline, suggesting that nonnoxious skin stimulation activates NBM projecting to the parietal cortex (Hotta et al., 2014). Innocuous brushing of the hindlimb induces an increase in the BOLD signal of the contralateral parietal cortex, over the hindlimb somatosensory area, as well as the contralateral NBM. The latency of the BOLD response by NBM is earlier than that in the parietal cortex. This result is consistent with the finding that an increase in parietal rCBF induced by electrical stimulation of NBM starts a few seconds after the onset of stimulation (Biesold et al., 1989; Hotta et al., 2011). Because an increase in the BOLD signal usually correlates with an increase in neuronal activities (Lee et al., 2010; Logothetis et al., 2001), the

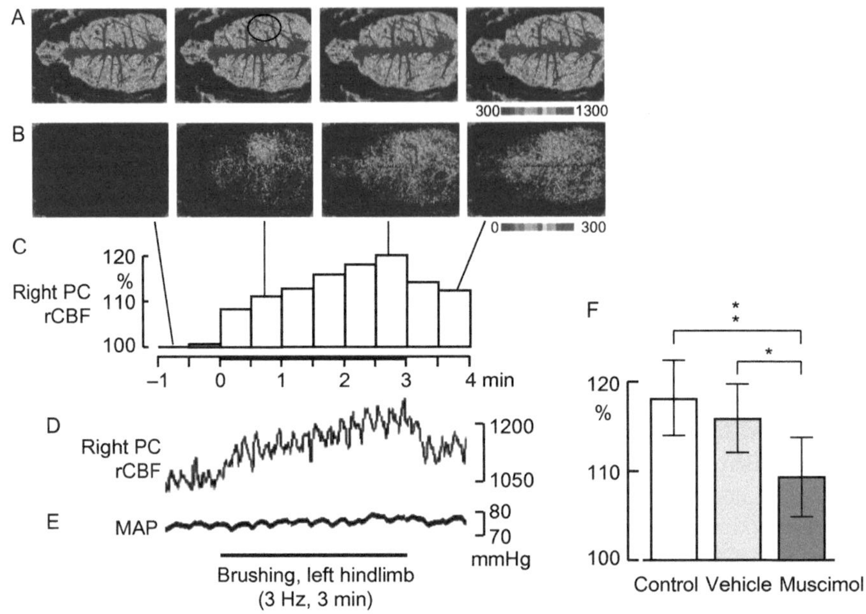

FIG. 5

Spatiotemporal changes in regional cerebral blood flow (rCBF) evoked by innocuous brushing (3 Hz, 3 min). Individual example showing rCBF variations in the right parietal cortex induced by innocuous brushing of the contralateral left hindlimb. (A) Averaged signal over selected intervals of 30 s. (B) Differential signal change taken from (A) when subtracting the baseline signal from subsequent images. (C) Percentage signal change in the right parietal cortex averaged every 30 s during the 5-min trial (data extracted from the region of interest (ROI) indicated by the *black circle* in A). (D) rCBF signal from the same ROI sampled at 1 Hz and temporally smoothed with a time constant of 3 s. (E) Mean arterial pressure (MAP) during the 5-min trial. (F) Effect of vehicle and muscimol injection into the right part of the nucleus basalis of Meynert on peak rCBF changes evoked by innocuous 3-Hz brushing. Muscimol significantly decreased rCBF in the right part of the parietal cortex when the left hindlimb was stimulated. $*p<0.05$ and $**p<0.01$.

Modified from Piché, M., Uchida, S., Hara, S., Aikawa, Y., Hotta, H., 2010. Modulation of somatosensory-evoked cortical blood flow changes by GABAergic inhibition of the nucleus basalis of Meynert in urethane-anaesthetized rats. J. Physiol. 588, 2163–2171.

result supports the assumption stated earlier that NBM neurons are activated by nonnoxious cutaneous stimulation.

Changes in the BOLD signal are significantly greater in NBM in the contralateral than in the ipsilateral side of the stimulated hindlimb, suggesting that nonnoxious skin stimulation predominantly activates the contralateral NBM. Each cortical area of the primary somatosensory (SI), secondary somatosensory (SII), and primary motor (MI) receives cholinergic afferents from neurons widely distributed throughout NBM, and each NBM neuron projects to a restricted cortical area without significant collateralization to adjacent subdivisions within the SI or to areas of the SI and SII or SI and MI (Baskerville et al., 1993). However, the ascending neural pathways from cutaneous mechanoreceptors to the contralateral NBM are unknown. Future studies are required to identify the specific neural pathway from the limbs to the contralateral NBM and, in particular, to determine whether these results can be generalized to other stimulus-evoked alterations of rCBF, such as those elicited by auditory and visual stimulation.

4 DIFFERENCES IN rCBF RESPONSES BETWEEN DIFFERENT ANIMAL SPECIES

The differences among animal species in the neurogenic regulation of the parenchymal rCBF are very important to consider, particularly in reference to understanding the data acquired using imaging techniques of the human brain.

4.1 NBM VASODILATIVE SYSTEM

In many species, the basal forebrain comprises a population of large cholinergic neurons that send axons to the entire cortex (Nieuwenhuys et al., 2008). Vasodilative responses in the cerebral cortex to the stimulation of cholinergic nucleus in the basal forebrain were first shown using rats (see the review of Sato and Sato, 1992) and subsequently in cats (Hotta et al., 2007b) and mice (Hotta et al., 2011). Many aspects of the rCBF response during stimulation of the basal forebrain, independent of changes in blood pressure, are similar among these species, including their magnitudes and kinetics as well as their stimulus strength dependence (Fig. 6), and laterality with responses produced predominantly in the ipsilateral cortex. In contrast, species variations exist in the regulation of cortical rCBF.

In rats, mAChRs and nAChRs may be involved to a similar extent in the cortical rCBF response induced by NBM stimulation, because administration of atropine (a muscarinic cholinergic blocker) reduces the response to 40% of the control, and further administration of mecamylamine (a nicotinic cholinergic blocker) largely abolishes the response (Biesold et al., 1989). In mice, it is interesting that most of the change in rCBF induced by low-intensity stimulation (twice the threshold intensity) is nearly abolished by atropine; however, the addition of mecamylamine is required to reduce changes caused by high-intensity stimuli (three times threshold intensity; Hotta et al., 2011). Moreover, the level of cortical desynchronization in the cortical neurons of mice induced by stimulation of the basal forebrain affects the contribution of nAChRs as follows: VIP-positive and layer I interneurons are preferentially activated through nAChRs during strong cortical desynchronization (Alitto and Dan, 2013).

In rodents, focal stimulation of the unilateral region of NBM increases rCBF in broad areas, including the frontal, parietal, and occipital lobes of cortices ipsilateral to the stimulated NBM (Adachi et al., 1990a; Hotta et al., 2011; see Fig. 3). However, in cats, focal stimulation of NBM increases rCBF in restricted areas of the ipsilateral

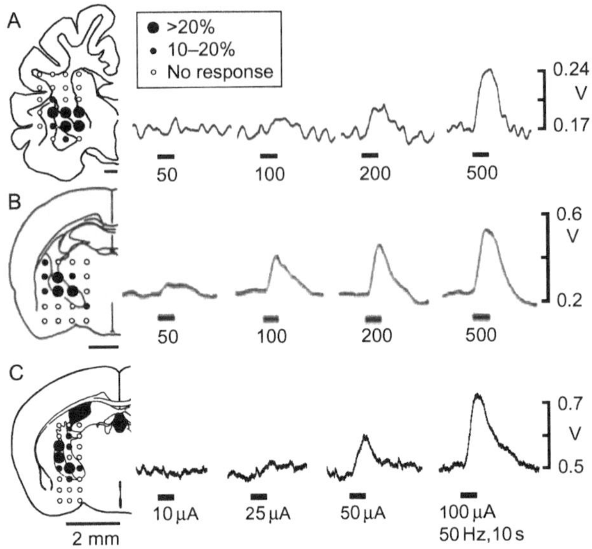

FIG. 6

Changes in regional cerebral blood flow (rCBF) in the parietal cortex ipsilateral to the stimulated nucleus basalis of Meynert in a cat (A), a rat (B), and a mouse (C). *Left*: Diagram of a coronal section of the brain showing the relationship between stimulating sites and rCBF responses in the parietal cortex induced by electrical stimulation of the brain. The *filled circles* indicate stimulation sites that increase rCBF, and sizes of circles represent the magnitude of the responses. *Open circles* indicate no significant response. *Right*: Sample recordings of rCBF obtained using laser Doppler flowmetry.

Modified from Hotta, H., Uchida, S., Shiba, K., 2007b. Cerebral cortical blood flow response during basal forebrain stimulation in cats. Neuroreport. 18, 809–812; Biesold, D., Inanami, O., Sato, A., Sato, Y., 1989. Stimulation of the nucleus basalis of Meynert increases cerebral cortical blood flow in rats. Neurosci. Lett. 98, 39–44; Hotta, H., Uchida, S., Kagitani, F., Maruyama, N., 2011. Control of cerebral cortical blood flow by stimulation of basal forebrain cholinergic areas in mice. J. Physiol. Sci. 61, 201–209.

cortex (Hotta et al., 2007b). The response of rCBF in the ipsilateral primary somatosensory cortex induced by a focal electrical stimulation of the unilateral basal forebrain (Hotta et al., 2007b) is greatest when the tip of the electrode is located within the area containing basal forebrain neurons projecting to the primary somatosensory cortex (Barstad and Bear, 1990). The results suggest that the topography of basal forebrain neurons that induce vasodilation in each cortical area in cats is more differentiated than that of rodents. Such a species difference is consistent with findings that NBM becomes progressively larger and more conspicuous with increasing cerebralization, reaching its greatest development in primates (Nieuwenhuys et al., 2008). In primates, the cholinergic cells projecting to the entire cortex are subdivided into five groups in monkeys and six in humans, according to the topography of their projections (Liu et al., 2015).

4.2 EFFECT OF NOXIOUS SOMATIC STIMULATION

There are differences between anesthetized rats and cats in rCBF changes induced by noxious somatosensory stimuli. When central nervous system of rats (Adachi et al., 1990c) or cats (Sakiyama et al., 1998) is intact, a noxious stimulus applied to the unilateral forelimb increases the rCBF bilaterally in numerous areas of the brain in association with increased systemic arterial pressure. However, there is a significant difference in rCBF if the secondary effect of increased blood pressure is eliminated by cutting the spinal cord at the upper thoracic level to block the neural connection between the sympathetic nervous system and afferent information from the forelimb. In anesthetized rats, bilateral widespread increases in rCBF are preserved (Adachi et al., 1990c). However, in cats, an increase in rCBF is observed in unilateral, restricted areas of primary somatosensory cortex, representing the forelimb area (Fig. 7A–F; Hotta et al., 2005).

The widespread increase in cortical rCBF independent of changes in blood pressure of rats may be attributed, in part, to the activation of intracranial cholinergic vasodilative fibers that originate in NBM and project to widespread areas of the cortex. For example, noxious forepaw stimulation applied to each side of the forelimb generally excites NBM neurons projecting to the cortex (Akaishi et al., 1990). In cats, the ipsilateral increase in rCBF following somatosensory noxious stimulation of a forelimb in spinal cord-intact animals (Sakiyama et al., 1998) is caused entirely by the increase in systemic blood pressure, while the contralateral effect is likely caused, in part, by active vasodilation. Considering that basal forebrain-mediated vasodilative responses are observed in rats and cats as described earlier, whereas somatically induced vasodilative responses differ, these findings suggest the following possibilities: (1) basal forebrain vasodilative neurons may not be activated in cats by noxious somatosensory stimulation at least under such experimental conditions or (2) in contrast to the rats' basal forebrain which apparently diffusely receives bilateral noxious somatosensory input, the cats' basal forebrain may selectively receive contralateral noxious somatosensory input.

4.3 EFFECT OF NONNOXIOUS SOMATIC STIMULATION

Nonnoxious somatic stimulation increases cortical rCBF, predominantly in the contralateral parietal cortex, over the representation of the stimulated site, and across different species, including humans (Hagen and Pardo, 2002; Stringer et al., 2014). In anesthetized rats, brushing of the hindlimb at 3 Hz for 3 min effectively increases cortical rCBF, particularly in the contralateral parietal cortex, over the representation of the hindlimb (Piché et al., 2010, urethane), as reported by studies using electrical stimulation of a paw (Durduran et al., 2004; Dunn et al., 2005, α-chloralose; Royl et al., 2006, α-chloralose and urethane). Further, widespread modifications of cortical rCBF, including the parietal, occipital, and frontal cortexes, are simultaneously observed (Piché et al., 2010). Interestingly, variations of rCBF in the parietal cortex peak with shorter latency at a higher stimulus frequency (3 Hz)

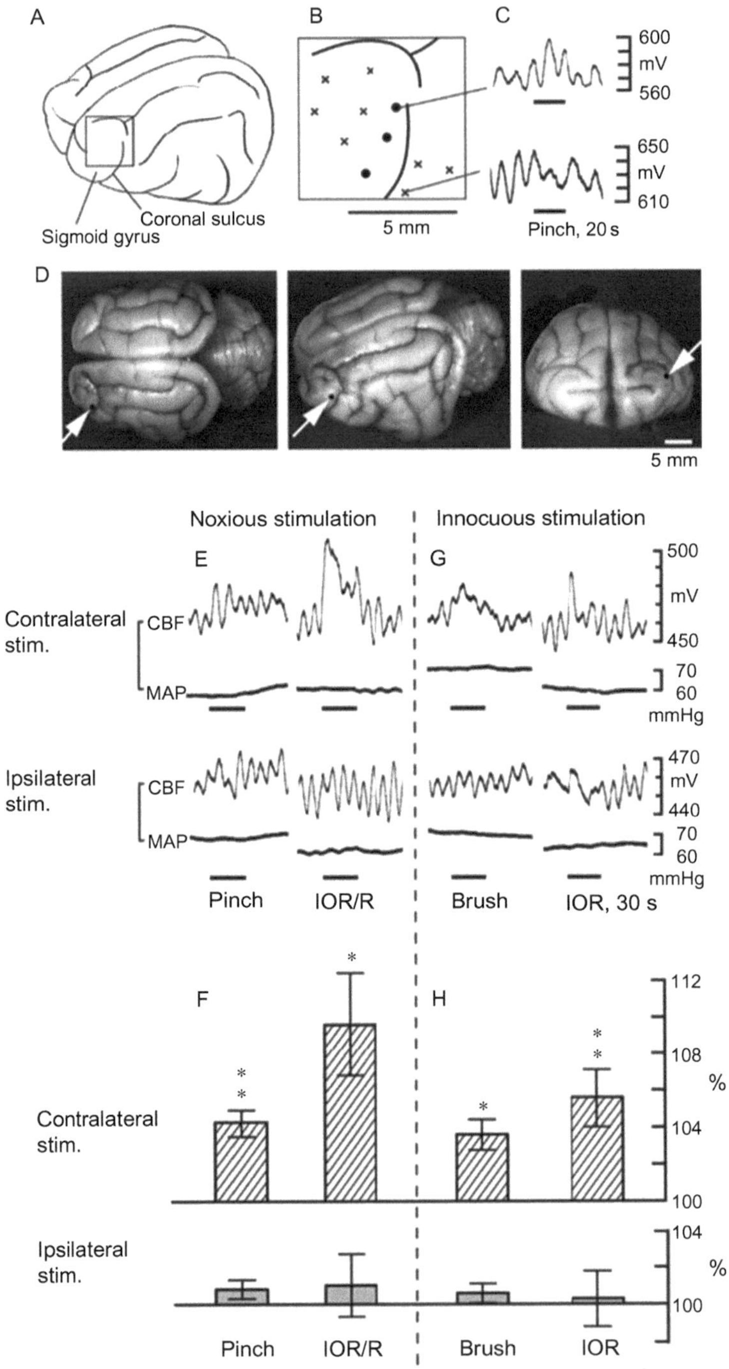

FIG. 7

See legend on opposite page.

and with higher amplitudes induced by prolonged stimulation at a lower stimulus frequency (1 Hz), reaching a 60% increase after 15 min. These characteristics are consistent with the temporal summation of hemodynamic responses associated with each brush stroke. During electrical stimulation of the hindpaw, the rCBF response in the parietal cortex is influenced by the duration and frequency of the stimulus (Ureshi et al., 2004, α-chloralose).

Innocuous mechanical stimulation of the skin, such as brushing at 1 Hz for 15–20 s, is ineffective in modulating NBM responses (Akaishi et al., 1990, urethane) or cortical rCBF (Adachi et al., 1990c, halothane). The same innocuous skin stimulation applied for 10–15 min significantly increases ACh release (Kurosawa et al., 1992, halothane) and rCBF (Piché et al., 2010, urethane) in the parietal cortex. These results indicate that the duration of the stimulus and temporal summation may represent critical factors required to activate cholinergic vasodilative fibers originating in NBM in response to innocuous stimulation. In cats, rCBF increases within several seconds in response to nonnoxious brushing and joint rotation applied to the forelimb (1 Hz, 20–30 s) contralateral but not ipsilateral to the rCBF on the recording side (Hotta et al., 2005, α-chloralose and urethane) (Fig. 7G and H). These results indicate that in rats, the duration of a stimulus and temporal summation are likely critical factors that augment rCBF in response to innocuous stimulation, although such a summation may not be required for the response of cats. However, because anesthesia profoundly affects the temporal dynamics of vascular responses to varying extents, depending on the type of anesthesia and dose (Masamoto and Kanno, 2012), such a difference between rats and cats may be explained by these factors.

FIG. 7—CONT'D

The effects of cutaneous and articular stimulation of a forelimb on cerebral blood flow in the forelimb area of the primary somatosensory cortex of anesthetized cats with a transected spinal cord at the T1 level. (A–C) The effects of noxious pinching on specific, topographical areas within the contralateral cerebral blood flow (CBF) recording region. (A) Cat brain. (B) Enlarged view of the left primary somatosensory cortex. *Filled circles* indicate an increased response, and *crosses* indicate no response. (C) Sample recordings of CBF in response to pinching the right forepaw. (D) Three different views of the cat brain. A *black circle* indicates the recording site that acquired the data in (E) and (G). (E and G) Sample recordings of cortical CBF and mean arterial pressure (MAP) after application of a noxious (E) or innocuous (G) stimulus to the right (*upper panel*, contralateral stim.) or left (*bottom panel*, ipsilateral stim.) forelimb. (F and H) Summary of the CBF responses to noxious (F) and innocuous (H) stimulation of the forelimb contralateral (*upper graph*) and ipsilateral (*bottom graph*) to the recording side. The mean values of CBF and MAP during the stimulation are expressed as the percentage of the prestimulus control value. Each *column* and *vertical bar* indicates the mean ± standard error of the mean. $*p<0.05$ and $**p<0.01$ relative to prestimulus control values using a paired *t*-test. Articular stimulation was applied by the rhythmic inward–outward rotation of an elbow joint from the midposition either within the physiological working range (innocuous; IOR) or against definite resistance of joint structures (noxious; IOR/R) with each half-cycle of 1 s (Hotta et al., 2005).

5 ROLE OF NEUROGENIC VASODILATION: NEUROPROTECTION

In patients with Alzheimer's disease accompanying dementia, degeneration of the cells in the basal forebrain is quite striking (Arendt et al., 1985; Mann et al., 1986; Terry et al., 1981; Whitehouse et al., 1982). Cholinergic neurons originating in NBM and septal complex of the basal forebrain project to the cerebral cortex and hippocampus, respectively. Therefore, there is a possible link between the cholinergic system and cognitive mechanism. The cholinergic system induces vasodilation in the cerebral cortex and hippocampus (Sato and Sato, 1992). These vasodilative responses are independent of systemic blood pressure and regional cerebral glucose metabolism.

What is the physiological relevance of these vasodilatory responses? Neurons in the cerebral cortex and hippocampus are quite vulnerable to ischemia (Kirino, 1982; Pulsinelli et al., 1982). The late death of neurons after transient ischemia, termed "delayed neuronal death," occurs in the hippocampus and cerebral cortex (Kirino, 1982; Pulsinelli et al., 1982). From these findings, it can be hypothesized that increases in cortical and hippocampal rCBF induced by the activation of the cholinergic neural vasodilative system that originates in the basal forebrain protects against the delayed death of cortical and hippocampal neurons caused by ischemia in the cerebral cortex and hippocampus. Such a possibility is clarified by studies of two rat models of ischemia described later (Hotta et al., 2002; Kagitani et al., 2000).

5.1 ANTIISCHEMIC EFFECT OF NICOTINIC STIMULATION IN THE HIPPOCAMPUS

A cholinergic neural vasodilative response in the cerebral cortex and hippocampus, independent of metabolic vasodilation, is induced by activating nAChRs via cholinergic inputs from NBM and septal complex in the basal forebrain and projecting to the cortex and hippocampus, respectively (Cao et al., 1989; Nakajima et al., 2003; Sato and Sato, 1992). Kagitani et al. (2000) conducted a study in rats to determine whether rCBF in the hippocampus (Hpc-rCBF) increases after stimulation of the nAChRs. This study found that the injection of nicotine (i.v.) protects hippocampal neurons from delayed death after the administration of transient ischemia. Hpc-rCBF was measured using a laser Doppler flowmeter. During 2 min intervals of transient occlusion of bilateral carotid arteries for 6 min as well as ischemia caused by permanent ligation of bilateral vertebral arteries, Hpc-rCBF decreases to approximately 16% of the preocclusion level, and 5 or 7 days after occlusion, delayed neuronal death occurs in approximately 70% of the CA1 hippocampal neurons. Further, Hpc-rCBF increases as a function of nicotine (30–100 μg/kg i.v.), independent of mean arterial pressure. Nicotine administered 5 min before occlusion significantly attenuates the occlusion-induced decrease in Hpc-rCBF. The delayed death of the CA1 hippocampal neurons after transient occlusion is attenuated by pretreatment with nicotine to approximately 50% (Fig. 8). The results indicate that the

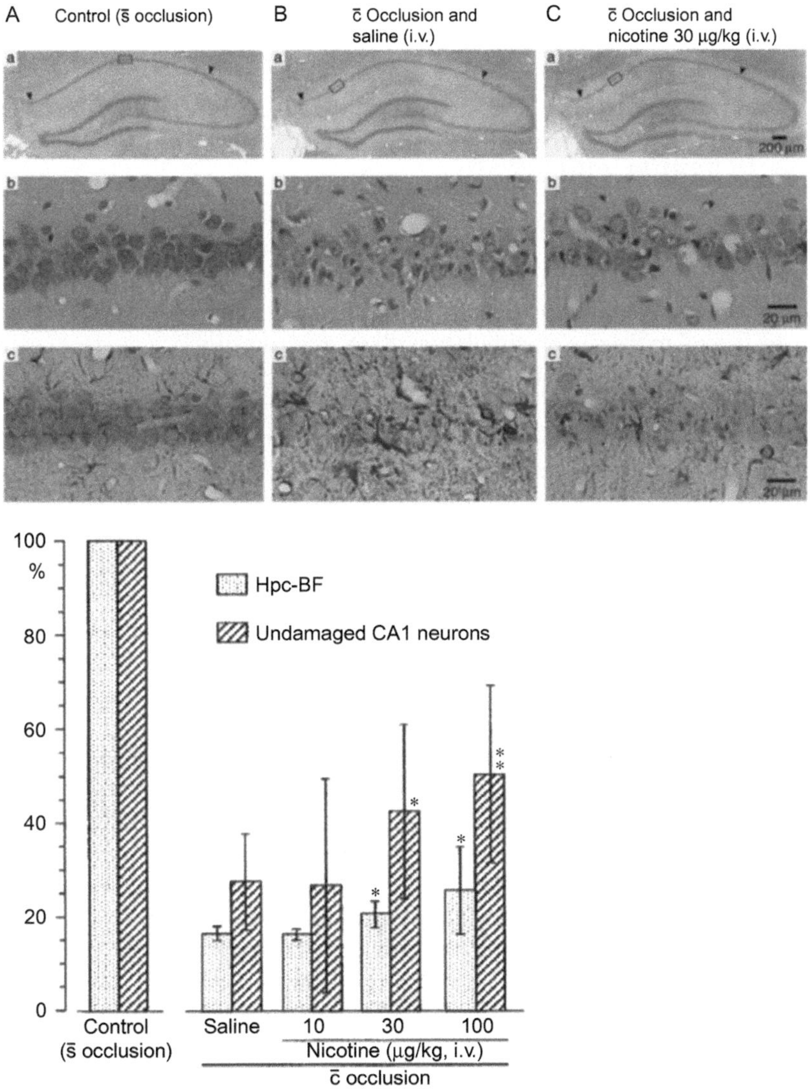

FIG. 8

The effect of nicotine on neuronal damage in the hippocampal CA1 region after occlusion. Representative histology of coronal sections (6 μm thick) of the hippocampal CA1 region in a normal control rat (Aa–c) and in two other rats (Ba–c and Ca–c) prepared on day 5 after an intermittent transient occlusion for 6 min after treatment with saline (B) or nicotine (30 μg/kg, C) and stained with hematoxylin and eosin (a and b) and glial fibrillary acidic protein (c). *Arrowheads* show medial and lateral borders of the CA1 region. *Lower graph*: Summary of the effects of nicotine on regional cerebral blood flow and delayed neuronal death after occlusion in the rat hippocampus. Hippocampus blood flow during occlusion (*stippled column*) and numbers of CA1 neurons undamaged on day 5 after occlusion (*hatched column*) are expressed as a percentage of the control value (Kagitani et al., 2000).

nAChR stimulation-induced increases in Hpc-rCBF protect hippocampal neurons from ischemia-induced delayed death.

5.2 STIMULATION OF NBM INDUCES AN ANTIISCHEMIC EFFECT IN THE CEREBRAL CORTEX

A study was conducted to determine whether an increase in the rCBF in the cortex produced by electrical stimulation of the cholinergic vasodilative system of NBM protects cortical neurons from delayed death. Because the vasodilative response caused by stimulation of the unilateral NBM is elicited predominantly in the cerebral cortex ipsilateral to the site of stimulation (Sato and Sato, 1992), a model of cerebral ischemia was developed that induces a moderate delayed neuronal death in the ipsilateral cerebral cortex. A laser Doppler flowmeter was used to measure rCBF, and delayed neuronal death of the cerebral cortex was induced by occlusions of the unilateral common carotid artery at intervals of 5 s for 60 min. The histology of the cortical hemisphere was analyzed at three different coronal levels. In control rats without occlusion, 6000–8000 intact and 9–19 damaged neurons were detected, respectively, in the cortical hemisphere at each coronal level. During the occlusions, rCBF ipsilateral to the occluded artery decreases by 13–32% of the preocclusion level. Five days after the occlusions are induced, the numbers of damaged neurons increase to 75–181. Regional differences in the degree of delayed neuronal death occur on the fifth day after the occlusions, which correlates with the magnitude of the decrease in rCBF during the occlusions (Hotta et al., 2002) (Fig. 9). The regional differences may be explained by the morphological features of arterial anastomosis in the rat cerebral arterial circle (Brown, 1966).

Repetitive electrical stimulation is delivered to NBM ipsilateral to the artery occluded, starting 5 min before the occlusions and ends approximately at the time the occlusions are terminated. The increase in rCBF induced by stimulation of NBM prevents the occlusion-induced decrease in rCBF in the three cortices. The delayed death of the cortical neurons previously observed after the occlusions is barely detectable in all cortices when NBM is stimulated. The increase in the numbers of damaged neurons following the occlusions correlates with the decrease in rCBF during the occlusions without or with stimulation of NBM. When rCBF is $<100\%$ of the control during the occlusions without or with stimulation of NBM, the numbers of damaged neurons correlate with the decrease in rCBF. Thus, neuronal damage is more extensive when rCBF decreases (Fig. 9). In contrast, when rCBF is $>100\%$ of the control during occlusions with stimulation of NBM, the numbers of damaged neurons do not correlate with the increase in rCBF and remain at approximately the control level (Fig. 9). The results suggest that activation of the vasodilative system that originates in NBM protects cortical neurons against ischemia-induced delayed death by preventing a decrease in rCBF that is distributed throughout the cortices (Hotta et al., 2002).

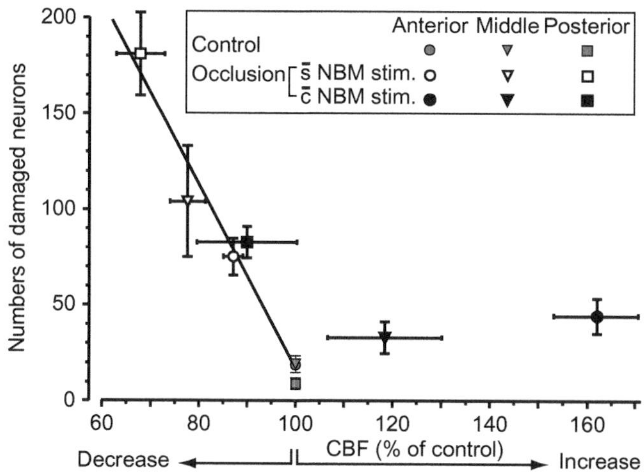

FIG. 9

Effect of nucleus basalis of Meynert (NBM) stimulation on neuronal damage in the cortex after occlusion. Relationship between regional cerebral blood flow (rCBF) during occlusion (expressed as percentage of the preocclusion control rCBF) and numbers of damaged neurons in one coronal section (6 μm thick) of the left cortices of control rats (*gray symbols*) and in rats on the fifth day after intermittent occlusions for 60 min without (*white symbols*) and with (*black symbols*) NBM stimulation. Each point represents mean ± standard error of the mean. Linear regression analysis of data rCBF <100%, $y = -4.8x + 496$ ($r^2 = 0.96$, $p < 0.0001$; Hotta et al., 2002).

5.3 STIMULATION OF NBM INCREASES THE SECRETION OF NGF

Neurotrophic factors such as NGF are neuroprotective. In the cortex, NGF-like immunoreactivity localizes to the pyramidal and nonpyramidal neurons in II/III and V/VI layers (Nishio et al., 1994; Senut et al., 1990). To assess the possibility that stimulation of NBM may act on cortical neurons to enhance the release of NGF, which acts as an endogenous neuroprotective factor in the cerebral cortex (Cheng and Mattson, 1991; Culmsee et al., 1999; Shimohama et al., 1993), a microdialysis method was developed for collecting cortical extracellular NGF, and the concentrations of NGF are determined using a highly sensitive enzyme-linked immunosorbent assay (Hotta et al., 2007a). In rats anesthetized with halothane, the concentration of NGF in the cortical perfusate under resting conditions is stable throughout the 12-h collection period. Focal stimulation of NBM (1 s on/2 s off, 50 Hz, 200 μA for 100 min) significantly increases the initial concentration of NGF (approximately 20 pg/mL) by a factor of 1.7 5 h after stimulation, and NGF concentration increases during the entire course of experiment. The increase in the concentration in cortical extracellular NGF occurs on the ipsilateral side of the cortex that receives cholinergic projections from the stimulated NBM. When NBM contralateral to the site of measurements is stimulated, an NGF response is not elicited. NGF response is abolished by the nicotinic blocker mecamylamine (20 mg/kg i.v.) but is unaffected by the muscarinic blocker atropine (5 or 25 mg/kg i.v.) (Hotta et al., 2009a).

Because ACh release increases during stimulation of NBM (Kurosawa et al., 1989), the delayed and prolonged increase in cortical extracellular NGF, first detectable hours after basal forebrain stimulation, strongly suggests that cholinergic inputs initiate the synthesis and perhaps the subsequent secretion of NGF. NGF-like immunoreactivity localizes exclusively to neurons in the healthy adult cerebral cortex (Nishio et al., 1994), although reactive astrocytes in the cerebral cortex observed under pathological conditions, such as ischemia, show NGF-like immunoreactivity (Lee et al., 1996). Transcripts of nAChRs in the parietal cortex are mainly detected in layers II/III and V pyramidal neurons but not in astrocytes that express glial fibrillary acidic protein (Birtsch et al., 1997). The detection of NGF-like immunoreactivity in the parietal cortex ipsilateral to the site of NBM stimulation is not altered when NGF secretion increases (Hotta et al., 2009a). This finding suggests that the increased NGF concentration in the cortical extracellular fluid may be caused by increases in the expression and secretion of NGF by neurons that normally produce NGF but not by recruitment of other cells such as astrocytes.

NGF increases the survival of cholinergic neuron after experimental injury and maintains and regulates the phenotype of uninjured cholinergic neurons (Rattray, 2001). In the normal adult cerebral cortex, endogenous NGF in the cortical extracellular fluid contributes to the maintenance of cortical cholinergic boutons (Debeir et al., 1999). Because NGF is secreted following activation of the basal forebrain cholinergic system, it is reasonable to suggest that the cortical levels of NGF are physiologically matched to cholinergic synapse activity during conditions that activate the cholinergic system, such as walking (Kimura et al., 1994; Nakajima et al., 2003), processing of sensory stimuli (Hotta et al., 2014; Kurosawa et al., 1992; Uchida and Kagitani, 2015; Uchida et al., 2000a), and learning (Pepeu and Giovannini, 2004).

Although stimulating NBM increases cortical rCBF, the magnitudes of NGF responses do not correlate with the changes in rCBF in experiments using AChR blockers. Although mecamylamine and atropine reduce the stimulus-dependent changes in cortical rCBF during stimulation of NBM (Biesold et al., 1989; Hotta et al., 2009a), only mecamylamine prevents increases in cortical NGF levels (Hotta et al., 2009a). The vasodilation and increase in NGF secretion may act cooperatively to protect neurons in the cerebral cortex. Thus, activation of NBM may protect the cerebral cortex, not only by vasodilation during activation but also by a prolonged increase in cortical extracellular NGF concentrations after activation. In contrast, transport of insulin-like growth factor-1 from the circulation into the cortex is increased associated with an increase in rCBF (Nishijima et al., 2010; see Nishijima, 2016, in this volume). The concentrations of such circulating neurotrophic factors that cross the BBB may increase during stimulation of NBM caused by the increase in rCBF.

6 THE EFFECT OF AGING ON THE NEURAL REGULATION OF rCBF

The brain is vulnerable to even a transient disturbance of its blood supply, and various mechanisms maintain CBF to support sufficient levels of brain activity. There is increasing evidence that vascular risk factors such as aging, hypertension, diabetes mellitus, and obesity impair cognition, although the underlying mechanisms are unknown (Kitagawa, 2010: Tarantini et al., 2015). Declines in regulatory functions that maintain CBF as well as aging of blood vessels might increase the risk of cerebrovascular accidents and contribute to neurodegenerative diseases of the elderly.

Although total CBF under resting conditions is not significantly affected by age, an age-related decrease in rCBF occurs in the limbic system and a part of the association cortex in healthy subjects aged 30–85 years (Martin et al., 1991). Orthostatic hypotension and postprandial hypotension frequently occur in elderly people because of an impaired arterial baroreceptor reflex (Hotta and Uchida, 2010). These changes in the baroreceptor reflex in aged people increase the risk of repeated episodes of cerebral ischemia during every hypotensive episode, because the lower limit of autoregulation shifts to the high blood pressure range with age (Chillon and Baumbach, 2002). Further, the adjustment of rCBF to neuronal activity via a neurovascular coupling mechanism is impaired during aging and pathophysiologic conditions. According to studies using fMRI, there is a decrease in the BOLD response of the elderly, for example, in the primary visual cortex in response to light stimulus (D'Esposito et al., 2003).

6.1 AGING OF NEUROVASCULAR COUPLING

It is not clear whether the age-related decrease in the BOLD response is caused by a reduction of neural activation or functional deterioration of neurovascular coupling mechanisms. However, there is evidence that neurovascular coupling is impaired in the elderly (Fabiani et al., 2014; Stefanova et al., 2013; Topcuoglu et al., 2009; Zaletel et al., 2005), which likely contributes to a significant age-related decline in higher cortical function, including cognition (Sorond et al., 2013) and coordination of gait (Sorond et al., 2011). Therapeutic interventions that improve neurovascular coupling in elderly patients may improve a range of age-related neurological deficits (Toth et al., 2014).

Inhibition of the synthesis of vasodilative mediators involved in neurovascular coupling using inhibitors of epoxygenase, NOS, and COX decreases neurovascular coupling of mice by >60%, which mimicks aging and significantly impairs spatial working and recognition memory as well as motor coordination (Tarantini et al., 2015). Blood pressure and basal rCBF (measured using arterial spin-labeling perfusion MRI) as well as neuronal responses such as evoked potential responses in the barrel cortex and basic synaptic transmission parameters are unaffected, suggesting that selective disruption of neurovascular coupling is associated with significant impairment of cognitive and sensorimotor function (Tarantini et al., 2015).

6.2 AGING OF THE CHOLINERGIC VASODILATIVE SYSTEM OF THE BASAL FOREBRAIN

In the basal forebrain, the numbers of cholinergic neurons projecting to the neocortex and hippocampus are significantly reduced in patients with Alzheimer's-type dementia (Arendt et al., 1985; McGeer et al., 1984; Whitehouse et al., 1982). The extent of cognitive decline and the decrease in the number of basal forebrain neurons correlate in Alzheimer's-type dementia and other diseases (Schliebs and Arendt, 2011). The rCBF decreases with the progression of the symptoms of dementia, particularly in the hippocampus and neocortex (Johnson et al., 2005; Petrella et al., 2003; Rodriguez et al., 2000). Cognitive function in normal elderly subjects marginally declines in association with an age-related decline in the number and size of basal forebrain cholinergic neurons (Grothe et al., 2013; McGeer et al., 1984).

In old rats, the increased response of the cortical rCBF induced by stimulation of NBM declines (Lacombe et al., 1997; Uchida et al., 2000b), mainly because of a decline in the function of nAChRs (Uchida et al., 1997) (Fig. 10B). The number of intracerebral nAChRs decreases remarkably in the elderly in humans (Nordberg et al., 1992). These declines can be exacerbated by the degeneration of cholinergic neurons in the basal forebrain that occurs in patients with Alzheimer's disease (Whitehouse et al., 1982). The cholinergic basal forebrain controls other functions as well as rCBF in the neocortex and hippocampus. Activation of mAChRs in cortical regions modulates the synaptic and firing properties of neurons that significantly alter the responsiveness of neurons. One example of this modulation that is dramatically evident in walking rats is the large-amplitude EEG oscillation (theta rhythm; Stewart and Fox, 1990). Further, activation of nAChRs in the cortex increases NGF secretion in the cortex (Hotta et al., 2009a). Impairments in basal forebrain cholinergic projections and in cortical cholinergic receptors that regulate rCBF and NGF secretion and modulate neuronal activity undoubtedly contribute to age-related cognitive impairments. In rats, the activity of this cholinergic vasodilative system declines with age, mainly because of the age-related decline in nAChR activity (Uchida et al., 1997). However, mAChR activity and release of ACh into the extracellular space in the cortex (Fig. 10C) are maintained during aging (Sato et al., 2002; Uchida et al., 2000b). NGF secretion was measured in adult (4–6 months) and aged (29–31 months) rats anesthetized with halothane that were administered unilateral electrical stimulation of NBM (Hotta et al., 2009a). The basal levels of extracellular NGF in the parietal cortex of the older rats were significantly higher than that in adult rats. However, stimulation of NBM does not induce significant changes in the concentrations of cortical extracellular NGF (Fig. 10A). In aged rats, cortical rCBF

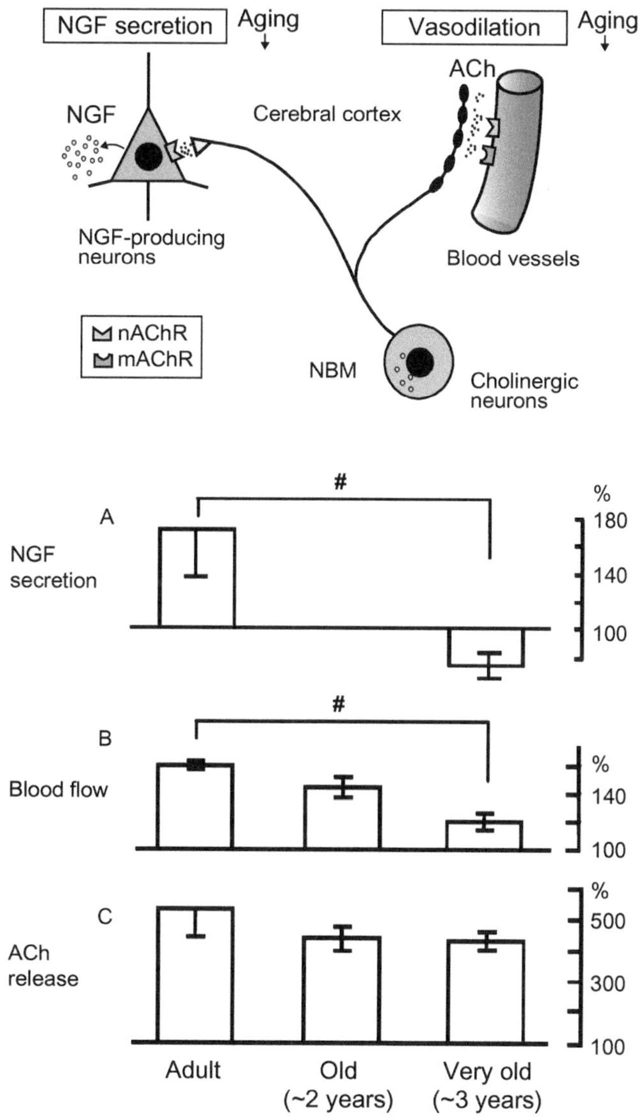

FIG. 10

Age-related changes in nerve growth factor (NGF) secretion (A), regional cerebral blood flow (B), and acetylcholine release (C) in the rat cortex induced by nucleus basalis of Meynert (NBM) stimulation. *Upper inset*: Mechanisms by which NGF secretion and vasodilation in the cerebral cortex occur in response to NBM stimulation and the consequences of aging (Hotta et al., 2009a).

Data from Uchida, S., Suzuki, A., Kagitani, F., Hotta, H., 2000b. Effects of age on cholinergic vasodilation of cortical cerebral blood vessels in rats. Neurosci. Lett. 294, 109–112; Hotta, H., Kagitani, F., Kondo, M., Uchida, S., 2009a. Basal forebrain stimulation induces NGF secretion in ipsilateral parietal cortex via nicotinic receptor activation in adult, but not aged rats. Neurosci. Res. 63, 122–128.

increases 25% during NBM stimulation, but the magnitude of the response is reduced compared with adult rats (55%). The absence of an NGF response in aged rats suggests a decline in the number or activity of cortical nAChRs. The failure of the physiological pairing between the activity of the cholinergic system and NGF secretion may contribute to the age-related decline in synaptic plasticity. Interestingly, higher basal levels of extracellular NGF in aged animals (Hotta et al., 2009a) suggest the action of a compensatory mechanism to maintain cortical NGF levels and thereby preserve cholinergic neurons.

7 CONCLUSION

Accumulating evidence suggests that complex mechanisms, which involve various neural components comprising local cortical circuits, regulate parenchymal arterioles. Local excitatory pyramidal neurons that represent 80% of the cells in the cortex are a major source of spikes and may be a candidate mediator of neurovascular coupling. However, the remaining 20% of cells, which include GABAergic inhibitory interneurons, can suppress excitatory neurons and induce vasodilation, independent of the activity of pyramidal neurons. Such local neuronal mechanisms of regulation of the local arteriolar tone may represent basic components of neurovascular coupling, which act in broad areas of the brain. Further, specific neural vasodilative mechanisms must act in the neocortex and hippocampus, because cholinergic afferent fibers from the basal forebrain projecting to the neocortex and hippocampus exert strong vasodilative activity. These mechanisms contribute, in part, to the vascular response of activated cortical areas, possibly supporting sufficient blood flow to the activated loci to protect against neuronal death. The vasodilation system originating in the basal forebrain may contribute to the responses detected using brain imaging techniques, such as fMRI, that measure changes in rCBF. A cholinergic vasodilative system may contribute to one of the major factors for a specific feature of neurovascular associations in the cortex and hippocampus.

A relationship may exist among layer specificity of the arteriolar response to stimulation of NBM, neuronal expression of NGF, and cholinergic terminal density. It is well known by investigators in the field of neuropathology that neurons in layers II/III in the cerebral cortex are the most vulnerable to transient ischemia and neurodegenerative diseases. It is therefore of interest to consider the significance of layer-specific dilation of the penetrating arteriole that is induced by the activation of NBM. Recent advances in analytical techniques show that neurogenic control of parenchymal arterioles plays an important physiological role in cerebral circulation that maintains brain function. In numerous species, the cholinergic neurons of the basal forebrain send axons to the entire cortex and may possess vasodilative functions. Because NBM becomes progressively larger and more conspicuous with increasing cerebralization, reaching its greatest complexity in humans, future studies are required to explore the possibility that the highly complex vasodilative neural circuit of the human brain is regulated by the NBM.

ACKNOWLEDGMENT

The author thanks Dr. H. Suzuki for help with the illustration.

REFERENCES

Adachi, T., Biesold, D., Inanami, O., Sato, A., 1990a. Stimulation of the nucleus basalis of Meynert and substantia innominata produces widespread increases in cerebral blood flow in the frontal, parietal and occipital cortices. Brain Res. 514, 163–166.

Adachi, T., Inanami, O., Ohno, K., Sato, A., 1990b. Responses of regional cerebral blood flow following focal electrical stimulation of the nucleus basalis of Meynert and the medial septum using the [14C] iodoantipyrine method in rats. Neurosci. Lett. 112, 263–268.

Adachi, T., Meguro, K., Sato, A., Sato, Y., 1990c. Cutaneous stimulation regulates blood flow in cerebral cortex in anesthetized rats. Neuroreport 1, 41–44.

Adachi, T., Baramidze, D.G., Sato, A., 1992a. Stimulation of the nucleus basalis of Meynert increases cortical cerebral blood flow without influencing diameter of the pial artery in rats. Neurosci. Lett. 143, 173–176.

Adachi, T., Inanami, O., Sato, A., 1992b. Nitric oxide (NO) is involved in increased cerebral cortical blood flow following stimulation of the nucleus basalis of Meynert in anesthetized rats. Neurosci. Lett. 139, 201–204.

Akaishi, T., Kimura, A., Sato, A., Suzuki, A., 1990. Responses of neurons in the nucleus basalis of Meynert to various afferent stimuli in rats. Neuroreport 1, 37–39.

Alitto, H.J., Dan, Y., 2013. Cell-type-specific modulation of neocortical activity by basal forebrain input. Front. Syst. Neurosci. 6, 79.

Anenberg, E., Chan, A.W., Xie, Y., LeDue, J.M., Murphy, T.H., 2015. Optogenetic stimulation of GABA neurons can decrease local neuronal activity while increasing cortical blood flow. J. Cereb. Blood Flow Metab. 35, 1579–1586.

Arendt, T., Bigl, V., Tennstedt, A., Arendt, A., 1985. Neuronal loss in different parts of the nucleus basalis is related to neuritic plaque formation in cortical target areas in Alzheimer's disease. Neuroscience 14, 1–14.

Barstad, K.E., Bear, M.F., 1990. Basal forebrain projections to somatosensory cortex in the cat. J. Neurophysiol. 64, 1223–1232.

Baskerville, K.A., Chang, H.T., Herron, P., 1993. Topography of cholinergic afferents from the nucleus basalis of Meynert to representational areas of sensorimotor cortices in the rat. J. Comp. Neurol. 335, 552–562.

Benagiano, V., Virgintino, D., Rizzi, A., Errede, M., Bertossi, M., Troccoli, V., Roncali, L., Ambrosi, G., 2000. Cholinergic nerve fibres associated with the microvessels of the human cerebral cortex: a study based on monoclonal immunocytochemistry for choline acetyltrasferase. Eur. J. Histochem. 44, 165–169.

Biesold, D., Inanami, O., Sato, A., Sato, Y., 1989. Stimulation of the nucleus basalis of Meynert increases cerebral cortical blood flow in rats. Neurosci. Lett. 98, 39–44.

Birtsch, C., Wevers, A., Traber, J., Maelicke, A., Bloch, W., Schröder, H., 1997. Expression of alpha 4–1 and alpha 5 nicotinic cholinoceptor mRNA in the aging rat cerebral cortex. Neurobiol. Aging 18, 335–342.

Brown, J.O., 1966. The morphology of circulus arteriosus cerebri in rats. Anat. Rec. 156, 99–106.

Cao, W.H., Inanami, O., Sato, A., Sato, Y., 1989. Stimulation of the septal complex increases local cerebral blood flow in the hippocampus in anesthetized rats. Neurosci. Lett. 107, 135–140.

Cauli, B., Tong, X.K., Rancillac, A., Serluca, N., Lambolez, B., Rossier, J., Hamel, E., 2004. Cortical GABA interneurons in neurovascular coupling: relays for subcortical vasoactive pathways. J. Neurosci. 24, 8940–8949.

Cheng, B., Mattson, M.P., 1991. NGF and bFGF protect rat hippocampal and human cortical neurons against hypoglycemic damage by stabilizing calcium homeostasis. Neuron 7, 1031–1041.

Chillon, J.M., Baumbach, G.L., 2002. Autoregulation: arterial and intracranial pressure. In: Edvinsson, L., Krause, D.N. (Eds.), Cerebral Blood Flow and Metabolism. Lippincott Williams & Wilkins, Philadelphia, pp. 395–412.

Clifford, P.M., Siu, G., Kosciuk, M., Levin, E.C., Venkataraman, V., D'Andrea, M.R., Nagele, R.G., 2008. Alpha7 nicotinic acetylcholine receptor expression by vascular smooth muscle cells facilitates the deposition of Abeta peptides and promotes cerebrovascular amyloid angiopathy. Brain Res. 1234, 158–171.

Culmsee, C., Semkova, I., Krieglstein, J., 1999. NGF mediates the neuroprotective effect of the α2-adrenoceptor agonist clenbuterol in vitro and in vivo: evidence from an NGF-antisense study. Neurochem. Int. 35, 47–57.

D'Esposito, M., Deouell, L.Y., Gazzaley, A., 2003. Alterations in the BOLD fMRI signal with ageing and disease: a challenge for neuroimaging. Nat. Rev. Neurosci. 4, 863–872.

De Lima, A.D., Singer, W., 1986. Cholinergic innervation of the cat striate cortex: a choline acetyltransferase immunocytochemical analysis. J. Comp. Neurol. 250, 324–338.

Debeir, T., Saragovi, H.U., Cuello, A.C., 1999. A nerve growth factor mimetic TrkA antagonist causes withdrawal of cortical cholinergic boutons in the adult rat. Proc. Natl. Acad. Sci. U.S.A. 96, 4067–4072.

DeFelipe, J., López-Cruz, P.L., Benavides-Piccione, R., Bielza, C., Larrañaga, P., Anderson, S., Burkhalter, A., Cauli, B., Fairén, A., Feldmeyer, D., Fishell, G., Fitzpatrick, D., Freund, T.F., González-Burgos, G., Hestrin, S., Hill, S., Hof, P.R., Huang, J., Jones, E.G., Kawaguchi, Y., Kisvárday, Z., Kubota, Y., Lewis, D.A., Marín, O., Markram, H., McBain, C.J., Meyer, H.S., Monyer, H., Nelson, S.B., Rockland, K., Rossier, J., Rubenstein, J.L., Rudy, B., Scanziani, M., Shepherd, G.M., Sherwood, C.C., Staiger, J.F., Tamás, G., Thomson, A., Wang, Y., Yuste, R., Ascoli, G.A., 2013. New insights into the classification and nomenclature of cortical GABAergic interneurons. Nat. Rev. Neurosci. 14, 202–216.

Descarries, L., Gisiger, V., Steriade, M., 1997. Diffuse transmission by acetylcholine in the CNS. Prog. Neurobiol. 53, 603–625.

Dunn, A.K., Devor, A., Dale, A.M., Boas, D.A., 2005. Spatial extent of oxygen metabolism and hemodynamic changes during functional activation of the rat somatosensory cortex. Neuroimage 27, 279–290.

Durduran, T., Burnett, M.G., Yu, G., Zhou, C., Furuya, D., Yodh, A.G., Detre, J.A., Greenberg, J.H., 2004. Spatiotemporal quantification of cerebral blood flow during functional activation in rat somatosensory cortex using laser-speckle flowmetry. J. Cereb. Blood Flow Metab. 24, 518–525.

Elhusseiny, A., Hamel, E., 2000. Muscarinic—but not nicotinic—acetylcholine receptors mediate a nitric oxide-dependent dilation in brain cortical arterioles: a possible role for the M5 receptor subtype. J. Cereb. Blood Flow Metab. 20, 298–305.

Fabiani, M., Gordon, B.A., Maclin, E.L., Pearson, M.A., Brumback-Peltz, C.R., Low, K.A., McAuley, E., Sutton, B.P., Kramer, A.F., Gratton, G., 2014. Neurovascular coupling in normal aging: a combined optical, ERP and fMRI study. Neuroimage 85, 592–607.

Fukuda, M., Poplawsky, A.J., Kim, S.-G., 2016. Submillimeter-resolution fMRI: Toward understanding local neural processing. New horizons in neurovascular coupling: a bridge between brain circulation and neural plasticity. Prog. Brain Res. 225, 123–152.

Fukuyama, H., Ouchi, Y., Matsuzaki, S., Ogawa, M., Yamauchi, H., Nagahama, Y., Kimura, J., Yonekura, Y., Shibasaki, H., Tsukada, H., 1996. Focal cortical blood flow activation is regulated by intrinsic cortical cholinergic neurons. Neuroimage 3, 195–201.

Goadsby, P.J., Edvinsson, L., 2002. Neurovascular control of the cerebral circulation. In: Edvinsson, L., Krause, D.N. (Eds.), Cerebral Blood Flow and Metabolism. Lippincott Williams & Wilkins, Philadelphia, pp. 395–412.

Grothe, M., Heinsen, H., Teipel, S., 2013. Longitudinal measures of cholinergic forebrain atrophy in the transition from healthy aging to Alzheimer's disease. Neurobiol. Aging 34, 1210–1220.

Hagen, M.C., Pardo, J.V., 2002. PET studies of somatosensory processing of light touch. Behav. Brain Res. 135, 133–140.

Hallström, A., Sato, A., Sato, Y., Ungerstedt, U., 1990. Effect of stimulation of the nucleus basalis of Meynert on blood flow and extracellular lactate in the cerebral cortex with special reference to the effect of noxious stimulation of skin and hypoxia. Neurosci. Lett. 116, 227–232.

Hillman, E.M.C., 2014. Coupling mechanism and significance of the BOLD signal: a status report. Annu. Rev. Neurosci. 37, 161–181.

Hotta, H., Uchida, S., 2010. Aging of the autonomic nervous system and possible improvements in autonomic activity using somatic afferent stimulation. Geriatr. Gerontol. Int. 10 (Suppl. 1), S127–S136.

Hotta, H., Uchida, S., Kagitani, F., 2002. Effects of stimulating the nucleus basalis of Meynert on blood flow and delayed neuronal death following transient ischemia in the rat cerebral cortex. Jpn. J. Physiol. 52, 383–393.

Hotta, H., Kanai, C., Uchida, S., Kanda, K., 2004. Stimulation of the nucleus basalis of Meynert increases diameter of the parenchymal blood vessels in the rat cerebral cortex. Neurosci. Lett. 358, 103–106.

Hotta, H., Sato, A., Schmidt, R.F., Suzuki, A., 2005. Cerebral regional cortical blood flow response during joint stimulation in cats. Neuroreport 16, 1693–1695.

Hotta, H., Uchida, S., Kagitani, F., 2007a. Stimulation of the nucleus basalis of Meynert produces an increase in the extracellular release of nerve growth factor in the rat cerebral cortex. J. Physiol. Sci. 57, 383–387.

Hotta, H., Uchida, S., Shiba, K., 2007b. Cerebral cortical blood flow response during basal forebrain stimulation in cats. Neuroreport 18, 809–812.

Hotta, H., Kagitani, F., Kondo, M., Uchida, S., 2009a. Basal forebrain stimulation induces NGF secretion in ipsilateral parietal cortex via nicotinic receptor activation in adult, but not aged rats. Neurosci. Res. 63, 122–128.

Hotta, H., Lazar, J., Orman, R., Koizumi, K., Shiba, K., Kamran, H., Stewart, M., 2009b. Vagus nerve stimulation-induced bradyarrhythmias in rats. Auton. Neurosci. 151, 98–105.

Hotta, H., Uchida, S., Kagitani, F., Maruyama, N., 2011. Control of cerebral cortical blood flow by stimulation of basal forebrain cholinergic areas in mice. J. Physiol. Sci. 61, 201–209.

Hotta, H., Masamoto, K., Uchida, S., Sekiguchi, Y., Takuwa, H., Kawaguchi, H., Shigemoto, K., Sudo, R., Tanishita, K., Ito, H., Kanno, I., 2013. Layer-specific dilation of penetrating arteries induced by stimulation of the nucleus basalis of Meynert in the mouse frontal cortex. J. Cereb. Blood Flow Metab. 33, 1440–1447.
Hotta, H., Watanabe, N., Piché, M., Hara, S., Yokawa, T., Uchida, S., 2014. Non-noxious skin stimulation activates the nucleus basalis of Meynert and promotes NGF secretion in the parietal cortex via nicotinic ACh receptors. J. Physiol. Sci. 64 (4), 253–260.
Itoh, Y., Suzuki, N., 2012. Control of brain capillary blood flow. J. Cereb. Blood Flow Metab. 32, 1167–1176.
Ji, L., Zhou, J., Zafar, R., Kantorovich, S., Jiang, R., Carney, P.R., Jiang, H., 2012. Cortical neurovascular coupling driven by stimulation of channelrhodopsin-2. PLoS One 7, e46607.
Johnson, N.A., Jahng, G.H., Weiner, M.W., Miller, B.L., Chui, H.C., Jagust, W.J., Gorno-Tempini, M.L., Schuff, N., 2005. Pattern of cerebral hypoperfusion in Alzheimer disease and mild cognitive impairment measured with arterial spin-labeling MR imaging: initial experience. Radiology 234, 851–859.
Kagitani, F., Uchida, S., Hotta, H., Sato, A., 2000. Effects of nicotine on blood flow and delayed neuronal death following intermittent transient ischemia in rat hippocampus. Jpn. J. Physiol. 50, 585–595.
Kalaria, R.N., Homayoun, P., Whitehouse, P.J., 1994. Nicotinic cholinergic receptors associated with mammalian cerebral vessels. J. Auton. Nerv. Syst. 49 (Suppl.), S3–S7.
Kimura, A., Sato, A., Takano, Y., 1990. Stimulation of the nucleus basalis of Meynert does not influence glucose utilization of the cerebral cortex in anesthetized rats. Neurosci. Lett. 119, 101–104.
Kimura, A., Okada, K., Sato, A., Suzuki, H., 1994. Regional cerebral blood flow in the frontal, parietal and occipital cortices increases independently of systemic arterial pressure during slow walking in conscious rats. Neurosci. Res. 20, 309–315.
Kirino, T., 1982. Delayed neuronal death in the gerbil hippocampus following ischemia. Brain Res. 239, 57–69.
Kitagawa, K., 2010. Cerebral blood flow measurement by PET in hypertensive subjects as a marker of cognitive decline. J. Alzheimers Dis. 20, 855–859.
Kocharyan, A., Fernandes, P., Tong, X.K., Vaucher, E., Hamel, E., 2008. Specific subtypes of cortical GABA interneurons contribute to the neurovascular coupling response to basal forebrain stimulation. J. Cereb. Blood Flow Metab. 28, 221–231.
Kubota, Y., Shigematsu, N., Karube, F., Sekigawa, A., Kato, S., Yamaguchi, N., Hirai, Y., Morishima, M., Kawaguchi, Y., 2011. Selective coexpression of multiple chemical markers defines discrete populations of neocortical GABAergic neurons. Cereb. Cortex 21, 1803–1817.
Kurosawa, M., Sato, A., Sato, Y., 1989. Stimulation of the nucleus basalis of Meynert increases acetylcholine release in the cerebral cortex in rats. Neurosci. Lett. 98, 45–50.
Kurosawa, M., Sato, A., Sato, Y., 1992. Cutaneous mechanical sensory stimulation increases extracellular acetylcholine release in cerebral cortex in anesthetized rats. Neurochem. Int. 21, 423–427.
Lacombe, P., Sercombe, R., Verrecchia, C., Philipson, V., MacKenzie, E.T., Seylaz, J., 1989. Cortical blood flow increases induced by stimulation of the substantia innominata in the unanesthetized rat. Brain Res. 491, 1–14.
Lacombe, P., Sercombe, R., Vaucher, E., Seylaz, J., 1997. Reduced cortical vasodilatory response to stimulation of the nucleus basalis of Meynert in the aged rat and evidence for a control of the cerebral circulation. Ann. N.Y. Acad. Sci. 826, 410–415.

Lacroix, A., Toussay, X., Anenberg, E., Lecrux, C., Ferreirós, N., Karagiannis, A., Plaisier, F., Chausson, P., Jarlier, F., Burgess, S.A., Hillman, E.M., Tegeder, I., Murphy, T.H., Hamel, E., Cauli, B., 2015. COX-2-derived prostaglandin E2 produced by pyramidal neurons contributes to neurovascular coupling in the rodent cerebral cortex. J. Neurosci. 35 (34), 11791–11810.

Lauritzen, M., Mathiesen, C., Schaefer, K., Thomsen, K.J., 2012. Neuronal inhibition and excitation, and the dichotomic control of brain hemodynamic and oxygen responses. Neuroimage 62, 1040–1050.

Lecrux, C., Toussay, X., Kocharyan, A., Fernandes, P., Neupane, S., Lévesque, M., Plaisier, F., Shmuel, A., Cauli, B., Hamel, E., 2011. Pyramidal neurons are "neurogenic hubs" in the neurovascular coupling response to whisker stimulation. J. Neurosci. 31, 9836–9847.

Lecrux, C., Kocharyan, A., Sandoe, C.H., Tong, X.K., Hamel, E., 2012. Pyramidal cells and cytochrome P450 epoxygenase products in the neurovascular coupling response to basal forebrain cholinergic input. J. Cereb. Blood Flow Metab. 32, 896–906.

Lee, S.H., Dan, Y., 2012. Neuromodulation of brain states. Neuron 76, 209–222.

Lee, T.H., Kato, H., Kogure, K., Itoyama, Y., 1996. Temporal profile of nerve growth factor-like immunoreactivity after transient focal cerebral ischemia in rats. Brain Res. 713, 199–210.

Lee, J.H., Durand, R., Gradinaru, V., Zhang, F., Goshen, I., Kim, D.S., Fenno, L.E., Ramakrishnan, C., Deisseroth, K., 2010. Global and local fMRI signals driven by neurons defined optogenetically by type and wiring. Nature 465, 788–792.

Lehmann, J., Nagy, J.I., Atmadja, S., Fibiger, H.C., 1980. The nucleus basalis magnocellularis: the origin of a cholinergic projection to the neocortex of the rat. Neuroscience 5, 1161–1174.

Linville, D.G., Williams, S., Arneric, S.P., 1993. Basal forebrain control of cortical cerebral blood flow is independent of local cortical neurons. Brain Res. 622, 26–34.

Liu, A.K., Chang, R.C., Pearce, R.K., Gentleman, S.M., 2015. Nucleus basalis of Meynert revisited: anatomy, history and differential involvement in Alzheimer's and Parkinson's disease. Acta Neuropathol. 129, 527–540.

Logothetis, N.K., Pauls, J., Augath, M., Trinath, T., Oeltermann, A., 2001. Neurophysiological investigation of the basis of the fMRI signal. Nature 412, 150–157.

Luiten, P.G.M., Spencer Jr., D.G., Traber, J., Gaykema, R.P.A., 1985. The pattern of cortical projections from the intermediate parts of the magnocellular nucleus basalis in the rat demonstrated by tracing with *Phaseolus vulgaris*-leucoagglutinin. Neurosci. Lett. 57, 137–142.

Luiten, P.G.M., Gaykema, R.P.A., Traber, J., Spencer Jr., D.G., 1987. Cortical projection patterns of magnocellular basal nucleus subdivisions as revealed by anterogradely transported *Phaseolus vulgaris* leucoagglutinin. Brain Res. 413, 229–250.

Mann, D.M.A., Yates, P.O., Marcyniuk, B., 1986. A comparison of nerve cell loss in cortical and subcortical structures in Alzheimer's disease. J. Neurol. Neurosurg. Psychiatry 49, 310–312.

Martin, A.J., Friston, K.J., Colebatch, J.G., Frackowiak, R.S.J., 1991. Decreases in regional cerebral blood flow with normal aging. J. Cereb. Blood Flow Metab. 11, 684–689.

Masamoto, K., Kanno, I., 2012. Anesthesia and the quantitative evaluation of neurovascular coupling. J. Cereb. Blood Flow Metab. 32, 1233–1247.

McGeer, P.L., McGeer, E.G., Suzuki, J., Dolman, C.E., Nagai, T., 1984. Aging, Alzheimer's disease, and the cholinergic system of the basal forebrain. Neurology 34, 741–745.

Metherate, R., Cox, C.L., Ashe, J.H., 1992. Cellular bases of neocortical activation: modulation of neural oscillations by the nucleus basalis and endogenous acetylcholine. J. Neurosci. 12, 4701–4711.

Moro, V., Kacem, K., Springhetti, V., Seylaz, J., Lasbennes, F., 1995. Microvessels isolated from brain: localization of muscarinic sites by radioligand binding and immunofluorescent techniques. J. Cereb. Blood Flow Metab. 15, 1082–1092.

Nakajima, K., Uchida, S., Suzuki, A., Hotta, H., Aikawa, Y., 2003. The effect of walking on regional blood flow and acetylcholine in the hippocampus in conscious rats. Auton. Neurosci. 103, 83–92.

Nakao, Y., Gotoh, J., Kuang, T.Y., Cohen, D.M., Pettigrew, K.D., Sokoloff, L., 1999. Cerebral blood flow responses to somatosensory stimulation are unaffected by scopolamine in unanesthetized rat. J. Pharmacol. Exp. Ther. 290, 929–934.

Nieuwenhuys, R., Voogd, J., van Huijzen, C., 2008. The Human Central Nervous System, fourth ed. Springer, Berlin.

Nishijima, T., Torres-Aleman, I., Soya, H., 2016. Exercise and cerebrovascular plasticity. New horizons in neurovascular coupling: a bridge between brain circulation and neural plasticity. Prog. Brain Res. 225, 243–268.

Nishijima, T., Piriz, J., Duflot, S., Fernandez, A.M., Gaitan, G., Gomez-Pinedo, U., Verdugo, J.M., Leroy, F., Soya, H., Nuñez, A., Torres-Aleman, I., 2010. Neuronal activity drives localized blood–brain-barrier transport of serum insulin-like growth factor-I into the CNS. Neuron 67, 834–846.

Nishio, T., Furukawa, S., Akiguchi, I., Oka, N., Ohnishi, K., Tomimoto, H., Nakamura, S., Kimura, J., 1994. Cellular localization of nerve growth factor-like immunoreactivity in adult rat brain: quantitative and immunohistochemical study. Neuroscience 60, 67–84.

Nordberg, A., Alafuzoff, I., Winblad, B., 1992. Nicotinic and muscarinic subtypes in the human brain: changes with aging and dementia. J. Neurosci. Res. 31, 103–111.

Nuriya, M., Hirase, H., 2016. Involvement of astrocytes in neurovascular communication. New horizons in neurovascular coupling: a bridge between brain circulation and neural plasticity. Prog. Brain Res. 225, 41–62.

Ogawa, S., Lee, T.M., Kay, A.R., Tank, D.W., 1990. Brain magnetic resonance imaging with contrast dependent on blood oxygenation. Proc. Natl. Acad. Sci. U.S.A. 87, 9868–9872.

Ogawa, M., Magata, Y., Ouchi, Y., Fukuyama, H., Yamauchi, H., Kimura, J., Yonekura, Y., Konishi, J., 1994. Scopolamine abolishes cerebral blood flow response to somatosensory stimulation in anesthetized cats: PET study. Brain Res. 650, 249–252.

Pepeu, G., Giovannini, M.G., 2004. Changes in acetylcholine extracellular levels during cognitive processes. Learn. Mem. 11, 21–27.

Peters, A., Jones, E.G., 1984. Cerebral cortex. Cellular Components of the Cerebral Cortex, vol. 1. Plenum Press, New York.

Petrella, J.R., Coleman, R.E., Doraiswamy, P.M., 2003. Neuroimaging and early diagnosis of Alzheimer disease: a look to the future. Radiology 226, 315–336.

Piché, M., Uchida, S., Hara, S., Aikawa, Y., Hotta, H., 2010. Modulation of somatosensory-evoked cortical blood flow changes by GABAergic inhibition of the nucleus basalis of Meynert in urethane-anaesthetized rats. J. Physiol. 588, 2163–2171.

Pulsinelli, W.A., Brierley, J.B., Plum, F., 1982. Temporal profile of neuronal damage in a model of transient forebrain ischemia. Ann. Neurol. 11, 491–498.

Raszkiewicz, J.L., Linville, D.G., Kerwin Jr., J.F., Wagenaar, F., Arneric, S.P., 1992. Nitric oxide synthase is critical in mediating basal forebrain regulation of cortical cerebral circulation. J. Neurosci. Res. 33, 129–135.

Rattray, M., 2001. Is there nicotinic modulation of nerve growth factor? Implications for cholinergic therapies in Alzheimer's disease. Biol. Psychiatry 49, 185–193.

Rodriguez, G., Vitali, P., Calvini, P., Bordoni, C., Girtler, N., Taddei, G., Mariani, G., Nobili, F., 2000. Hippocampal perfusion in mild Alzheimer's disease. Psychiatry Res. 100, 65–74.

Royl, G., Leithner, C., Sellien, H., Müller, J.P., Megow, D., Offenhauser, N., Steinbrink, J., Kohl-Bareis, M., Dirnagl, U., Lindauer, U., 2006. Functional imaging with laser speckle contrast analysis: vascular compartment analysis and correlation with laser Doppler flowmetry and somatosensory evoked potentials. Brain Res. 1121, 95–103.

Saito, T., Sakamoto, K., Koizumi, K., Stewart, M., 2006. Repeatable focal seizure suppression: a rat preparation to study consequences of seizure activity based on urethane anesthesia and reversible carotid artery occlusion. J. Neurosci. Methods 155, 241–250.

Sakiyama, Y., Sato, A., Senda, M., Ishikawa, K., Toyama, H., Schmidt, R.F., 1998. Positron emission tomography reveals changes in global and regional cerebral blood flow during noxious stimulation of normal and inflamed elbow joints in anesthetized cats. Exp. Brain Res. 118, 439–446.

Sato, A., Sato, Y., 1992. Regulation of regional cerebral blood flow by cholinergic fibers originating in the basal forebrain. Neurosci. Res. 14, 242–274.

Sato, A., Sato, Y., 1995. Cholinergic neural regulation of regional cerebral blood flow. Alzheimer Dis. Assoc. Disord. 9, 28–38.

Sato, A., Sato, Y., Uchida, S., 1994. Blood flow in the sciatic nerve is regulated by vasoconstrictive and vasodilative nerve fibers originating from the ventral and dorsal roots of the spinal nerves. Neurosci. Res. 21, 125–133.

Sato, A., Sato, Y., Uchida, S., 2002. Regulation of cerebral cortical blood flow by the basal forebrain cholinergic fibers and aging. Auton. Neurosci. 96, 13–19.

Schliebs, R., Arendt, T., 2011. The cholinergic system in aging and neuronal degeneration. Behav. Brain Res. 221, 555–563.

Senut, M.C., Lamour, Y., Lee, J., Brachet, P., Dicou, E., 1990. Neuronal localization of the nerve growth factor precursor-like immunoreactivity in the rat brain. Int. J. Dev. Neurosci. 8, 65–80.

Shimohama, S., Ogawa, N., Tamura, Y., Akaike, A., Tsukahara, T., Iwata, H., Kimura, J., 1993. Protective effect of nerve growth factor against glutamate-induced neurotoxicity in cultured cortical neurons. Brain Res. 632, 296–302.

Sorond, F.A., Kiely, D.K., Galica, A., Moscufo, N., Serrador, J.M., Iloputaife, I., Egorova, S., Dell'Oglio, E., Meier, D., Newton, E., Milberg, W.P., Guttmann, C.G., Lipsitz, L.A., 2011. Neurovascular coupling is impaired in slow walkers: the MOBILIZE Boston Study. Ann. Neurol. 70, 213–220.

Sorond, F.A., Hurwitz, S., Salat, D.H., Greve, D.N., Fisher, N.D.L., 2013. Neurovascular coupling, cerebral white matter integrity, and response to cocoa in older people. Neurology 81, 904–909.

Stefanova, I., Stephan, T., Becker-Bense, S., Dera, T., Brandt, T., Dieterich, M., 2013. Age-related changes of blood-oxygen-level-dependent signal dynamics during optokinetic stimulation. Neurobiol. Aging 34, 2277–2286.

Stewart, M., Fox, S.E., 1990. Do septal neurons pace the hippocampal theta rhythm? Trends Neurosci. 13, 163–168.

Stringer, E.A., Qiao, P.G., Friedman, R.M., Holroyd, L., Newton, A.T., Gore, J.C., Chen, L.M., 2014. Distinct fine-scale fMRI activation patterns of contra- and ipsilateral somatosensory areas 3b and 1 in humans. Hum. Brain Mapp. 35, 4841–4857.

Takata, N., Nagai, T., Ozawa, K., Oe, Y., Mikoshiba, K., Hirase, H., 2013. Cerebral blood flow modulation by Basal forebrain or whisker stimulation can occur independently of large cytosolic Ca^{2+} signaling in astrocytes. PLoS One 8, e66525.

Tarantini, S., Hertelendy, P., Tucsek, Z., Valcarcel-Ares, M.N., Smith, N., Menyhart, A., Farkas, E., Hodges, E.L., Towner, R., Deak, F., Sonntag, W.E., Csiszar, A., Ungvari, Z., Toth, P., 2015. Pharmacologically-induced neurovascular uncoupling is associated with cognitive impairment in mice. J. Cereb. Blood Flow Metab. 35, 1871–1881.

Terry, R.D., Peck, A., DeTeresa, R., Schechter, R., Horoupian, D.S., 1981. Some morphometric aspects of the brain in senile dementia of the Alzheimer type. Ann. Neurol. 10, 184–192.

Topcuoglu, M.A., Aydin, H., Saka, E., 2009. Occipital cortex activation studied with simultaneous recordings of functional transcranial Doppler ultrasound (fTCD) and visual evoked potential (VEP) in cognitively normal human subjects: effect of healthy aging. Neurosci. Lett. 452, 17–22.

Toth, P., Tarantini, S., Tucsek, Z., Ashpole, N.M., Sosnowska, D., Gautam, T., Ballabh, P., Koller, A., Sonntag, W.E., Csiszar, A., Ungvari, Z., 2014. Resveratrol treatment rescues neurovascular coupling in aged mice: role of improved cerebromicrovascular endothelial function and downregulation of NADPH oxidase. Am. J. Physiol. Heart Circ. Physiol. 306, H299–H308.

Tsukada, H., Kakiuchi, T., Ando, I., Shizuno, H., Nakanishi, S., Ouchi, Y., 1997. Regulation of cerebral blood flow response to somatosensory stimulation through the cholinergic system: a positron emission tomography study in unanesthetized monkeys. Brain Res. 749, 10–17.

Uchida, S., Kagitani, F., 2015. Effect of acupuncture-like stimulation on cortical cerebral blood flow in aged rats. J. Physiol. Sci. 65, 67–75.

Uchida, S., Kagitani, F., Nakayama, H., Sato, A., 1997. Effect of stimulation of nicotinic cholinergic receptors on cortical cerebral blood flow and changes in the effect during aging in anesthetized rats. Neurosci. Lett. 228, 203–206.

Uchida, S., Kagitani, F., Suzuki, A., Aikawa, Y., 2000a. Effect of acupuncture-like stimulation on cortical cerebral blood flow in anesthetized rats. Jpn. J. Physiol. 50, 495–507.

Uchida, S., Suzuki, A., Kagitani, F., Hotta, H., 2000b. Effects of age on cholinergic vasodilation of cortical cerebral blood vessels in rats. Neurosci. Lett. 294, 109–112.

Uematsu, M., Hirai, Y., Karube, F., Ebihara, S., Kato, M., Abe, K., Obata, K., Yoshida, S., Hirabayashi, M., Yanagawa, Y., Kawaguchi, Y., 2008. Quantitative chemical composition of cortical GABAergic neurons revealed in transgenic venus-expressing rats. Cereb. Cortex 18, 315–330.

Umbriaco, D., Watkins, K.C., Descarries, L., Cozzari, C., Hartman, B.K., 1994. Ultrastructural and morphometric features of the acetylcholine innervation in adult rat parietal cortex: an electron microscopic study in serial sections. J. Comp. Neurol. 348, 351–373.

Ureshi, M., Matsuura, T., Kanno, I., 2004. Stimulus frequency dependence of the linear relationship between local cerebral blood flow and field potential evoked by activation of rat somatosensory cortex. Neurosci. Res. 48, 147–153.

Van Der Zee, E.A., De Jong, G.I., Strosberg, A.D., Luiten, P.G., 1993. Muscarinic acetylcholine receptor-expression in astrocytes in the cortex of young and aged rats. Glia 8, 42–50.

Vaucher, E., Hamel, E., 1995. Cholinergic basal forebrain neurons project to cortical microvessels in the rat: electron microscopic study with anterogradely transported

Phaseolus vulgaris leucoagglutinin and choline acetyltransferase immunocytochemistry. J. Neurosci. 15, 7427–7441.

Vaucher, E., Borredon, J., Bonvento, G., Seylaz, J., Lacombe, P., 1997. Autoradiographic evidence for flow-metabolism uncoupling during stimulation of the nucleus basalis of Meynert in the conscious rat. J. Cereb. Blood Flow Metab. 17, 686–694.

Whitehouse, P.J., Price, D.L., Struble, R.G., Clark, A.W., Coyle, J.T., DeLong, M.R., 1982. Alzheimer's disease and senile dementia: loss of neurons in the basal forebrain. Science 215, 1237–1239.

Wu, H., Williams, J., Nathans, J., 2014. Complete morphologies of basal forebrain cholinergic neurons in the mouse. eLife 3, e02444.

Yamada, K., 2016. Vascular potassium channels in NVC. New horizons in neurovascular coupling: a bridge between brain circulation and neural plasticity. Prog. Brain Res. 225, 63–73.

Zaletel, M., Strucl, M., Pretnar-Oblak, J., Zvan, B., 2005. Age-related changes in the relationship between visual evoked potentials and visually evoked cerebral blood flow velocity response. Funct. Neurol. 20, 115–120.

Zhang, F., Xu, S., Iadecola, C., 1995. Role of nitric oxide and acetylcholine in neocortical hyperemia elicited by basal forebrain stimulation: evidence for an involvement of endothelial nitric oxide. Neuroscience 69, 1195–1204.

CHAPTER

Involvement of astrocytes in neurovascular communication

2

M. Nuriya*, H. Hirase†,1

*Keio University, Shinjuku, Tokyo, Japan

†RIKEN Brain Science Institute, Wako, Saitama, Japan

[1]Corresponding author: Tel.: +81-48-4676918; Fax: +81-48-4679652, e-mail address: hirase@brain.riken.jp

Abstract

The vascular interface of the brain is distinct from that of the peripheral tissue in that astrocytes, the most numerous glial cell type in the gray matter, cover the vasculature with their endfeet. This morphological feature of the gliovascular junction has prompted neuroscientists to suggest possible functional roles of astrocytes including astrocytic modulation of the vasculature. Additionally, astrocytes develop an intricate morphology that intimately apposes neuronal synapses, making them an ideal cellular mediator of neurovascular coupling. In this article, we first introduce the classical anatomical and physiological findings that led to the proposal of various gliovascular interaction models. Next, we touch on the technological advances in the past few decades that enabled investigations and evaluations of neuro-glio-vascular interactions *in situ*. We then review recent experimental findings on the roles of astrocytes in neurovascular coupling from the viewpoints of intra- and intercellular signalings in astrocytes.

Keywords

Astrocyte, Calcium, Neurovascular coupling, Endfeet, Gap junction, Potassium, AQP4

1 INTRODUCTION

1.1 HISTORICAL PERSPECTIVES

Plato (428–348 BC) arguably put forward the fundamental notion that the *psyche* (the mind) is in the brain. In the second century AD, when the territorial spread of the Roman Empire was at its peak, the Greek physician Galen (also known as Claudius Galenus) supported this idea by adding the concept of the *pneuma* (the spirit). He noted that the vital *pneuma* comes from the arterial system and is transformed to the psychic *pneuma* in the brain (Lloyd, 2007). While the importance of cerebral blood flow (CBF) has been recognized, it was not until about a century ago that

Progress in Brain Research, Volume 225, ISSN 0079-6123, http://dx.doi.org/10.1016/bs.pbr.2016.02.001

the CBF was discovered to be dynamically regulated by local brain activity. This phenomenon is referred to as functional hyperemia and was first described in humans by Mosso (1880). Roy and Sherrington (1890) independently investigated functional hyperemia in dogs and proposed that the metabolic processes triggered by local neural activity produce chemicals that change the vessel diameter.

Incidentally, the discovery of glia, nonneuronal cells in the brain, occurred at nearly the same time. Virchow is considered the first person to describe glia in the central nervous system. Virchow regarded glia as glue (Nervenkitt) that connects neural tissues (Kettenmann and Verkhratsky, 2008). The taxonomy and morphology of glial cells was investigated in the late 19th and early 20th centuries by multiple independent scientists including Koelliker, Deiters, Golgi, Cajal, and Río-Hortega, to name a few. Cajal's gold impregnation staining method enabled visualization of the somata and major branching processes of astrocytes and, thus, revealed that astrocytes extend their endfeet to the nearby vasculature. Golgi's silver staining method labeled the astrocyte morphology more extensively and showed that astrocytes have a bushy appearance with microprocesses that surround the synaptic structures. From a morphological viewpoint, it was clear that astrocyte is an interface that interacts with synaptic and vascular networks. Despite this morphological configuration, functional hyperemia has been considered to be mediated by direct coupling of neurons to the vasculature (see Chapter 1) with vasodilators such as carbon dioxide, potassium (K^+), and nitric oxide. However, with the sophistication of brain activity instrumentation, glial involvement, in particular that of astrocytes, has increasingly gained recognition.

1.2 TECHNICAL ADVANCEMENTS

Hemodynamics in the brain has been investigated for over a century using various techniques (see Chapters 4–6). Several technical advances in recent decades have critically contributed to our understanding of neuro-glio-vascular coupling. Arguably the most important of those was the application of the two-photon imaging technique (Denk et al., 1990). Two-photon imaging uses infrared light to excite fluorescent dyes and proteins in the brain tissue as deep as 1 mm. The use of ultrashort pulse lasers allows a two-photon excitation at the focus that achieves the horizontal spatial resolution of a micrometer. Recent imaging advances have been accelerated by the ever-increasing repertoire of fluorescent molecules, enabling live imaging of CBF together with morphological and functional changes of neurovascular units. Notably, super-resolution techniques such as stimulated emission depletion (STED) microscopy have come of age and submicron imaging of astrocytes became feasible with brain slices (Panatier et al., 2014). These high-resolution functional imaging techniques are commercially available and used to observe neurovascular units in living animals with unprecedented spatial resolution and noninvasiveness.

Another significant methodological leap is in molecular genetics. Because of the highly intermingled anatomical layout and the expressions of common receptors among different cell types, the pharmacological dissection of astrocytic function is often challenging. Molecular genetics approaches that introduce functional alternations to a target cell population have become an essential tool for hypothesis-driven experiments. For instance, using loss-of-function and gain-of-function approaches, one can test the causal relationship between a biological phenomenon and the presence of a gene/protein in the target cell population. Cell-type specific manipulation of gene expression is primarily performed in two ways: first, site-specific recombinase technology (eg, Cre-LoxP or Flp-FRT) allows genome excision at targeted sites. Second, the tetracycline-controlled transcriptional activation system allows reversible manipulation of gene transcription. Cell-type specificity of these manipulations is achieved by incorporating cell-type specific promoters or modifying the gene locus that has specific expression in the target cell type (ie, knock-in). With the advent of genome editing techniques such as CRISPR/Cas9 and TALEN systems, genomic manipulations of mice are becoming a matter of a few months. Moreover, the use of recombinant viruses provides further spatial and temporal confinement of gene manipulation. These techniques are readily applicable to astrocytes in rodents.

Advances in molecular genetics also enable gain-of-function experiments with relatively noninvasive target cell activation. Namely, opto- or pharmacogenetic molecules can be expressed in a cell-type selective manner, allowing dynamic and specific manipulations (for a review, Xie et al., 2015). For example, in animals expressing channelrodopsin-2 variants in astrocytes, blue light can directly and noninvasively activate the astrocytes (Masamoto et al., 2015; Perea et al., 2014; Sasaki et al., 2012). In the case of pharmacogenetics, systemic ligand administration activates foreign receptors expressed in a target cell population. Pharmacogenetic receptors are typically G protein-coupled receptors (GPCRs) such as designer receptors exclusively activated by designer drugs (DREADDs) that are activated exclusively by the synthetic ligand clozapine-*N*-oxide (CNO). Astrocyte pharmacogenetics has started to be employed in *in vivo* experiments (Agulhon et al., 2013; Bonder and McCarthy, 2014).

For applications of certain molecules at a high spatiotemporal resolution, photolysis of caged compounds (ie, uncaging) became available. Uncaging has been used popularly in the synaptic physiology community since it provides a means to stimulate single synapses, for instance, by glutamate uncaging. For astrocytes, this technology is effectively used to uncage calcium (Ca^{2+}) to elevate intracellular Ca^{2+} levels and investigate the role of Ca^{2+} dynamics at various scales. Of note, a recent paper by Wang et al. (2013) reported that spurious glutamate secretion occurs with the photolysis process. Perhaps in the near future, focal activation of receptors by some optogenetic means will be a viable option for subcellular manipulations. While we must be careful about the correctness of molecular genetic manipulations since these techniques sometimes suffer from unintended leaky expression (Fujita et al., 2014), the imaging and genetic manipulation advancements continuously provide a potent means that accelerates our understanding of neuro-glio-vascular coupling as discussed in the following sections.

2 ASTROCYTIC CALCIUM DYNAMICS AND VASCULAR CONTROL

2.1 CALCIUM ELEVATIONS IN ASTROCYTES

Since astrocytes do not express significant amounts of voltage-gated ion channels, the membrane potential fluctuations of an astrocyte are usually within a few millivolts from the resting membrane potential of below −80 mV (Amzica and Massimini, 2002; Mishima et al., 2007; Mishima and Hirase, 2010). While voltage-based signals (eg, action potentials) are unlikely means of cellular communications for astrocytes, intracellular elevations of Ca^{2+} have long been hypothesized to reflect the output of astrocytes.

Astrocytic Ca^{2+} elevation was first demonstrated in cultured preparations (Pearce et al., 1986). Subsequently, it was found that the elevation of Ca^{2+} was mediated by the Ca^{2+} release from internal stores (Ahmed et al., 1990; Cornell-Bell et al., 1990; Glaum et al., 1990; Jensen and Chiu, 1990) such as the endoplasmic reticulum and mitochondria. Moreover, the activation of GPCRs, such as metabotropic glutamate receptors and P2Y purinergic receptors, was demonstrated to trigger large and long-lasting Ca^{2+} elevations (Helen et al., 1992; Nakahara et al., 2002). In addition to Ca^{2+} elevations in single astrocytes, Ca^{2+} elevations have been observed to propagate across neighboring cells in cultured preparations (Cornell-Bell et al., 1990; Dani et al., 1992; Nedergaard, 1994). The mechanism of propagation has been shown to be both communication through gap junctions and secretion of bioactive molecules from astrocytes, which will be discussed further in the next section. Numerous studies reported similar Ca^{2+} elevations in acute slices of the cortex, hippocampus, cerebellum, and other parts of the brain. Importantly, astrocytic Ca^{2+} elevations were observed in living rodent brains (Hirase et al., 2004b; Stosiek et al., 2003).

Astrocytic Ca^{2+} elevations are observed in various spatial scales. For example, large-amplitude cytosolic Ca^{2+} elevations occur simultaneously in both the soma and branching processes. Ca^{2+} elevations can also occur in a discrete focalized spot within a process or in an expanded continuous region of a process (Di Castro et al., 2011; Fiacco and McCarthy, 2004). Somatic and expanded Ca^{2+} elevations are critically dependent on the IP_3 type 2 receptor (IP_3R2), the Ca^{2+} channel localized in the endoplasmic reticulum (Petravicz et al., 2008). On the other hand, a small proportion of focal events are still observed in the IP_3R2 knockout mouse brain (Di Castro et al., 2011; Kanemaru et al., 2014; Srinivasan et al., 2015). Interestingly, some of the focal events have been shown to be mediated by the influx of extracellular Ca^{2+}. Focal Ca^{2+} events are more frequently observed with membrane-bound genetic Ca^{2+} sensors and one of the transient receptor potential cation channels, TRPA1, has been proposed to be involved in such focalized events (Shigetomi et al., 2013).

Earlier *in vivo* studies showed that nearby astrocytes tend to have synchronized Ca^{2+} elevations in the anesthetized rodent cortex (Hirase et al., 2004b; Takata and Hirase, 2008). In awake mice, astrocytes in the somatosensory cortex hindlimb area are reported to elevate Ca^{2+} levels with running behavior (Dombeck et al., 2007).

Moreover, recent studies report that synchronized astrocytic Ca^{2+} elevations over a wide region of the cortex are prevalent in unanesthetized mice (Ding et al., 2013; Paukert et al., 2014). Large-scale synchronized Ca^{2+} elevations are also observed in cerebellar Bergmann glia, a specialized form of astrocytes that enwrap cerebellar Purkinje cell synapses, at times of motor initiation (Nimmerjahn et al., 2009). Interestingly, radially expanding transglial Ca^{2+} waves are observed in the cerebellum, although their functional role is not yet known (Hoogland et al., 2009).

2.2 EXTRACELLULAR CHANGES BY ASTROCYTIC CALCIUM ELEVATIONS

While intracellular Ca^{2+} elevation is a remarkable means by which astrocytes respond to extracellular stimuli, the functional significance of the Ca^{2+} elevation remains under active investigation. Gliotransmission, the process in which astrocytes secrete bioactive molecules, has been proposed as the main consequence of intracellular Ca^{2+} elevation. Since glutamate and adenosine triphosphate (ATP) were discovered to be gliotransmitters of astrocytes (Guthrie et al., 1999; Parpura et al., 1994), many other molecules have been described as gliotransmitters, including amino acids (eg, GABA, D-serine), peptides (eg, S100B), and cytokines (eg, TNFα) (Volterra and Meldolesi, 2005). Gliotransmission can occur by channel or vesicular release. Vesicular release of gliotransmitters can occur in a Ca^{2+}-dependent manner (for review, Montana et al., 2006). For instance, interference of the SNARE proteins that mediate vesicle fusion has been shown to compromise purinergic gliotransmission (Pascual et al., 2005). On the other hand, gliotransmission by channel release can occur independently of intracellular Ca^{2+} concentration. However, the Best1 channel, which mediates astrocytic gliotransmission of glutamate and GABA, has been reported to operate in a Ca^{2+}-dependent manner (Woo et al., 2012), suggesting that some channel-mediated gliotransmission depends on intracellular Ca^{2+}.

In recent years, astrocytic Ca^{2+} elevations has been described to modulate the extracellular K^+ concentration (Wang et al., 2012). According to Wang et al. (2012), the intracellular elevation of Ca^{2+} by GPCR activation leads to increases of Na^+,Ca^{2+} exchanger (NCX) activity, which in turn elevates Na^+,K^+-ATPase activity. The net effect is a decrease in extracellular K^+ concentration by a few hundred nanomolar with a time course of minutes. Remarkably, such K^+ changes are shown to influence synaptic transmission and plasticity. Since K^+ concentration is the most influential component of the Goldman–Hodgkin–Katz equation to determine resting membrane potential, the extracellular K^+ modulation by astrocytic Ca^{2+} is supposed to have influences on other cell types, including vascular cells.

2.3 ASTROCYTE–VASCULAR INTERACTIONS OBSERVED IN ACUTE BRAIN SLICES

GPCR-triggered astrocytic modulation of vascular tone has been hypothesized by Harder et al. (1998) and first demonstrated in acute rat cortical slice preparations by Zonta et al. (2003). This initial study reported vessel dilation after stimulation of neural activity and subsequent astrocytic Ca^{2+} elevation. It should be noted that most of the experiments were performed in a condition in which vessels were constricted by default by the blocking of nitric oxide (NO) formation. Pharmacological dissection, although not specific to astrocytes, showed the importance of group I metabotropic glutamate receptors (ie, mGluR1 and mGluR5) activation and prostaglandin-forming cyclooxygenase (COX) enzyme activity.

Interestingly, a study in the mouse cortex reported that astrocytic Ca^{2+} elevations cause constriction in adjacent vessels (Mulligan and MacVicar, 2004). Aside from the species difference, it should be noted that the experiments were performed without NO inhibitor pretreatment, so the vessels were presumably dilated by default. This study performed photolysis of caged Ca^{2+} compounds to elevate Ca^{2+} concentrations specifically in astrocytes. The study suggested Ca^{2+}-dependent synthesis of arachidonic acid (AA) in astrocytes and transfer of the AA to smooth muscle cells. Thereafter, the smooth muscle cells produce 20-hydroxyeicosatetraeonic acid (20-HETE) from the AA and constrict the vessels.

Together, these studies suggest the involvement of astrocytic Ca^{2+} signaling in blood vessel diameter regulation (Fig. 1). They also suggest that the initial condition of the experiment could play a significant role in the direction of astrocyte-mediated vessel diameter modulation. Additionally, a later study demonstrated that the oxygenation level of the perfusion solution is one of the key determinants of the polarity of astrocytic Ca^{2+}-mediated vessel diameter regulation (Gordon et al., 2008). Furthermore, the degree of astrocytic Ca^{2+} elevation was reported to influence the direction of vascular tone (Girouard et al., 2010); normal Ca^{2+} elevations lead to vasodilation, whereas high Ca^{2+} elevations (>500 nM) lead to vasoconstriction. While these studies gained insight into how astrocytic Ca^{2+} elevations may modulate local CBF, *in vivo* studies are desired to ultimately verify the astrocytic modulation of local CBF.

2.4 ASTROCYTE–VASCULAR INTERACTIONS OBSERVED WITH *IN VIVO* EXPERIMENTS

In vivo experiments serve as powerful support to validate observations as well as hypotheses reported in acute slice studies since they incorporate the parameters crucial to understanding neuro-glio-vascular interactions such as blood pressure, flow shear stress, body temperature, blood sugar level, and gas tension. Another advantage is that sensory input can be used as a means to provide physiological stimulation. With the advent of two-photon microscopy, direct monitoring of the cerebral capillaries and their blood flow has become feasible (Kleinfeld et al., 1998).

Initial reports that assessed the role of astrocytic Ca^{2+} elevation in local CBF regulation using *in vivo* preparations appeared to support astrocytic involvement. For instance, Ca^{2+} uncaging in astrocyte endfeet were shown to dilate the contacting arteries in the anesthetized mouse cortex, suggesting a causal relationship between astrocytic Ca^{2+} elevation and CBF regulation (Takano et al., 2006). This astrocyte–

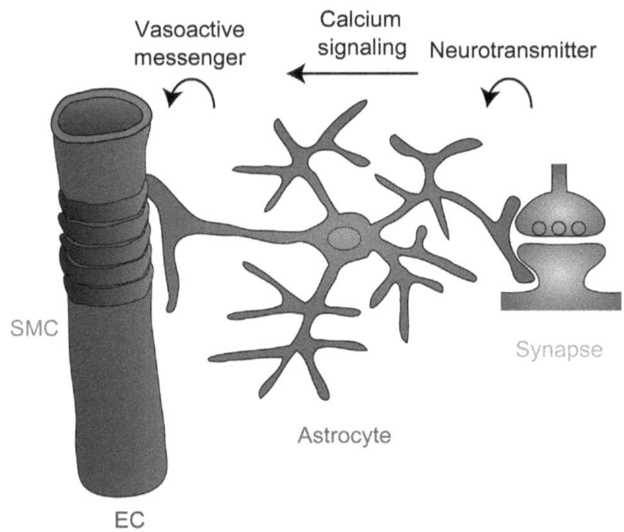

FIG. 1

Controversial involvement of astrocytic Ca^{2+} elevation in vascular tone regulation. Neurotransmitters from synaptic activity signal the astrocyte to elevate its Ca^{2+} levels. Ca^{2+}-induced Ca^{2+} release occurs within the astrocyte and vasomodulators such as arachidonic acid, prostaglandin, and epoxyeicosatrienoic acid are generated. These vasomodulators are then received by vascular SMCs that induce vessel constriction or dilation. This circuit has gained some experimental support, yet other experiments challenge its validity. In addition, this model suggests somewhat a specialized role of the class of astrocytes that contact arterioles, while the majority of astrocytes are associated with the capillaries. *EC*, endothelial cell; *SMC*, smooth muscle cell.

vascular interaction depended on prostaglandin E2 and its metabolizing enzyme COX-1, which was shown to be expressed in high amounts in astrocytic endfeet contacting the penetrating cortical arteries. Another study investigating the mouse olfactory bulb (Petzold et al., 2008) demonstrated that astrocytic Ca^{2+} elevation accompanies odor-induced hyperemia in the corresponding glomerulus. Subsequently, they demonstrated that the mGluR5 antagonist MPEP blocks this Ca^{2+} elevation and halves the functional hyperemia. Moreover, they also showed that COX-1 is involved in the expression of the hyperemia. As the vast majority of the sensory afferents in the mammalian brain are glutamatergic, and sensory-driven astrocytic Ca^{2+} elevations are mGluR5-dependent (Petzold et al., 2008; Wang et al., 2006), glutamate-mediated neuron–astrocyte–vasculature coupling appears to be an attractive model that explains functional hyperemia, the basis of signals obtained using functional magnetic resonance imaging (Fig. 1). On the other hand, synchronized astrocytic Ca^{2+} elevations are correlated with decreased CBF in the hippocampus of living mice, although the downstream pathway has yet to be investigated (Kuga et al., 2011).

Controversy on the functional relevance of astrocytic Ca^{2+} elevations in CBF has arisen as IP_3R2 knockout mice became available. As mentioned earlier, IP_3R2 is predominantly expressed in brain astrocytes and is the major Ca^{2+} channel for the endoplasmic reticulum. Deficiency of IP_3R2 results in the diminishment of large cytosolic Ca^{2+} elevations in astrocytes (Petravicz et al., 2008; Takata et al., 2011); as such, this animal provides an opportunity to test the causal relationship between CBF regulation and astrocytic Ca^{2+} elevations. Three independent studies reported that IP_3R2 knockout mice retain normal hyperemia induced by hindlimb, whisker, or visual stimulation (Bonder and McCarthy, 2014; Nizar et al., 2013; Takata et al., 2013). Similar results were attained in mouse fMRI experiments (Jego et al., 2014). Moreover, the CBF modulation induced by basal forebrain stimulation is not affected in IP_3R2 knockout mice either (Takata et al., 2013). Bonder and McCarthy (2014) additionally performed a gain-of-function experiment in which Gq-GPCR signals are activated exclusively in visual cortical astrocytes by a pharmacogenetical means, producing negative results for astrocytic Ca^{2+}-mediated CBF modulation. Very recently, an acute cortical study using a genetic Ca^{2+} indicator demonstrated that electrical stimulation can change vascular tone without inducing astrocytic Ca^{2+} elevations in the endfeet (Institoris et al., 2015). These experiments pose a challenge to the astrocyte–Ca^{2+}–CBF model.

2.5 ALTERNATIVE PERSPECTIVES TO ASTROCYTIC CALCIUM AND VASCULAR TONE

A recent paper by Rosenegger et al. (2015) suggests an alternative way that astrocytic Ca^{2+} levels are related to vascular tone; in rat cortical slices, they found that the basal Ca^{2+} level of an astrocyte is a determinant of the contacting vascular tone. Depleting the basal intracellular Ca^{2+} in an astrocyte resulted in constriction of the contacting arteriole. This phenomenon was not observed under COX-1 blockade. In the same study, COX-1 blockade in the awake mouse cortex tonically constricted the arterioles, although sensory-induced hyperemia was not affected. Such a mechanism might account for the slow increase of CBF in a bicuculline-induced hyperactivity focus (Hirase et al., 2004a) since neural activity affects basal astrocytic Ca^{2+} levels (Jackson and Robinson, 2015). According to Zheng et al. (2015), the basal level of *in vivo* astrocytic Ca^{2+} is on average around 100 nM with a distribution of 50–200 nM. Future studies should identify the degree of basal Ca^{2+} decrease by which significant arteriole constriction is induced and if such a Ca^{2+} level reduction is physiologically feasible.

While large somatic Ca^{2+} elevations in astrocytes are almost completely diminished in IP_3R2 knockout mice, Ca^{2+} elevations in astrocytic microprocesses are observed with a reduced frequency in these mice. Using a transgenic mouse that expresses the highly sensitive genetic Ca^{2+} indicator protein YC-Nano50, Kanemaru et al. (2014) reported that *in vivo* Ca^{2+} event frequency of astrocytic microprocesses is reduced to one quarter with IP_3R2 gene deletion. Moreover, the study reported that spontaneous Ca^{2+} elevations in astrocytic endfeet are extremely rare in both wild-type and IP_3R2 knockout mouse. On the other hand, using the virally expressed GCaMP6f Ca^{2+} indicator, Srinivasan et al. (2015) reported only a marginal decline in microprocess Ca^{2+} event frequency. Moreover, this study reported that Ca^{2+} fluctuations in astrocytic endfeet persist in IP_3R2 knockout mice. Despite the IP_3R2 knockout mouse studies, recent *in vivo* studies show correlated activities of endfoot Ca^{2+} activity and CBF modulations (Lind et al., 2013; Otsu et al., 2015). It remains possible that the persistence of functional hyperemia in IP_3R2 knockout mice is due to the residual Ca^{2+} elevations in astrocytic microprocesses and endfeet.

Causality between astrocytic Ca^{2+} elevation and CBF changes has been supposed by comparison of their time courses. In most studies, sensory-driven astrocytic Ca^{2+} takes at least a few seconds, similar to the time course of hemodynamic changes. In some studies, Ca^{2+} responses from astrocytic branches or microprocesses are reported to be as fast as that of neurons, which occur within a second (Dombeck et al., 2007; Jourdain et al., 2007; Lind et al., 2013; Winship et al., 2007). However, one must be aware that the depth resolution of a two-photon microscope can be as large as a few micrometers in suboptimal conditions such as *in vivo* observation with a beam underfill regime. Therefore, two-photon imaging of mixed labeling of neuronal and astrocytic processes with Ca^{2+} sensors could include the signals above or below the intended depth. We must also be aware of the caveat recently reported by Wang et al. (2013); ultraviolet flash photolysis of caged compounds in astrocytes causes a spurious glutamate release and hence distinguished from Ca^{2+} elevations caused by GPCR activation. Finally, high-intensity two-photon excitation light itself can induce cellular activation (Hirase et al., 2002; Kuga et al., 2011; Wang et al., 2006), and TRP channels are gated by temperature, which might increase at the focal spot when a slow scan is performed.

3 METABOLIC NEUROVASCULAR COUPLING MEDIATED BY ASTROCYTES

Although the active role of astrocytes in CBF modulation remains controversial as far as the intracellular Ca^{2+} elevation is concerned, astrocytic involvement in neurovascular coupling may be effectively addressed from the viewpoints of metabolites and ion flow. For example, it has been nearly two decades since the lactate shuttle hypothesis was proposed for energy substrate transfer between CBF and neurons (for a review, Magistretti et al., 1999). In the lactate shuttle hypothesis, astrocytes synthesize lactate via the glycolysis of glucose and/or glucose-6-phosphate from CBF and glycogenolysis, respectively. Astrocytic lactate is then released to the extracellular space and taken up by neurons through monocarboxylate transporters. Many experiments lend credence to this model (for a review, Bélanger et al., 2011), although some concerns have been raised recently (Lundgaard et al., 2015).

An important feature of astrocytes in the metabolic neurovascular coupling is that neighboring astrocytes are physically connected with gap junctions to form analog networks called astrocyte syncytium (Giaume et al., 2010). Gap junctions allow communication of the connected astrocytes by exchanges of small molecules approximately <1000 Da. This form of signaling allows astrocytes to communicate in analog fashion over distances by diffusion and a chemical gradient. Indeed, using mouse hippocampal slices, a study demonstrated that glucose is transported over distances to support neuronal activity via gap junction-mediated astrocytic syncytium (Rouach et al., 2008). Importantly, the spread of glucose was shown to be neuronal (synaptic) activity dependent. Such an activity-dependent supply of glucose matches the proposed role of astrocytes in neurovascular coupling, namely, astrocytes as a mediator of neuronal energy metabolism. Accordingly, astrocytes constitute a special highway that connects the neurons with the vasculature. Furthermore, as the efficiency of gap junctional communications is regulated in physiological and pathophysiological conditions (Giaume and Liu, 2012), astrocytes may play a key role in regulating neurovascular coupling through changes in gap junctional coupling. While the effect exerted by a single astrocyte on the vasculature could be small, it is tempting to postulate that a mass action of the astrocytic syncytium may be capable of exerting visible effects on the vasculature (Fig. 2).

4 K^+ AND WATER REDISTRIBUTION THROUGH ASTROCYTES

In addition to metabolites, astrocytes are highly capable of sensing and handling extracellular ionic changes since they express rich amounts of various ion channels and transporters in their highly ramified membranes (Verkhratsky and Nedergaard, 2014). Indeed, a role of astrocyte has been proposed as to redistribute neuronal activity-induced extracellular K^+ elevations to the vasculature (Newman et al., 1984). This "potassium siphoning" was originally proposed to occur by an action of K^+-induced depolarization and high K^+ channel expression at the endfeet to dilate neighboring arterioles (Paulson and Newman, 1987). Although this model requires revision as neither simple astrocytic depolarization nor Kir4.1 knockout changed the responses of the retinal arterioles (Metea et al., 2007), extracellular K^+ is still considered a strong candidate for vascular tone control (see Chapter 3). For instance, astrocytic endfeet express BK-type K^+ channels, which are activated by intracellular Ca^{2+} elevation and cause K^+ efflux. Such an extracellular K^+ efflux, possibly with additional messengers, has been shown to hyperpolarize smooth muscle cells and dilate arterioles (Filosa et al., 2006). Combined with the astrocytic Ca^{2+}-dependent uptake of extracellular K^+ (Wang et al., 2012), this pathway could provide a major mechanism that redistributes excess extracellular K^+ from the neuropil to the perivascular space (Fig. 3). Importantly, extracellular K^+ increase elevates astrocytic cyclic adenosine monophosphate (cAMP) levels, which in turn promotes lactate generation by glycolysis (Choi et al., 2012), suggesting a functional coupling among neural activity, astrocyte metabolism, and vascular tone.

In considering ionic environment, one cannot neglect the effect of water homeostasis in astrocytes. Previous studies have revealed that extracellular volume in the

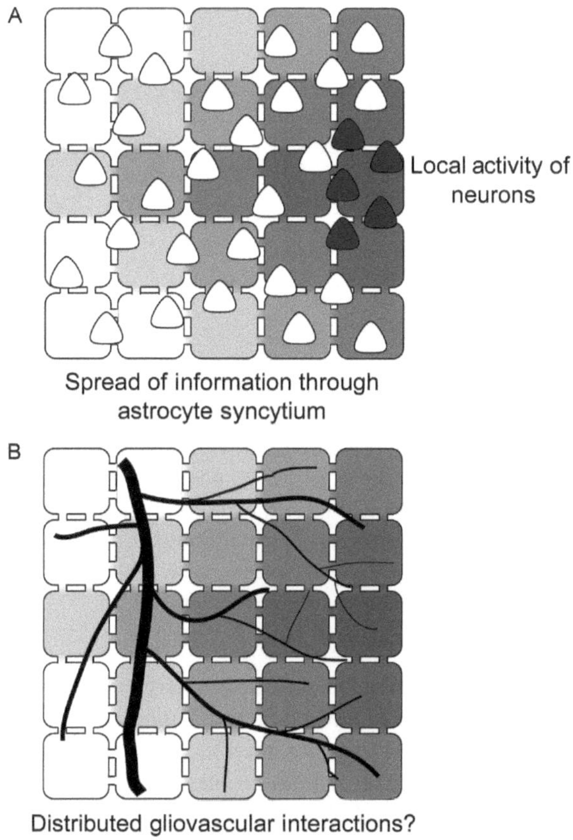

FIG. 2

Astrocyte syncytium in neurovascular coupling. Neighboring astrocytes are connected by gap junctions and form a large cellular assembly called the astrocyte syncytium (*squares*) that is overlaid on the neuronal network (*triangles*). Due to the analog nature of the gap junction coupling, information of a single astrocyte (eg, induced by local neuronal activity shown in *yellow* (dark gray in the print version)) passively spread for distances over the syncytium (A), which then induce distributed gliovascular interactions with associated vasculatures (B). Gap junction efficacy is regulated by the surrounding neural activity and other means like pH, which possibly provides a basis for a prioritized metabolic supply from the cerebral blood flow to the region of high neural activity.

brain parenchyma can change dramatically under both physiological and pathological conditions (Syková and Nicholson, 2008), which are inevitably accompanied by water redistributions. Astrocytes are considered the main cell type involved in water homeostasis since they are the major cells that express water channel aquaporins in the brain (Papadopoulos and Verkman, 2013). Astrocytes express aquaporin 4 (AQP4) that is specifically enriched at the endfeet, suggesting a functional role with

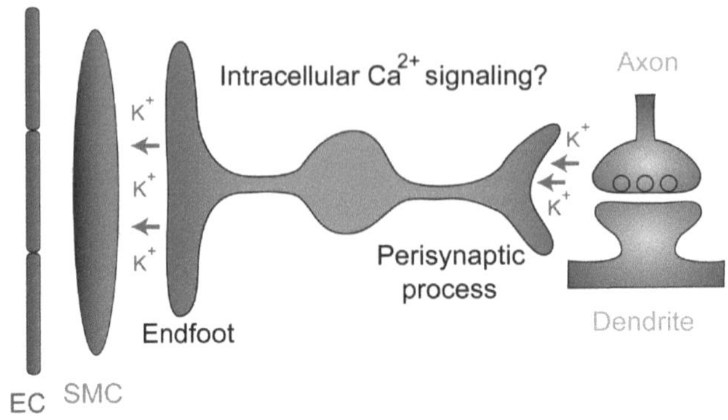

FIG. 3

Extracellular K^+ regulation in the neuro-glio-vascular circuit. Synaptic activity elevates the extracellular K^+ level by action potential-associated K^+ efflux from the axon and synaptic *N*-methyl-D-aspartate receptor (NMDAR). Excess K^+ enters the astrocyte and is redistributed to the extracellular space at the vascular end. A recent work suggests that astrocytic Ca^{2+} elevations promote Na–K–ATPase activity; hence, K^+ influx occurs in the astrocyte. Other studies showed the recruitment of BK-type K^+ channels with astrocytic Ca^{2+} elevation. These pathways provide a potential mechanism for K^+ siphoning.

the vasculature (Amiry-Moghaddam et al., 2004). Interestingly, AQP4 proteins at the endfeet are arranged in a highly ordered structure called orthogonal arrays of particles (OAP). OAP are considered to be tethered to the dystrophin–protein complex, which is tightly linked to the basal lamina around the blood vessels (Waite et al., 2012). The water permeability of AQP4 is regulated by its phosphorylation and other metal ion-dependent mechanisms (for a review, Yukutake and Yasui, 2010), although the significance of such regulations in live animals has yet to be examined in the brain. The regulation of water flux and homeostasis is proposed to impact extracellular concentrations of ions and chemical messengers. In particular, extracellular K^+ clearance and astrocytic K^+ buffering are compromised in AQP4 knockout mice (Binder et al., 2006; Strohschein et al., 2011). Also, the extracellular space increases in AQP4 knockout mice (Papadopoulos and Verkman, 2005; Yao et al., 2008), possibly due to diminished water flux through the astrocytes.

The significance of water flux by AQP4 was recently highlighted in a proposed mechanism of interstitial fluid exchange by a series of work from Nedergaard's group. By injecting fluorescent dyes into the cisterna magna, a reservoir of cerebrospinal fluid (CSF), they observed that the fluorescent dyes in the cortical parenchyma spread from the arterial perivascular space to the venular side (Iliff et al., 2012). In the same study, this flow system was shown to accelerate the clearance of extracellular soluble proteins via AQP4. Accordingly, they proposed that AQP4 mediates convective water flow through astrocytes that promotes the turnover of interstitial fluid and washout of extracellular protein wastes. Drawing an analogy to the lymphatic system available in other body parts, this flow system is termed the glymphatic system. The subsequent work showed that the driving force of the fluid transport to the interstitial space is the arterial pulsation (Iliff et al., 2013). While the initial study by Iliff et al. (2012) was performed in anesthetized mice, a later study using unanesthetized mice revealed that the glymphatic system is more active during sleep than awake periods (Xie et al., 2013). As the glymphatic system has been shown effective for the clearance of protein wastes and neurotoxic products including amyloid-β, this observation suggested a new biological significance of sleep. Furthermore, noradrenaline, a potent messenger that act on astrocytes *in vivo* (Bekar et al., 2008; Ding et al., 2013; Paukert et al., 2014), is shown to be the main neuromodulator that controls the glymphatic system (Xie et al., 2013). Although the exact mechanism of how noradrenaline controls the glymphatic system is not yet clear, the extracellular volume change associated with noradrenaline is thought to be a potent mechanism (Jessen et al., 2015).

5 STRUCTURAL SPECIALIZATION OF ASTROCYTIC ENDFEET

It is often assumed that metabolites and other molecules inside an astrocyte diffuse freely. Given the intricate morphology of astrocytes, this picture is undoubtedly over-simplified and chemical signaling inside astrocytes need to be considered in the context of their complex structures. Diffusion rates in various parts of an astrocyte were investigated by the imaging of uncaged fluorescent molecules or fluorescence recovery after photobleaching. As a result, the diffusion rate was found to be an order of magnitude lower in the endfoot than in the soma or other processes (Fig. 4; Nuriya and Yasui, 2013). In fact, the fluorescent molecules were retained for tens of seconds inside the endfeet. This observation leads to the idea that the endfoot is a diffusion-limited subcellular compartment. Considering the function of endfeet as the interface to intake metabolites from the CBF, the slow diffusion hints to a specialized biochemical process in this compartment.

In this respect, it is noteworthy that astrocyte endfeet are enriched with functional proteins that enable specific biochemical processes at the gliovascular interface. These include the highly enriched water channel AQP4 and the potassium channel Kir4.1 (Bragg et al., 2006; Nagelhus et al., 2004) as well as more widely localized proteins such as the TRPV4 cation channel (Benfenati et al., 2007). By leveraging this structurally and functionally specialized interface, astrocytes can communicate with vascular structures in a highly specialized manner. For instance, a recent study reported that blood vessels affects astrocyte physiology via pressure-mediated signaling sensed by TRPV4 (Kim et al., 2015). In this case, vascular flow/pressure in the arterioles directly acts on biochemical pathways presumably at the endfeet, which could in turn initiate interactions with vascular cells. Intriguingly, functional interactions between AQP4 and TRPV4 have been reported (Benfenati et al., 2011; Jo et al., 2015; Mola et al., 2015), suggesting the potential convergence of Ca^{2+} signaling and water/ion homeostasis at the endfeet. In this way, the structural specializations of astrocytes support dynamic bidirectional signaling between astrocytes and blood vessels.

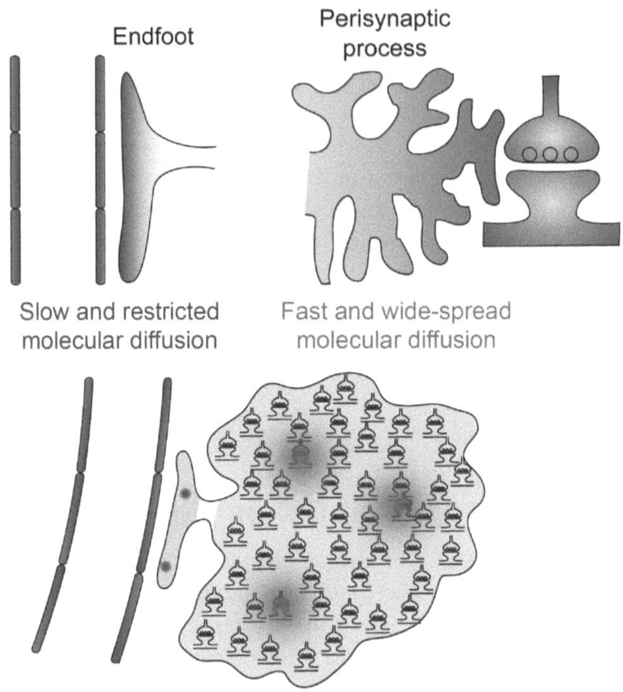

FIG. 4

Two modes of molecular diffusion inside an astrocyte. The diffusion of molecules is spatially limited and distinctively slower in the endfoot, which effectively creates specialized biochemical compartments within it. In contrast, diffusion is fast and widespread in the rest of the processes that surround the synapses within the astrocyte domain. Such free diffusion can buffer rapid and large changes in the extracellular environment.

In sharp contrast to the endfeet, Nuriya and Yasui (2013) found that molecular diffusions are much faster in the rest of astrocytes, including the fine processes that have mutual communication with tens of thousands of axons and dendritic spines as tripartite synapses (Halassa et al., 2007; Ogata and Kosaka, 2002). Interestingly, this differential diffusion pattern between endfeet and other parts of the astrocytes was observed with biologically innate fluorescent molecules of different sizes as well as the bioactive glucose analog 2-NBDG, suggesting that this principle applies to a wide variety of molecules in astrocytes. In the context of astrocytic metabolic coupling to neurons and the vasculature, it suggests that astrocytes can integrate dynamically changing information from large numbers of synapses, and convey this information to blood vessels (Fig. 4). The idea of astrocytes as spatial and temporal integrators can be further extended to the gap junction-mediated astrocytic syncytium, which also provides the regulation capacity. Taken together, astrocytes capitalize on their complex morphology as well as multicellular syncytium to achieve efficient and dynamic neurovascular coupling (Fig. 2).

6 FUTURE PERSPECTIVES: IN RELATION TO NEUROPATHOLOGY

In the last few decades, the field of neurology has changed dramatically owing to better understanding of the molecular pathophysiology of neurons. As the genomic information of humans as well as other organisms has become widely available, numerous genetic risk factors have been identified for different diseases and attempts have been made to explain the pathology from the perspective of neuronal changes resulting from these genetic factors. Recently, however, increasing numbers of neurological diseases have also been linked to pathological changes of astrocytes.

Morphological changes of astrocytes have been recognized for a long time in various pathological conditions including stroke and Alzheimer's disease. These observations were largely considered hallmarks of astrocytic responses to neuronal dysfunction. In fact, astrocytes do respond to pathological signals and become reactive, presumably with an initial purpose of recovering the damaged tissue. However, reactive astrocytes can result in aggravation in later stages. For example, reduced Kir4.1 expression is reported in mouse models of Huntington's disease (Tong et al., 2014). Of particular interest in light of neurovascular coupling are the recent studies reporting that vascular contacts made by astrocytic endfeet are compromised in glioma and Alzheimer's disease models, resulting in impaired astrocyte-Ca^{2+}-mediated CBF regulation (Kimbrough et al., 2015; Watkins et al., 2014). Moreover, alternations in AQP4 level or localization are frequently observed in many neurological disorders (Badaut et al., 2014). As age-dependent AQP4 decreases are reported to be related to immune responses involving the complement cascade (Soto et al., 2015), pathological changes may include glial cell types other than astrocytes, such as microglia. All of these changes affect neuro-glio-vascular coupling and potentially the CSF drainage system as discussed earlier.

In most cases, we face a chicken-and-egg problem; does astrocyte pathophysiology lead to neuronal and vascular malfunctions or vice versa. Refinement of disease models with noninvasive observations and manipulations will likely accelerate the assessment of the cause and mechanism. Given the intricate and mutual interactions among neurons, glia, and the vasculature, it is also conceivable that all the three components interdependently contribute to the development of a neurological condition. Either way, as astrocytes lie closer to CBF than neurons, astrocytes can be a potent first-order target for systemic drug treatment for neurological and psychiatric disorders.

ACKNOWLEDGMENTS

This work was supported by the RIKEN Brain Science Institute, KAKENHI grants (23115522, 23657106, 26117520, 26650052), and HFSP (RGP0036/2014). We thank members of the laboratory for comments on earlier versions of the manuscript. The authors declare no competing financial interests.

REFERENCES

Agulhon, C., Boyt, K.M., Xie, A.X., Friocourt, F., Roth, B.L., McCarthy, K.D., 2013. Modulation of the autonomic nervous system and behaviour by acute glial cell Gq protein-coupled receptor activation *in vivo*. J. Physiol. 591, 5599–5609.

Ahmed, Z., Lewis, C.A., Faber, D.S., 1990. Glutamate stimulates release of Ca^{2+} from internal stores in astroglia. Brain Res. 516, 165–169.

Amiry-Moghaddam, M., Frydenlund, D.S., Ottersen, O.P., 2004. Anchoring of aquaporin-4 in brain: molecular mechanisms and implications for the physiology and pathophysiology of water transport. Neuroscience 129, 999–1010.

Amzica, F., Massimini, M., 2002. Glial and neuronal interactions during slow wave and paroxysmal activities in the neocortex. Cereb. Cortex 12, 1101–1113.

Badaut, J., Fukuda, A.M., Jullienne, A., Petry, K.G., 2014. Aquaporin and brain diseases. Biochim. Biophys. Acta 1840, 1554–1565.

Bekar, L.K., He, W., Nedergaard, M., 2008. Locus coeruleus alpha-adrenergic-mediated activation of cortical astrocytes *in vivo*. Cereb. Cortex 18, 2789–2795.

Bélanger, M., Allaman, I., Magistretti, P.J., 2011. Brain energy metabolism: focus on astrocyte-neuron metabolic cooperation. Cell Metab. 14, 724–738.

Benfenati, V., Amiry-Moghaddam, M., Caprini, M., Mylonakou, M.N., Rapisarda, C., Ottersen, O.P., Ferroni, S., 2007. Expression and functional characterization of transient receptor potential vanilloid-related channel 4 (TRPV4) in rat cortical astrocytes. Neuroscience 148, 876–892.

Benfenati, V., Caprini, M., Dovizio, M., Mylonakou, M.N., Ferroni, S., Ottersen, O.P., Amiry-Moghaddam, M., 2011. An aquaporin-4/transient receptor potential vanilloid 4 (AQP4/TRPV4) complex is essential for cell-volume control in astrocytes. Proc. Natl. Acad. Sci. U. S. A. 108, 2563–2568.

Binder, D.K., Yao, X., Zador, Z., Sick, T.J., Verkman, A.S., Manley, G.T., 2006. Increased seizure duration and slowed potassium kinetics in mice lacking aquaporin-4 water channels. Glia 53, 631–636.

Bonder, D.E., McCarthy, K.D., 2014. Astrocytic Gq-GPCR-linked IP3R-dependent Ca^{2+} signaling does not mediate neurovascular coupling in mouse visual cortex *in vivo*. J. Neurosci. 34, 13139–13150.

Bragg, A.D., Amiry-Moghaddam, M., Ottersen, O.P., Adams, M.E., Froehner, S.C., 2006. Assembly of a perivascular astrocyte protein scaffold at the mammalian blood–brain barrier is dependent on alpha-syntrophin. Glia 53, 879–890.

Choi, H.B., Gordon, G.R.J., Zhou, N., Tai, C., Rungta, R.L., Martinez, J., Milner, T.A., Ryu, J.K., McLarnon, J.G., Tresguerres, M., Levin, L.R., Buck, J., MacVicar, B.A., 2012. Metabolic communication between astrocytes and neurons via bicarbonate-responsive soluble adenylyl cyclase. Neuron 75, 1094–1104.

Cornell-Bell, A.H., Finkbeiner, S.M., Cooper, M.S., Smith, S.J., 1990. Glutamate induces calcium waves in cultured astrocytes: long-range glial signaling. Science 247 (4941), 470–473.

Dani, J.W., Chernjavsky, A., Smith, S.J., 1992. Neuronal activity triggers calcium waves in hippocampal astrocyte networks. Neuron 8, 429–440.

Denk, W., Strickler, J.H., Webb, W.W., 1990. Two-photon laser scanning fluorescence microscopy. Science 248 (4951), 73–76.

Di Castro, M.A., Chuquet, J., Liaudet, N., Bhaukaurally, K., Santello, M., Bouvier, D., Tiret, P., Volterra, A., 2011. Local Ca^{2+} detection and modulation of synaptic release by astrocytes. Nat. Neurosci. 14, 1276–1284.

Ding, F., O'Donnell, J., Thrane, A.S., Zeppenfeld, D., Kang, H., Xie, L., Wang, F., Nedergaard, M., 2013. α1-Adrenergic receptors mediate coordinated Ca(2+) signaling of cortical astrocytes in awake, behaving mice. Cell Calcium 54, 387–394.

Dombeck, D.A., Khabbaz, A.N., Collman, F., Adelman, T.L., Tank, D.W., 2007. Imaging large-scale neural activity with cellular resolution in awake, mobile mice. Neuron 56, 43–57.

Fiacco, T.A., McCarthy, K.D., 2004. Intracellular astrocyte calcium waves *in situ* increase the frequency of spontaneous AMPA receptor currents in CA1 pyramidal neurons. J. Neurosci. 24, 722–732.

Filosa, J.A., Bonev, A.D., Straub, S.V., Meredith, A.L., Wilkerson, M.K., Aldrich, R.W., Nelson, M.T., 2006. Local potassium signaling couples neuronal activity to vasodilation in the brain. Nat. Neurosci. 9, 1397–1403.

Fujita, T., Chen, M.J., Li, B., Smith, N.A., Peng, W., Sun, W., Toner, M.J., Kress, B.T., Wang, L., Benraiss, A., Takano, T., Wang, S., Nedergaard, M., 2014. Neuronal transgene expression in dominant-negative SNARE mice. J. Neurosci. 34, 16594–16604.

Giaume, C., Koulakoff, A., Roux, L., Holcman, D., Rouach, N., 2010. Astroglial networks: a step further in neuroglial and gliovascular interactions. Nat. Rev. Neurosci. 11, 87–99.

Giaume, C., Liu, X., 2012. From a glial syncytium to a more restricted and specific glial networking. J. Physiol. Paris 106, 34–39.

Girouard, H., Bonev, A.D., Hannah, R.M., Meredith, A., Aldrich, R.W., Nelson, M.T., 2010. Astrocytic endfoot Ca^{2+} and BK channels determine both arteriolar dilation and constriction. Proc. Natl. Acad. Sci. U. S. A. 107, 3811–3816.

Glaum, S.R., Holzwarth, J.A., Miller, R.J., 1990. Glutamate receptors activate Ca^{2+} mobilization and Ca^{2+} influx into astrocytes. Proc. Natl. Acad. Sci. U. S. A. 87, 3454–3458.

Gordon, G.R., Choi, H.B., Rungta, R.L., Ellis-Davies, G.C., MacVicar, B.A., 2008. Brain metabolism dictates the polarity of astrocyte control over arterioles. Nature 456, 745–749.

Guthrie, P.B., Knappenberger, J., Segal, M., Bennett, M.V., Charles, A.C., Kater, S.B., 1999. ATP released from astrocytes mediates glial calcium waves. J. Neurosci. 19, 520–528.

Halassa, M.M., Fellin, T., Takano, H., Dong, J.H., Haydon, P.G., 2007. Synaptic islands defined by the territory of a single astrocyte. J. Neurosci. 27, 6473–6477.

Harder, D.R., Alkayed, N.J., Lange, A.R., Gebremedhin, D., Roman, R.J., 1998. Functional hyperemia in the brain: hypothesis for astrocyte-derived vasodilator metabolites. Stroke 29, 229–234.

Helen, C., Kastritsis, C., Salm, A.K., McCarthy, K., 1992. Stimulation of the P 2Y purinergic receptor on type 1 astroglia results in inositol phosphate formation and calcium mobilization. J. Neurochem. 58, 1277–1284.

Hirase, H., Creso, J., Buzsaki, G., 2004a. Capillary level imaging of local cerebral blood flow in bicuculline-induced epileptic foci. Neuroscience 128, 209–216.

Hirase, H., Nikolenko, V., Goldberg, J.H., Yuste, R., 2002. Multiphoton stimulation of neurons. J. Neurobiol. 51, 237–247.

Hirase, H., Qian, L., Bartho, P., Buzsaki, G., 2004b. Calcium dynamics of cortical astrocytic networks *in vivo*. PLoS Biol. 2, E96.

Hoogland, T.M., Kuhn, B., Gobel, W., Huang, W., Nakai, J., Helmchen, F., Flint, J., Wang, S.S., 2009. Radially expanding transglial calcium waves in the intact cerebellum. Proc. Natl. Acad. Sci. U. S. A. 106, 3496–3501.

Iliff, J.J., Wang, M., Liao, Y., Plogg, B.A., Peng, W., Gundersen, G.A., Benveniste, H., Vates, G.E., Deane, R., Goldman, S.A., Nagelhus, E.A., Nedergaard, M., 2012. A paravascular pathway facilitates CSF flow through the brain parenchyma and the clearance of interstitial solutes, including amyloid β. Sci. Transl. Med. 4, 147ra111.

Iliff, J.J., Wang, M., Zeppenfeld, D.M., Venkataraman, A., Plog, B.A., Liao, Y., Deane, R., Nedergaard, M., 2013. Cerebral arterial pulsation drives paravascular CSF-interstitial fluid exchange in the murine brain. J. Neurosci. 33, 18190–18199.

Institoris, Á., Rosenegger, D.G., Gordon, G.R., 2015. Arteriole dilation to synaptic activation that is sub-threshold to astrocyte endfoot Ca(2+) transients. J. Cereb. Blood Flow Metab. 35, 1411–1415.

Jackson, J.G., Robinson, M.B., 2015. Reciprocal regulation of mitochondrial dynamics and calcium signaling in astrocyte processes. J. Neurosci. 35, 15199–15213.

Jego, P., Pacheco-Torres, J., Araque, A., Canals, S., 2014. Functional MRI in mice lacking IP3-dependent calcium signaling in astrocytes. J. Cereb. Blood Flow Metab. 34, 1599–1603.

Jensen, A.M., Chiu, S.Y., 1990. Fluorescence measurement of changes in intracellular calcium induced by excitatory amino acids in cultured cortical astrocytes. J. Neurosci. 10, 1165–1175.

Jessen, N.A., Munk, A.S.F., Lundgaard, I., Nedergaard, M., 2015. The glymphatic system: a beginner's guide. Neurochem. Res. 40 (12), 2583–2599.

Jo, A.O., Ryskamp, D.A., Phuong, T.T.T., Verkman, A.S., Yarishkin, O., MacAulay, N., Križaj, D., 2015. TRPV4 and AQP4 channels synergistically regulate cell volume and calcium homeostasis in retinal Müller glia. J. Neurosci. 35, 13525–13537.

Jourdain, P., Bergersen, L.H., Bhaukaurally, K., Bezzi, P., Santello, M., Domercq, M., Matute, C., Tonello, F., Gundersen, V., Volterra, A., 2007. Glutamate exocytosis from astrocytes controls synaptic strength. Nat. Neurosci. 10, 331–339.

Kanemaru, K., Sekiya, H., Xu, M., Satoh, K., Kitajima, N., Yoshida, K., Okubo, Y., Sasaki, T., Moritoh, S., Hasuwa, H., Mimura, M., Horikawa, K., Matsui, K., Nagai, T., Iino, M., Tanaka, K.F., 2014. *In vivo* visualization of subtle, transient, and local activity of astrocytes using an ultrasensitive Ca(2+) indicator. Cell Rep. 8, 311–318.

Kettenmann, H., Verkhratsky, A., 2008. Neuroglia: the 150 years after. Trends Neurosci. 31, 653–659.

Kim, K.J., Iddings, J.A., Stern, J.E., Blanco, V.M., Croom, D., Kirov, S.A., Filosa, J.A., 2015. Astrocyte contributions to flow/pressure-evoked parenchymal arteriole vasoconstriction. J. Neurosci. 35, 8245–8257.

Kimbrough, I.F., Robel, S., Roberson, E.D., Sontheimer, H., 2015. Vascular amyloidosis impairs the gliovascular unit in a mouse model of Alzheimer's disease. Brain 138, 3716–3733.

Kleinfeld, D., Mitra, P.P., Helmchen, F., Denk, W., 1998. Fluctuations and stimulus-induced changes in blood flow observed in individual capillaries in layers 2 through 4 of rat neocortex. Proc. Natl. Acad. Sci. U. S. A. 95, 15741–15746.

Kuga, N., Sasaki, T., Takahara, Y., Matsuki, N., Ikegaya, Y., 2011. Large-scale calcium waves traveling through astrocytic networks *in vivo*. J. Neurosci. 31, 2607–2614.

Lind, B.L., Brazhe, A.R., Jessen, S.B., Tan, F.C.C., Lauritzen, M.J., 2013. Rapid stimulus-evoked astrocyte Ca^{2+} elevations and hemodynamic responses in mouse somatosensory cortex *in vivo*. Proc. Natl. Acad. Sci. U. S. A. 110, E4678–E4687.

Lloyd, G., 2007. Pneuma between body and soul. J. R. Anthropol. Inst. 13, S135–S146.

Lundgaard, I., Li, B., Xie, L., Kang, H., Sanggaard, S., Haswell, J.D.R., Sun, W., Goldman, S., Blekot, S., Nielsen, M., Takano, T., Deane, R., Nedergaard, M., 2015. Direct neuronal glucose uptake heralds activity-dependent increases in cerebral metabolism. Nat. Commun. 6, 6807.

Magistretti, P.J., Pellerin, L., Rothman, D.L., Shulman, R.G., 1999. Energy on demand. Science 283 (5401), 496–497.

Masamoto, K., Unekawa, M., Watanabe, T., Toriumi, H., Takuwa, H., Kawaguchi, H., Kanno, I., Matsui, K., Tanaka, K.F., Tomita, Y., Suzuki, N., 2015. Unveiling astrocytic control of cerebral blood flow with optogenetics. Sci. Rep. 5, 11455.

Metea, M.R., Kofuji, P., Newman, E.A., 2007. Neurovascular coupling is not mediated by potassium siphoning from glial cells. J. Neurosci. 27, 2468–2471.

Mishima, T., Hirase, H., 2010. *In vivo* intracellular recording suggests that gray matter astrocytes in mature cerebral cortex and hippocampus are electrophysiologically homogeneous. J. Neurosci. 30, 3093–3100.

Mishima, T., Sakatani, S., Hirase, H., 2007. Intracellular labeling of single cortical astrocytes *in vivo*. J. Neurosci. Methods 166, 32–40.

Mola, M.G., Sparaneo, A., Gargano, C.D., Spray, D.C., Svelto, M., Frigeri, A., Scemes, E., Paola Nicchia, G., 2015. The speed of swelling kinetics modulates cell volume regulation and calcium signaling in astrocytes: a different point of view on the role of aquaporins. Glia 64 (1), 139–154.

Montana, V., Malarkey, E.B., Verderio, C., Matteoli, M., Parpura, V., 2006. Vesicular transmitter release from astrocytes. Glia 54, 700–715.

Mosso, A., 1880. Sulla circolazione del sangue nel cervello dell'uomo (On the circulation of blood in the human brain). R. Accad. Lincei 5, 237–358.

Mulligan, S.J., MacVicar, B.A., 2004. Calcium transients in astrocyte endfeet cause cerebrovascular constrictions. Nature 431, 195–199.

Nagelhus, E.A., Mathiisen, T.M., Ottersen, O.P., 2004. Aquaporin-4 in the central nervous system: cellular and subcellular distribution and coexpression with KIR4.1. Neuroscience 129, 905–913.

Nakahara, K., Okada, M., Nakanishi, S., 2002. The metabotropic glutamate receptor mGluR5 induces calcium oscillations in cultured astrocytes via protein kinase C phosphorylation. J. Neurochem. 69, 1467–1475.

Nedergaard, M., 1994. Direct signaling from astrocytes to neurons in cultures of mammalian brain cells. Science 263 (5154), 1768–1771.

Newman, E.A., Frambach, D.A., Odette, L.L., 1984. Control of extracellular potassium levels by retinal glial cell K+ siphoning. Science 225 (4667), 1174–1175.

Nimmerjahn, A., Mukamel, E.A., Schnitzer, M.J., 2009. Motor behavior activates Bergmann glial networks. Neuron 62, 400–412.

Nizar, K., et al., 2013. *In vivo* stimulus-induced vasodilation occurs without IP3 receptor activation and may precede astrocytic calcium increase. J. Neurosci. 33, 8411–8422.

Nuriya, M., Yasui, M., 2013. Endfeet serve as diffusion-limited subcellular compartments in astrocytes. J. Neurosci. 33, 3692–3698.

Ogata, K., Kosaka, T., 2002. Structural and quantitative analysis of astrocytes in the mouse hippocampus. Neuroscience 113, 221–233.

Otsu, Y., Couchman, K., Lyons, D.G., Collot, M., Agarwal, A., Mallet, J.-M., Pfrieger, F.W., Bergles, D.E., Charpak, S., 2015. Calcium dynamics in astrocyte processes during neurovascular coupling. Nat. Neurosci. 18, 210–218.

Panatier, A., Arizono, M., Nägerl, U.V., 2014. Dissecting tripartite synapses with STED microscopy. Philos. Trans. R. Soc. Lond. B Biol. Sci. 369, 20130597.

Papadopoulos, M.C., Verkman, A.S., 2005. Aquaporin-4 gene disruption in mice reduces brain swelling and mortality in pneumococcal meningitis. J. Biol. Chem. 280, 13906–13912.

Papadopoulos, M.C., Verkman, A.S., 2013. Aquaporin water channels in the nervous system. Nat. Rev. Neurosci. 14, 265–277.

Parpura, V., Basarsky, T.A., Liu, F., Jeftinija, K., Jeftinija, S., Haydon, P.G., 1994. Glutamate-mediated astrocyte-neuron signalling. Nature 369, 744–747.

Pascual, O., Casper, K.B., Kubera, C., Zhang, J., Revilla-Sanchez, R., Sul, J.Y., Takano, H., Moss, S.J., McCarthy, K., Haydon, P.G., 2005. Astrocytic purinergic signaling coordinates synaptic networks. Science 310 (5745), 113–116.

Paukert, M., Agarwal, A., Cha, J., Doze, V.A., Kang, J.U., Bergles, D.E., 2014. Norepinephrine controls astroglial responsiveness to local circuit activity. Neuron 82, 1263–1270.

Paulson, O.B., Newman, E.A., 1987. Does the release of potassium from astrocyte endfeet regulate cerebral blood flow? Science 237, 896–898.

Pearce, B., Albrecht, J., Morrow, C., Murphy, S., 1986. Astrocyte glutamate receptor activation promotes inositol phospholipid turnover and calcium flux. Neurosci. Lett. 72, 335–340.

Perea, G., Yang, A., Boyden, E.S., Sur, M., 2014. Optogenetic astrocyte activation modulates response selectivity of visual cortex neurons *in vivo*. Nat. Commun. 5, 3262.

Petravicz, J., Fiacco, T.A., McCarthy, K.D., 2008. Loss of IP3 receptor-dependent Ca^{2+} increases in hippocampal astrocytes does not affect baseline CA1 pyramidal neuron synaptic activity. J. Neurosci. 28, 4967–4973.

Petzold, G.C., Albeanu, D.F., Sato, T.F., Murthy, V.N., 2008. Coupling of neural activity to blood flow in olfactory glomeruli is mediated by astrocytic pathways. Neuron 58, 897–910.

Rosenegger, D.G., Tran, C.H.T., Wamsteeker Cusulin, J.I., Gordon, G.R., 2015. Tonic local brain blood flow control by astrocytes independent of phasic neurovascular coupling. J. Neurosci. 35, 13463–13474.

Rouach, N., Koulakoff, A., Abudara, V., Willecke, K., Giaume, C., 2008. Astroglial metabolic networks sustain hippocampal synaptic transmission. Science 322, 1551–1555.

Roy, C., Sherrington, C., 1890. On the regulation of the blood supply of the brain. J. Physiol. 11, 85–108.

Sasaki, T., Beppu, K., Tanaka, K.F., Fukazawa, Y., Shigemoto, R., Matsui, K., 2012. Application of an optogenetic byway for perturbing neuronal activity via glial photostimulation. Proc. Natl. Acad. Sci. U. S. A. 109, 20720–20725.

Shigetomi, E., Jackson-Weaver, O., Huckstepp, R.T., O'Dell, T.J., Khakh, B.S., 2013. TRPA1 channels are regulators of astrocyte basal calcium levels and long-term potentiation via constitutive D-serine release. J. Neurosci. 33, 10143–10153.

Soto, I., Graham, L.C., Richter, H.J., Simeone, S.N., Radell, J.E., Grabowska, W., Funkhouser, W.K., Howell, M.C., Howell, G.R., 2015. APOE stabilization by exercise prevents aging neurovascular dysfunction and complement induction. PLoS Biol. 13, e1002279.

Srinivasan, R., Huang, B.S., Venugopal, S., Johnston, A.D., Chai, H., Zeng, H., Golshani, P., Khakh, B.S., 2015. Ca^{2+} signaling in astrocytes from $Ip3r2^{-/-}$ mice in brain slices and during startle responses *in vivo*. Nat. Neurosci. 18, 708–717.

Stosiek, C., Garaschuk, O., Holthoff, K., Konnerth, A., 2003. *In vivo* two-photon calcium imaging of neuronal networks. Proc. Natl. Acad. Sci. U. S. A. 100, 7319–7324.

Strohschein, S., Hüttmann, K., Gabriel, S., Binder, D.K., Heinemann, U., Steinhäuser, C., 2011. Impact of aquaporin-4 channels on K^+ buffering and gap junction coupling in the hippocampus. Glia 59, 973–980.

Syková, E., Nicholson, C., 2008. Diffusion in brain extracellular space. Physiol. Rev. 88, 1277–1340.

Takano, T., Tian, G.F., Peng, W., Lou, N., Libionka, W., Han, X., Nedergaard, M., 2006. Astrocyte-mediated control of cerebral blood flow. Nat. Neurosci. 9, 260–267.

Takata, N., Hirase, H., 2008. Cortical layer 1 and layer 2/3 astrocytes exhibit distinct calcium dynamics *in vivo*. PLoS ONE 3, e2525.

Takata, N., Mishima, T., Hisatsune, C., Nagai, T., Ebisui, E., Mikoshiba, K., Hirase, H., 2011. Astrocyte calcium signaling transforms cholinergic modulation to cortical plasticity *in vivo*. J. Neurosci. 31, 18155–18165.

Takata, N., Nagai, T., Ozawa, K., Oe, Y., Mikoshiba, K., Hirase, H., 2013. Cerebral blood flow modulation by Basal forebrain or whisker stimulation can occur independently of large cytosolic Ca^{2+} signaling in astrocytes. PLoS ONE 8, e66525.

Tong, X., Ao, Y., Faas, G.C., Nwaobi, S.E., Xu, J., Haustein, M.D., Anderson, M.A., Mody, I., Olsen, M.L., Sofroniew, M.V., Khakh, B.S., 2014. Astrocyte Kir4.1 ion channel deficits contribute to neuronal dysfunction in Huntington's disease model mice. Nat. Neurosci. 17, 694–703.

Verkhratsky, A., Nedergaard, M., 2014. Astroglial cradle in the life of the synapse. Philos. Trans. R. Soc. B Biol. Sci. 369, 20130595.

Volterra, A., Meldolesi, J., 2005. Astrocytes, from brain glue to communication elements: the revolution continues. Nat. Rev. Neurosci. 6, 626–640.

Waite, A., Brown, S.C., Blake, D.J., 2012. The dystrophin-glycoprotein complex in brain development and disease. Trends Neurosci. 35, 487–496.

Wang, F., Smith, N.A., Xu, Q., Fujita, T., Baba, A., Matsuda, T., Takano, T., Bekar, L., Nedergaard, M., 2012. Astrocytes modulate neural network activity by Ca^{2+}-dependent uptake of extracellular K^+. Sci. Signal 5, ra26.

Wang, F., Smith, N.A., Xu, Q., Goldman, S., Peng, W., Huang, J.H., Takano, T., Nedergaard, M., 2013. Photolysis of caged Ca^{2+} but not receptor-mediated Ca^{2+} signaling triggers astrocytic glutamate release. J. Neurosci. 33, 17404–17412.

Wang, X., Lou, N., Xu, Q., Tian, G.F., Peng, W.G., Han, X., Kang, J., Takano, T., Nedergaard, M., 2006. Astrocytic Ca^{2+} signaling evoked by sensory stimulation *in vivo*. Nat. Neurosci. 9, 816–823.

Watkins, S., Robel, S., Kimbrough, I.F., Robert, S.M., Ellis-Davies, G., Sontheimer, H., 2014. Disruption of astrocyte-vascular coupling and the blood-brain barrier by invading glioma cells. Nat. Commun. 5, 4196.

Winship, I.R., Plaa, N., Murphy, T.H., 2007. Rapid astrocyte calcium signals correlate with neuronal activity and onset of the hemodynamic response *in vivo*. J. Neurosci. 27, 6268–6272.

Woo, D.H., Han, K.-S., Shim, J.W., Yoon, B.-E., Kim, E., Bae, J.Y., Oh, S.-J., Hwang, E.M., Marmorstein, A.D., Bae, Y.C., Park, J.-Y., Lee, C.J., 2012. TREK-1 and Best1 channels mediate fast and slow glutamate release in astrocytes upon GPCR activation. Cell 151, 25–40.

Xie, A.X., Petravicz, J., McCarthy, K.D., 2015. Molecular approaches for manipulating astrocytic signaling *in vivo*. Front. Cell. Neurosci. 9, 144.

Xie, L., Kang, H., Xu, Q., Chen, M.J., Liao, Y., Thiyagarajan, M., O'Donnell, J., Christensen, D.J., Nicholson, C., Iliff, J.J., Takano, T., Deane, R., Nedergaard, M., 2013. Sleep drives metabolite clearance from the adult brain. Science 342, 373–377.

Yao, X., Hrabetová, S., Nicholson, C., Manley, G.T., 2008. Aquaporin-4-deficient mice have increased extracellular space without tortuosity change. J. Neurosci. 28, 5460–5464.

Yukutake, Y., Yasui, M., 2010. Regulation of water permeability through aquaporin-4. Neuroscience 168, 885–891.

Zheng, K., Bard, L., Reynolds, J.P., King, C., Jensen, T.P., Gourine, A.V., Rusakov, D.A., 2015. Time-resolved imaging reveals heterogeneous landscapes of nanomolar Ca^{2+} in neurons and astroglia. Neuron 88, 277–288.

Zonta, M., Angulo, M.C., Gobbo, S., Rosengarten, B., Hossmann, K.A., Pozzan, T., Carmignoto, G., 2003. Neuron-to-astrocyte signaling is central to the dynamic control of brain microcirculation. Nat. Neurosci. 6, 43–50.

CHAPTER

Vascular potassium channels in NVC

3

K. Yamada[1]
Hirosaki University Graduate School of Medicine, Hirosaki, Aomori, Japan
[1]Corresponding author: Tel.: +81-172-395008; Fax: +81-172-395009,
e-mail address: kyamada@hirosaki-u.ac.jp

Abstract

It has long been proposed that the external potassium ion ($[K^+]_O$) works as a potent vasodilator in the dynamic regulation of local cerebral blood flow. Astrocytes may play a central role for producing K^+ outflow possibly through calcium-activated potassium channels on the end feet, responding to a rise in the intracellular Ca^{2+} concentration, which might well reflect local neuronal activity.

A mild elevation of $[K^+]_O$ in the end feet/vascular smooth muscle space could activate Na^+/K^+-ATPase concomitant with inwardly rectifying potassium (Kir) channels in vascular smooth muscle cells, leading to a hyperpolarization of vascular smooth muscle and relaxation of smooth muscle actin-positive vessels.

Also proposed notion is endothelial calcium-activated potassium channels and/or inwardly rectifying potassium channel-mediated hyperpolarization of vascular smooth muscle.

A larger elevation of $[K^+]_O$, which may occur pathophysiologically in such as spreading depression or stroke, can trigger a depolarization of vascular smooth muscle cells and vasoconstriction instead.

Keywords

Smooth muscle cell, Neurovascular coupling, Cerebral blood flow, Functional hyperemia, Astrocyte, BK channel, Inward rectifier, KATP channel

1 HYPERKALEMIA CAN PRODUCE RELAXATION OF VASCULAR SMOOTH MUSCLE CELLS

An increase in the external potassium ions (hyperkalemia) could lead to a relaxation of vascular smooth muscle cells and a dynamic increase in local cerebral blood flow (CBF) (for review, Atwell et al., 2010; Filosa and Iddings, 2013; Filosa et al., 2015; Hibino et al., 2010; Howarth, 2014; Longden and Nelson, 2015; MacVicar and Newman, 2015). Such potassium-mediated vascular response could be dose, time, and location dependent (Haddy, 1983; Horiuchi et al., 2002; Longden and Nelson, 2015; Nelson and Quayle, 1995).

In isolated blood vessels, vascular smooth muscle cells dilate when the extracellular K^+ concentration ($[K^+]_0$) is elevated from a hypothetical resting level (3–4 mM) to a moderately activated (postulated to be a physiological) level (for example, 5–6 mM up to 10 mM), provided if some vascular tone was maintained initially (Haddy, 2006; Horiuchi et al., 2002).

This vasodilation might occur in a biphasic manner, a transient constriction followed by dilation (Horiuchi et al., 2002). The dilation, but not the initial constriction, could be blocked by ouabain, suggesting an involvement of Na^+, K^+-ATPase (Fig. 1, also known as the sodium/potassium pump) in the hyperkalemia-induced vasodilation (Haddy, 2006; Horiuchi et al., 2002).

The biphasic (constriction to dilation) vascular response might be explained as follows. A moderate rise in $[K^+]_0$ would initially act as a driving force to the depolarized direction from the resting potential (E_{rest}) of vascular smooth muscle cells by the Nernst equation (the positive shift would be expected) (Nelson and Quayle, 1995). This could also lead to an activation of the electrogenic sodium/potassium pump (three Na^+ out for two K^+ in per cycle) on vascular smooth muscle cells regardless of vasoconstriction being detected actually (Horiuchi et al., 2002) or not (Haddy, 1983). It is reasonable to think that such pump activity would wane along with reduction in internal Na^+ and accumulation of internal K^+ (Longden and Nelson, 2015).

Concomitantly, it has been thought that inwardly rectifying potassium (Kir) channels in vascular smooth muscle cells are activated by a moderate rise in $[K^+]_0$ (Fig. 1; Atwell et al., 2010; Longden and Nelson, 2015; Nelson and Quayle, 1995). Ba^{2+}(100 μM)-sensitive Kir2.x subunit-containing channels have been hypothesized as the inwardly rectifying potassium channels (for review, Longden and Nelson, 2015; Zaritsky et al., 2000). The ouabain-sensitive hyperpolarization of vascular smooth muscle cells in response to the mild elevation in $[K^+]_0$ could well be due to an overcompensation of $[K^+]_0$ by the pump activity under $[K^+]_0$ supply by opening of inwardly rectifying potassium channels possibly Kir2.1 subunit-containing ones in vascular smooth muscle cells (Hibino et al., 2010; Longden and Nelson, 2015; Nelson and Quayle, 1995). Expression of inwardly rectifying potassium channels in endothelial cells has also been proposed; yet to be well established though, such dual expression system might work as a mutual reinforcer (Fig. 1; Longden and Nelson, 2015).

The activation of inwardly rectifying potassium channels would lead to an increase in potassium flow according to the Nernst equation. In vascular smooth muscle cells, if the inwardly rectifying potassium channels were activated, the net change in the membrane potential would be hyperpolarization (K^+ outward flow) during when the K^+ reversal potential E_K was in the range still more negative than E_{rest} of vascular smooth muscle cells (Hibino et al., 2010; Nelson and Quayle, 1995).

Nelson and colleagues discussed that the K^+ reversal potential E_K is −52 mV even when $[K^+]_0$ is 20 mM, and that it is more negative than E_{rest} of arterial smooth muscle which was estimated to be approximately −40 mV (Nelson and Quayle, 1995).

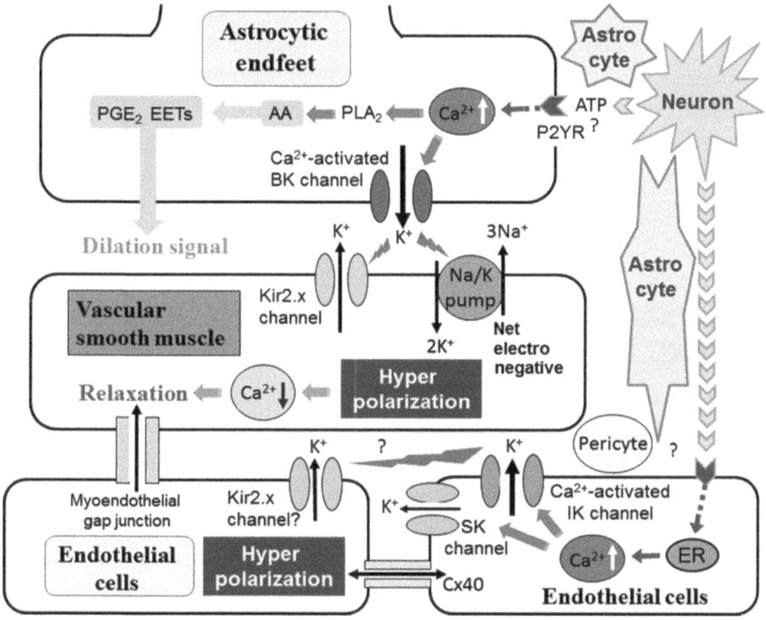

FIG. 1

Schematic representation of external potassium-mediated vasodilation. It has been postulated that an increase in the intracellular Ca^{2+} concentration in astrocytic end feet is triggered by neuronal and/or glial activity-dependent release of transmitters such as glutamate and/or adenosine triphosphate (ATP). This may activate large conductance Ca^{2+}-activated potassium (BK) channels on the end feet, leading to K^+ release to end feet/ vascular smooth muscle space. Astrocytic BK channel-induced increase in the external K^+ concentration, from the resting (~3 mM) to moderately elevated level (~5 up to 10 mM), may trigger concomitant activation of Na^+, K^+-ATPase and inwardly rectifying potassium (Kir) channels in smooth muscle actin-positive cells, leading to their hyperpolarization. Such hyperpolarization in the smooth muscle actin-positive cells would result in vasodilation due to voltage-dependent Ca^{2+} channel inactivation. A similar mechanism is believed to operate in the release of K^+ to vascular smooth muscle/endothelial cell space, but through activation of intermediate conductance (IK) and small conductance (SK) Ca^{2+}-activated potassium channels. In addition, it has been speculated that elevated external K^+ concentration may further activate Kir channels in endothelial cells, boosting hyperpolarization of smooth muscle actin-positive cells through myoendothelial gap junction, and hyperpolarization of endothelial cells. Alternatively, elevated Ca^{2+} in the astrocytic end feet may trigger phospholipase A_2 (PLA_2)-mediated liberation of arachidonic acid (AA). In a low pO_2 environment, a subsequent production of prostaglandin E2 (PGE_2) and/or epoxyeicosatrienoic acid (EET) may exert a vasodilating effect on smooth muscle actin-positive cells.

If K^+ taken up into vascular smooth muscle cells by sodium/potassium pump was compensated by K^+ outflow through the inwardly rectifying potassium channels to the extracellular space, the inwardly rectifying potassium channel could contribute to a prolonged hyperpolarization (Longden and Nelson, 2015). In any case, the hyperpolarization of vascular smooth muscle cells induce a decrease in $[Ca^{2+}]_i$ by such as a closure of voltage-gated calcium channels, leading to a relaxation of vascular smooth muscle cells (Filosa et al., 2006; Otsu et al., 2015).

In pathophysiological conditions, such as seizure, spreading depression, and stroke, opening of inwardly rectifying potassium channels might lead to a net depolarization of vascular smooth muscle cells due to a large increase in $[K^+]_0$ to the level whereby E_K exceeds (more positive than) E_{rest} (MacVicar and Newman, 2015).

2 A POSSIBLE SOURCE OF THE EXTRACELLULAR POTASSIUM IONS IN END FEET/VASCULAR SMOOTH MUSCLE SPACE

As a source for the extracellular potassium ions, perivascular astrocytes may be of particular importance due to their spatial arrangement that surrounds smooth muscle actin-expressing penetrating arterioles, smooth muscle actin-expressing precapillary microvessels, and that apposing to endothelial cells forming capillaries, and those forming venules ascending to the pial surface in the cerebral cortex (Hill et al., 2015; McCaslin et al., 2011; Otsu et al., 2015).

Providing that neuronal activity evokes intracellular Ca^{2+} increase in astrocytic end feet, large conductance, Ca^{2+}-activated K^+ (BK, also known as MaxiK or $K_{Ca}1.1$) channels that are expressed on astrocyte end feet may open, resulting in K^+ efflux from the end feet (Fig. 1). It has been widely accepted that BK channels potentially contribute to smooth muscle actin-mediated vasodilation and the activity-dependent increase in blood flow traditionally termed functional hyperemia (Filosa et al., 2006; Girouard et al., 2010; MacVicar and Newman, 2015; Nelson et al., 1995; Otsu et al., 2015; Takano et al., 2006).

Further, Girouard and colleagues have reported in mouse cortical brain slice superfused with a conventional Krebs–Ringer solution equilibrated with 95% O_2, that when intracellular Ca^{2+} concentration ($[Ca^{2+}]_i$) increased in a single astrocytic end foot from ~100 to ~300 nM by field (neuronal) stimulation or by uncaging Ca^{2+} in the end foot, microvessel adjacent to the end foot dilated by ~20%, whereas a larger $[Ca^{2+}]_i$ increase (to ~700 nM) caused vasoconstriction by ~30%, suggesting a bidirectional control of vascular diameter by $[Ca^{2+}]_i$ in astrocyte end feet (Girouard et al., 2010).

These authors have also reported that selective BK channel blocker, paxilline, reduced the vasodilation of the adjacent vessel in ~70% and 50% of cases tested in slice and in vivo, respectively, without affecting $[Ca^{2+}]_i$ increase in the end foot. These results implicate a BK channels-mediated as well as BK-independent vasodilation. Arachidonic acid (AA) and prostaglandin-mediated mechanism has been proposed for the latter BK-independent vasodilation (Girouard et al., 2010; Gordon et al., 2008; MacVicar and Newman, 2015). Furthermore, an interplay between AA-mediated metabolic regulation and potassium channel-mediated electrical regulation mechanisms for vasodilation has been proposed, wherein epoxyeicosatrienoic acids (EETs), which could be released from astrocytic end feet in response to glutamate, activate BK channels possibly in the end feet or in vascular smooth muscle, modulating the $[K^+]_O$ in end feet/vascular smooth muscle space (Filosa and Iddings, 2013; Higashimoto et al., 2010).

3 A POSSIBLE SOURCE OF THE EXTRACELLULAR POTASSIUM IONS IN VASCULAR SMOOTH MUSCLE/ENDOTHELIAL CELL SPACE

It has been reported that vascular endothelial cells express Ca^{2+}-activated small conductance (SK) and intermediate conductance (IM) potassium channels (Billaud et al., 2014). An opening of these Ca^{2+}-activated K^+ channels (K_{Ca}) would elevate $[K^+]_O$ in vascular smooth muscle/endothelial cell space along with $[Ca^{2+}]_i$ increase, leading to a hyperpolarization of vascular smooth muscle cells and a vasodilation in the mechanism similar to that discussed in the end feet–vascular smooth muscle interaction (Billaud et al., 2014; Filosa and Iddings, 2013).

An elevation in $[K^+]_O$ in vascular smooth muscle/endothelial cell space might also activate the inwardly rectifying potassium channels postulated to be expressed in endothelial cells such as Kir2.1, leading to endothelial cell-mediated hyperpolarization of vascular smooth muscle (Filosa and Iddings, 2013; McCarron and Halpern, 1990). A hyperpolarization of endothelial cells might be further transmitted to vascular smooth muscle cells via myoendothelial gap junctions (Sokoya et al., 2006).

Interestingly, it appears that the spatial distribution of IK and SK in endothelial cell membrane differs; IKs are mainly expressed at the point of contacts between vascular smooth muscle and endothelial cells, whereas SKs are predominantly expressed in endothelial cell to endothelial cell junctions (Billaud et al., 2014).

In addition, since endothelial cells are mutually connected with gap junctions such as Cx40, it is expected that hyperpolarization in one endothelial cell transfers rapidly to other endothelial cells (Billaud et al., 2014). An opening of inwardly rectifying potassium channels in vascular smooth muscle cells as well as in endothelial cells may provide robustness to the $[K^+]_O$-mediated vasodilation (Longden and Nelson, 2015). Mechanisms underlying neuronal and/or astrocytic control of endothelial cell membrane potential are still open to discussion.

4 OTHER POTASSIUM CHANNELS THAT COULD CRITICALLY CONTRIBUTE TO REGULATION OF NEUROVASCULAR COUPLING

Inwardly rectifying potassium channels such as Kir4.1 and/or Kir5.1 subunit-containing channels that are predominantly expressed in astrocytes have been proposed as candidates for the $[K^+]_O$ regulation mechanism (Bay and Butt, 2012; Hibino et al., 2010). It has been reported in particular that Kir4.1/Kir5.1 heteromer is expressed solely in the astrocytic end feet attached to blood vessels and to pia matter, whereas Kir4.1 homomer appears to distribute widely in astrocytic processes such as those surrounding synapse in the brain (Hibino et al., 2010). However, the function of these inwardly rectifying potassium channels located in the end feet for neurovascular coupling (NVC) has not been well established (Metea et al., 2007).

Another pathway, which does not involve endothelial cell hyperpolarization, proposed is a Ca^{2+}-dependent release of endothelium-derived hyperpolarizing factors (EDHFs) (Edwards et al., 2010). The term EDHF originally denotes the hypothetical endothelium-derived substrates responsible for hyperpolarization of "vascular smooth muscle" cells not associated with endothelium-derived relaxing factor (nitric oxide, NO) or prostacyclin (Edwards et al., 2010).

Analogously to in the end feet/vascular smooth muscle space, Ca^{2+}-dependent production of diffusible factors including such as EETs might well occur in vascular smooth muscle/endothelial cell space, and this process could well occur metabolically in the absence of hyperpolarization of endothelial cells (Edwards et al., 2010). For example, it has been proposed that EETs may activate BK or ATP-sensitive potassium (K_{ATP}) channels postulated to be expressed in vascular smooth muscle cells, from the fact that vascular smooth muscle hyperpolarization was blocked by inhibitors against BK (iberiotoxin) or K_{ATP} channels (glibenclamide).

K_{ATP} channels were first discovered in cardiac muscle by Noma in 1983, then were found in various other excitable cells including pancreatic B-cells (Aguilar-Bryan et al., 1998; Ashcroft et al., 1984; Inagaki et al., 1995; Seino, 1999), neurons (Amoroso et al., 1990; Ashford et al., 1988; Schiemann et al., 2012; Yamada and Inagaki, 2005; Yamada et al., 2001), and vascular cells (Flagg and Nichols, 2010; Kleppisch and Nelson, 1995; Quayle et al., 1997; Rosenblum, 2003; Standen et al., 1989; Wareing, 2014).

K_{ATP} channels couple intracellular ATP level, which reflects metabolic state of cells, to the cellular excitability by regulating K^+ flow by opening/closing of the channels (Aguilar-Bryan et al., 1998; Flagg and Nichols, 2010; Seino, 1999). K_{ATP} channels comprise a member of inward rectifier Kir6.x subunits (Kir6.1, Kir6.2, and Kir6.3) and the receptors for sulfonylureas (SUR1 and SUR2). Kir6.x subunits have two transmembrane domains forming the potassium-selective pore, whereas an SUR subunit, a member of ATP-binding cassette (ABC) protein superfamily, is the regulatory subunit having 17 transmembrane domains. Of divergent K_{ATP} channels, it has been thought that those consisting of Kir6.1 and SUR2 subunits (so-called vascular smooth muscle type) are central in control of vascular tone (Flagg and Nichols, 2010).

An activation of such vascular-type K_{ATP} channels would lead to a lowering of $[Ca^{2+}]_i$, leading to a vascular relaxation (Standen et al., 1989). However, it is still a challenging issue to elucidate detailed distribution of such vascular-type K_{ATP} channels, either in vascular smooth muscle cells and/or in other cells including astrocytic end feet and endothelial cells, as well as their function in the brain vasculature (Flagg and Nichols, 2010). RNA transcripts for the SUR2 gene (*ABCC9*) are found in endothelium (SUR2B), implicating the possibility of vascular-type K_{ATP} channels existing in both vascular smooth muscle cells and in nonsmooth muscle cells (Gendron et al., 2004).

It has been speculated that K_{ATP} channel-dependent communication by paracrine (endothelin) released from endothelial cells between endothelial cells and vascular smooth muscle cells (Malester et al., 2007). Numbers of questions remain unanswered; for example, what the ATP/ADP levels are in the brain vessel system wherein ATP might be a key intercellular messenger in NVC.

5 A CONSIDERATION ON AA-MEDIATED, PO_2-SENSITIVE VASOREGULATION

It has been proposed that under "low" pO_2, such as when acute slice was superfused with a solution equilibrated with 20% O_2 instead of conventionally used 95% O_2, an increase in $[Ca^{2+}]_i$ in the astrocyte end feet activates phospholipase A_2 (PLA_2) and AA production, which could be converted by cyclooxygenase (COX) to prostaglandin E_2 (PGE_2) in the end feet (Gordon et al., 2008; Takano et al., 2006). In such a "low" pO_2 condition, lactate may well be increased due to facilitated glycolysis, and since a rise in the extracellular lactate can attenuate reuptake function of prostaglandin transporters, a lactate/PG anion exchanger, smooth muscle relaxation could be then induced by an increase in PGE_2 in the extracellular space (Gordon et al., 2011; MacVicar and Newman, 2015). By contrast, under a conventional pO_2 condition such as when acute slice was superfused with a solution equilibrated with 95% O_2, Gordon and colleagues have suggested that an increase in $[Ca^{2+}]_i$ may produce 20-HETE from AA in the astrocyte end feet, leading to vasoconstriction (MacVicar and Newman, 2015).

It may be worth noting that neither blood flow-mediated supply of energy source nor removal of metabolites occurs in acute slice, wherein a solution equilibrated with 20% O_2 could well correspond, in our experience, to ~5–6% O_2 environment in vivo, which is a critical transition pO_2 for mice to survive for a short period (Yamada and Inagaki, 2005; Yamada et al., 2001). From these, it can be speculated that the "low" pO_2 condition might be a "hypoxic" rather than normal condition, although situation may critically depend on dynamic time course of hypoxia, slice thickness, temperature, and cellular activity.

6 RECENT TOPICS AND A FUTURE DIRECTION: PRECAPILLARY MICROVESSELS EXPRESSING SMOOTH MUSCLE ACTIN AND DIRECT NEURONAL GLUCOSE UPTAKE

A recent report using smooth muscle actin promoter-driven fluorescent protein transgenic mice has suggested that smooth muscle actin is expressed not only in cortical penetrating arterioles, but also in thinner (<10 μm) microvessels branching from the cortical penetrating arterioles, some of which were denoted as capillary. Furthermore, such smooth muscle actin-positive thin microvessels included those which were surrounded by "astrocytic" end feet, although the term "astrocyte" might include heterogeneous population of cells depending upon the marker and methods being used (Hill et al., 2015; McCaslin et al., 2011).

Considering these thin "precapillary" smooth muscle actin-expressing microvessels (<10 μm), possible relationships between smooth muscle actin-positive vessels and endothelial cells are expected to be main players in NVC, although a role of pericytes, which have contractile property, in the NVC is still a very interesting topic (Hall et al., 2014; Hill et al., 2015; Peppiatt et al., 2006).

The balance between brain activity and energy substrates such as glucose and oxygen in the regulation of vascular tone are pivotal for understanding NVC (MacVicar and Newman, 2015). Extracellular accumulation of lactate under neuronal activation is a key issue considering neuron activity-dependent regulation of vascular relaxation (Atwell et al., 2010; MacVicar and Newman, 2015). According to the astrocyte-to-neuron lactate shuttle hypothesis, glucose is taken up into astrocytes thereby converted to lactate, which is then taken up by neurons and further metabolized. Recently, this notion has been challenged by a direct neuronal glucose uptake (Lundgaard et al., 2015). A role of ATP in intercellular and intracellular events in the NVC under various physiological and pathophysiological states is still a critical issue to be elucidated.

Interestingly, a recent study proposed that a decrease in functional inwardly rectifying potassium channels in the smooth muscle membrane was caused by stress-induced glucocorticoid signaling, impairing the ability of the vessel to respond to local increase in extracellular potassium concentration during NVC, leading to suboptimal vasodilation and ultimately downregulation of local CBF in stressed condition (Longden et al., 2014). When considering brain functions in daily physiological conditions, a role of not only neurons/astrocytes but also of vascular cells should be a broad area to be explored.

7 SUMMARY

Mechanisms currently proposed for external potassium-triggered vasodilation in NVC can be summarized as follows (Fig. 1). A moderate elevation in $[K^+]_0$ can be produced via large conductance or intermediate/small conductance calcium-activated potassium channels found in astrocytic end feet and in endothelial cells, respectively. Such an increase in $[K^+]_0$ may trigger concomitant activation of Na^+, K^+-ATPase and inwardly rectifying potassium channels in smooth muscle actin-positive cells, leading to their membrane hyperpolarization. The hyperpolarization inactivates voltage-dependent Ca^{2+} channels, resulting in a relaxation of smooth muscle actin-positive cells (Atwell et al., 2010; Filosa and Iddings, 2013; Haddy, 2006; Longden and Nelson, 2015).

REFERENCES

Aguilar-Bryan, L., Clement 4th., J.P., Gonzalez, G., Kunjilwar, K., Babenko, A., Bryan, J., 1998. Toward understanding the assembly and structure of KATP channels. Physiol. Rev. 78, 227–245.

Amoroso, S., Schmid-Antomarchi, H., Fosset, M., Lazdunski, M., 1990. Glucose, sulfonylureas, and neurotransmitter release: role of ATP-sensitive K+ channels. Science 247, 852–854.

Ashcroft, F.M., Harrison, D.E., Ashcroft, S.J., 1984. Glucose induces closure of single potassium channels in isolated rat pancreatic beta-cells. Nature 312, 446–448.

Ashford, M.L., Sturgess, N.C., Trout, N.J., Gardner, N.J., Hales, C.N., 1988. Adenosine-5′-triphosphate-sensitive ion channels in neonatal rat cultured central neurones. Pflugers Arch. 412, 297–304.

Atwell, D., Buchan, A.M., Charpak, S., Lauritzen, M., MacVicar, B.A., Newman, E.A., 2010. Glial and neuronal control of brain blood flow. Nature 468, 232–243.

Bay, V., Butt, A.M., 2012. Relationship between glial potassium regulation and axon excitability: a role for glial Kir4.1 channels. Glia 60, 651–660.

Billaud, M., Lohman, A.W., Johnstone, S.R., Biwer, L.A., Mutchler, S., Isakson, B.E., 2014. Regulation of cellular communication by signaling microdomains in the blood vessel wall. Pharmacol. Rev. 66, 513–569.

Edwards, G., Feletou, M., Weston, A.H., 2010. Endothelium-derived hyperpolarizing factors and associated pathways: a synopsis. Pflugers Arch. 459, 863–879.

Filosa, J.A., Iddings, J.A., 2013. Astrocyte regulation of cerebral vascular tone. Am. J. Physiol. Heart Circ. Physiol. 305, H609–H619.

Filosa, J.A., Bonev, A.D., Straub, S.V., Meredith, A.L., Wilkerson, M.K., Aldrich, R.W., Nelson, M.T., 2006. Local potassium signaling couples neuronal activity to vasodilation in the brain. Nat. Neurosci. 9, 1397–1403.

Filosa, J.A., Morisson, H.W., Iddings, J.A., Du, W., Kim, K.J., 2015. Beyond neurovascular coupling, role of astrocytes in the regulation of vascular tone. Neuroscience. http://dx.doi.org/10.1016/j.neuroscience.2015.03.064.

Flagg, T.P., Nichols, C.G., 2010. Muscle K_{ATP} channels: recent insights to energy sensing and myoprotection. Physiol. Rev. 90, 799–829.

Gendron, M.E., Thorin, E., Perrault, L.P., 2004. Loss of endothelial K_{ATP} channel-dependent, NO-mediated dilation of endocardial resistance coronary arteries in pigs with left ventricular hypertrophy. Br. J. Pharmacol. 143, 285–291.

Girouard, H., Bonev, A.D., Hannah, R.M., Meredith, A., Aldrich, R.W., Nelson, M.T., 2010. Astrocyte endfoot Ca2+ and BK channels determine both arteriolar dilation and constriction. Proc. Natl. Acad. Sci. U. S. A. 107, 3811–3816.

Gordon, G.R.J., Choi, H.B., Rungta, R.I., Ellis-Davies, G.C.R., MacVicar, B.A., 2008. Brain metabolism dictates the polarity of astrocyte control over arterioles. Nature 456, 745–749.

Gordon, G.R.J., Howarth, C., MacVicar, B.A., 2011. Bidirectional control of arteriole diameter by astrocytes. Exp. Physiol. 96, 393–399.

Haddy, F.J., 1983. Potassium effects on contraction in arterial smooth muscle mediated by Na^+, K^+-ATPase. Fed. Proc. 42, 239–245.

Haddy, F.J., 2006. Role of potassium in regulating blood flow and blood pressure. Am. J. Physiol. Regul. Integr. Comp. Physiol. 290, R546–R552.

Hall, C.N., Reynell, C., Gesslein, B., Hamilton, N.B., Mishra, A., Sutherland, B.A., O'Farrell, F.M., Buchan, A.M., Lauritzen, M., Attwell, D., 2014. Capillary pericytes regulate cerebral blood flow in health and disease. Nature 508, 55–60.

Hibino, H., Inanobe, A., Furutani, K., Murakami, S., Findlay, I., Kurachi, Y., 2010. Inwardly rectifying potassium channels: their structure, function, and physiological roles. Physiol. Rev. 90, 291–366.

Higashimoto, H., Blanco, V.M., Tuniki, V.R., Falck, J.R., Filsa, J.A., 2010. Role of epoxyeicosatrienoic acids as autocrine metabolites in glutamate-mediated K+ signaling in perivascular astrocytes. Am. J. Physiol. Cell Physiol. 299, C1068–C1078.

Hill, R.A., Tong, L., Yuan, P., Murkinati, S., Gupta, S., Grutzendler, J., 2015. Regional blood flow in the normal and ischemic brain is controlled by arteriolar smooth muscle cell contractility and not by capillary pericytes. Neuron 87, 95–110.

Horiuchi, T., Dietrich, H.H., Hongo, K., Dacey Jr., R.G., 2002. Mechanism of extracellular K^+-induced local and conducted responses in cerebral penetrating arterioles. Stroke 33, 2692–2699.

Howarth, C., 2014. The contribution of astrocytes to the regulation of cerebral blood flow. Front. Neurosci. 8, 103. http://dx.doi.org/10.3389/fnins.2014.00103.

Inagaki, N., Gonoi, T., Clement 4th., J.P., Namba, N., Inazawa, J., Gonzalez, G., Aguilar-Bryan, L., Seino, S., Bryan, J., 1995. Reconstitution of IKATP: an inward rectifier subunit plus the sulfonylurea receptor. Science 270, 1166–1170.

Kleppisch, T., Nelson, M.T., 1995. ATP-sensitive K+ currents in cerebral arterial smooth muscle: pharmacological and hormonal modulation. Am. J. Physiol. 269, H1634–H1640.

Longden, T.A., Nelson, M.T., 2015. Vascular inward rectifier K+ channels as external K+ sensors in the control of cerebral blood flow. Microcirculation 22, 183–196.

Longden, T.A., Dabertrand, F., Hill-Eubanks, D.C., Hammack, S.E., Nelson, M.T., 2014. Stress-induced glucocorticoid signaling remodels neurovascular coupling through impairment of cerebrovascular inwardly rectifying K+ channel function. Proc. Natl. Acad. Sci. U. S. A. 111, 7462–7467.

Lundgaard, I., Li, B., Xie, L., Kang, H., Sanggaard, S., Haswell, J.D., Sun, W., Goldman, S., Blekot, S., Nielsen, M., Takano, T., Deane, R., Nedergaard, M., 2015. Direct neuronal glucose uptake Heralds activity-dependent increases in cerebral metabolism. Nat. Commun. http://dx.doi.org/10.1038/ncomms7807.

MacVicar, B.A., Newman, E.A., 2015. Astrocyte regulation of blood flow in the brain. Cold Spring Harb. Perspect. Biol. http://dx.doi.org/10.1101/cshperspect.a020388.

Malester, B., Tong, X., Ghiu, I., Kontogeorgis, A., Gutstein, D.E., Xu, J., Hendricks-Munoz, K.D., Coetzee, W.A., 2007. Transgenic expression of a dominant negative K_{ATP} channel subunit in the mouse endothelium: effects on coronary flow and endothelin-1 secretion. FASEB J. 21, 2162–2172.

McCarron, J.G., Halpern, W., 1990. Potassium dilates rat cerebral arteries by two independent mechanisms. Am. J. Physiol. Heart Circ. Physiol. 259, H902–H908.

McCaslin, A.F.H., Chen, B.R., Radosevich, A.J., Cauli, B., Hillman, E.M.C., 2011. In vivo 3D morphology of astrocyte-vasculature interactions in the somatosenosory cortex: implications for neurovascular coupling. J. Cereb. Blood Flow Metab. 31, 795–806.

Metea, M.R., Kofuji, P., Newman, E.A., 2007. Neurovascular coupling is not mediated by potassium siphoning from glial cells. J. Neurosci. 27, 2468–2471.

Nelson, M.T., Quayle, J.M., 1995. Physiological roles and properties of potassium channels in arterial smooth muscle. Am. J. Physiol. 268 (Cell Physiol. 37), C799–C822.

Nelson, M.T., Cheng, H., Rubart, M., Santana, L.F., Bonev, A.D., Knot, H.J., Lederer, W.J., 1995. Relaxation of arterial smooth muscle by calcium sparks. Science 270, 633–637.

Noma, A., 1983. ATP-regulated K+ channels in cardiac muscle. Nature 305, 147–148.

Otsu, Y., Couchman, K., Lyons, D.G., Collot, M., Agarwal, A., Mallet, J.-M., Pfrieger, F.W., Bergles, D.E., Charpak, S., 2015. Calcium dynamics in astrocyte processes during neurovascular coupling. Nat. Neurosci. 18, 210–218.

Peppiatt, C.M., Howarth, C., Mobbs, P., Atwell, D., 2006. Bidirectional control of CNS capillary diameter by pericytes. Nature 443, 700–704.

Quayle, J.M., Nelson, M.T., Standen, N.B., 1997. ATP-sensitive and inwardly rectifying potassium channels in smooth muscle. Physiol. Rev. 77, 1165–1232.

Rosenblum, W.I., 2003. ATP-sensitive potassium channels in the cerebral circulation. Stroke 34, 1547–1552.

Schiemann, J., Schlaudraff, F., Klose, V., Bingmer, M., Seino, S., Magill, P.J., Zaghloul, K.A., Schneider, G., Liss, B., Roeper, J., 2012. K-ATP channels in dopamine substantia nigra neurons control bursting and novelty-induced exploration. Nat. Neurosci. 15, 1272–1280.

Seino, S., 1999. ATP-sensitive potassium channels: a model of heteromultimeric potassium channel/receptor assemblies. Annu. Rev. Physiol. 61, 337–362.

Sokoya, E.M., Burns, A.R., Setiawan, C.T., Coleman, H.A., Parkington, H.C., Tare, M., 2006. Evidence for the involvement of myoendothelial gap junctions in EDHF-mediated relaxation in the rat middle cerebral artery. Am. J. Physiol. Heart Circ. Physiol. 291, H385–H393.

Standen, N.B., Quayle, J.M., Davies, N.W., Brayden, J.E., Huang, Y., Nelson, M.T., 1989. Hyperpolarizing vasodilators activate ATP-sensitive K^+ channels in arterial smooth muscle. Science 245, 177–180.

Takano, T., Tian, G.F., Peng, W., Lou, N., Libionka, W., Han, X., Nedergaard, M., 2006. Astrocyte-mediated control of cerebral blood flow. Nat. Neurosci. 9, 260–267.

Wareing, M., 2014. Oxygen sensitivity, potassium channels and regulation of placental vascular tone. Microcirculation 21, 58–66.

Yamada, K., Inagaki, N., 2005. Neuroprotection by KATP channels. J. Mol. Cell. Cardiol. 38, 945–949.

Yamada, K., Ji, J.J., Yuan, H., Miki, T., Sato, S., Horimoto, N., Shimizu, T., Seino, S., Inagaki, N., 2001. Protective role of ATP-sensitive potassium channels in hypoxia-induced generalized seizure. Science 292, 1543–1546.

Zaritsky, J.J., Eckman, D.M., Wellman, G.C., Nelson, M.T., Schwarz, T.L., 2000. Targeted disruption of Kir2.1 and Kir2.2 genes reveals the essential role of the inwardly rectifying K^+ current in K^+-mediated vasodilation. Circ. Res. 87, 160–166.

PART

Methodology for Measurements of Brain Circulation

II

CHAPTER

4

Bridging macroscopic and microscopic methods for the measurements of cerebral blood flow: Toward finding the determinants in maintaining the CBF homeostasis

I. Kanno[*,1], **K. Masamoto**[†]

**Molecular Imaging Center, National Institute of Radiological Sciences, Chiba, Japan*

[†]Brain Science Inspired Life Support Research Center, University of Electro-Communications, Tokyo, Japan

[1]Corresponding author: Tel.: +81-43-206-4716; Fax: +81-43-206-0819, e-mail address: kanno1009@gmail.com

Abstract

Methods exist to evaluate the cerebral blood flow (CBF) at both the macroscopic and microscopic spatial scales. These methods provide complementary information for understanding the mechanism in maintaining an adequate blood supply in response to neural demand. The macroscopic CBF assesses perfusion flow, which is usually measured using radioactive tracers, such as diffusible, nondiffusible, or microsphere. Each of them determines CBF based on indicator dilution principle or particle fraction principle under the assumption that CBF is steady state during the measurement. Macroscopic CBF therefore represents averaged CBF over a certain space and time domains. On the other hand, the microscopic CBF assesses bulk flow, usually measures using real-time microscopy. The method assesses hemodynamics of microvessels, ie, vascular dimensions and flow velocities of fluorescently labeled or nonlabeled RBC and plasma markers. The microscopic CBF continuously fluctuates in time and space. Smoothing out this heterogeneity may lead to underestimation in the macroscopic CBF. To link the two measurements, it is needed to introduce a common parameter which is measurable for the both methods, such as mean transit time. Additionally, applying the defined physiological and/or pharmacological perturbation may provide a good exercise to determine how the specific perturbations interfere the quantitative relationships between the macroscopic and microscopic CBF. Finally, bridging these two-scale methods potentially gives a further indication how the absolute CBF is regulated with respect to a specific type of the cerebrovascular tones or capillary flow velocities in the brain.

Progress in Brain Research, Volume 225, ISSN 0079-6123, http://dx.doi.org/10.1016/bs.pbr.2016.03.002

Keywords

Perfusion flow, Bulk flow, Quantitative measurement, Radioisotope tracer, Diffusible tracer, Nondiffusible tracer, Microsphere tracer, Fluorescent tracer, Heterogeneity, Neurovascular unit

The brain is an organ of the vessel. We believe cerebral blood flow (CBF) supplies oxygen and nutrition necessary to maintain brain function. Accurate measurement of CBF will help to determine whether this notion is correct. This chapter describes outline of CBF measurement and physiology, macroscopic and microscopic methods for measurements of CBF, and bridging these methodological features to find the determinants in maintaining the CBF.

1 OUTLINES OF CBF MEASUREMENT AND PHYSIOLOGY

1.1 HISTORY OF DEVELOPING THE TECHNIQUES FOR CBF MEASUREMENT

Kety and Schmidt first measured CBF using N_2O as an inert gas tracer. N_2O is freely diffusible into the brain tissue through blood–brain barrier (BBB). A subject was asked to continuously inhale N_2O for 10 min. The tracer is concentrated and reached to saturated level. During the inhalation the peripheral arterial blood and the jugular vein blood were sampled and the N_2O concentration was monitored. Using the arterial–vinous difference of N_2O concentration, mean transit time (MTT) of N_2O was calculated based on a Fick principle (1948). They obtained whole brain CBF as a reciprocal of MTT.

Lassen and Ingvar (1961) modified Kety's N_2O saturation principle by using a clearance principle with a bolus injection of the tracer into the carotid artery. This is mathematically equivalent to the differentiation of Kety's saturation principle and allowed to directly estimate MTT from the clearance curve of the tracer without the arterial blood sampling. This principle also allowed us to repeat measurements within a few or 10 min because of quick washout of these tracers. In addition, Lassen used the saline-solved radio isotope (RI) gas (^{133}Xe) instead of N_2O as a tracer (Lassen and Hoedt-Rasmussen, 1966). Use of the RI tracer enabled regional estimation of CBF and opened a subsequent era of brain imaging (Lassen et al., 1978).

Both saturation and clearance principles were extended into positron emission tomography (PET) techniques using ^{15}O (half-life $T_{1/2}=2$ min)-labeled gases to measure cerebral circulation and oxygen metabolism (Frackowiak et al., 1980; Kanno et al., 1987; Raichle et al., 1983). However, because PET requires large-scale facilities for cyclotron and chemical synthesis system, in most clinical circumstances, single-photon emission computed tomography (SPECT) becomes widely distributed and several RI tracers for CBF measurements are commercially available. Diffusible ^{133}Xe gas was used for absolute quantification of CBF until early 2000, and a variety of chemical microsphere tracers (see later) labeled with ^{99m}Tc ($T_{1/2}$=6 h) and ^{123}I ($T_{1/2}$=13 h) are used for relative quantification of CBF.

Magnetic resonance imaging (MRI) techniques made a new era for the brain imaging sciences. One of the widely used techniques is imaging of blood oxygenation level dependence (BOLD) found by Ogawa. In principle, the BOLD signal varies depending on the balances of blood volume, blood flow, and oxygen metabolism (Ogawa et al., 1992). Therefore, it is worthwhile to remind that BOLD signal differs from CBF in several aspects. Because of increased users of fMRI as an imaging tool for neurosciences, there are some confusing discussions that BOLD signal reflects CBF, or that BOLD signal reflects neural activity. Neither of them is true. As far as BOLD signal is used as a tool for mapping neural activities, it is no problem. However, when BOLD signal is used as a tool for pursuing physiological questions, we have to pay careful attention on its quantification. For example, the combination with hypercapnia condition, response of BOLD signal is limited due to a ceiling effect, that is, different from PET-CBF results (Shimosegawa et al., 1995).

Alternatively, a new MRI technique which measures the perfusion flow was subsequently developed (Detre et al., 1992). This MRI technique labels spin of water in arterial blood at the entry of the brain. Then, spin-labeled water diffuses into the brain tissue as a diffusible tracer (see later). Using two images with/without proton-spin water, quantitative CBF is calculated. The advantage of this technique, called arterial blood spin-labeling technique (Detre et al., 1992; Williams et al., 1992), is to use an endogenous tracer unlike an exogenous RI tracer used in PET.

1.2 DEFINITION OF BLOOD FLOW: BULK FLOW VS PERFUSION FLOW

When Kety first developed the quantitative method for measuring CBF, he defined CBF as blood flow per unit brain tissue. This means a perfusion flow that flows through a certain volume of the brain tissue (Fig. 1B). In contrast, a microscopic technique provides us a new concept of the blood flow which differs from the earlier

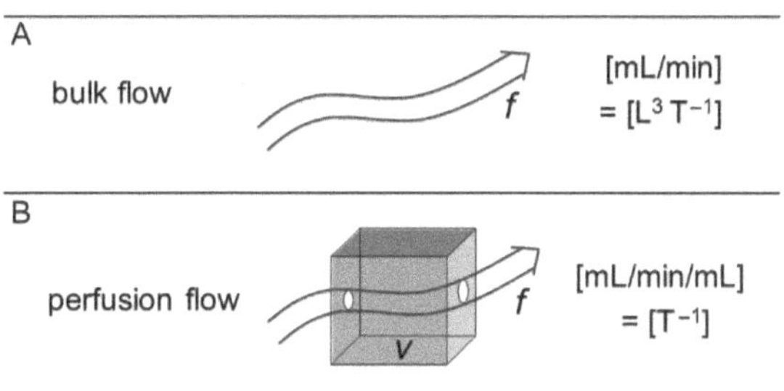

FIG. 1

Definitions for bulk flow and perfusion flow. (A) Bulk flow is defined as blood flow (*f*) flowing through vascular path. Dimension becomes volume per time. (B) Perfusion flow is defined as blood flow (*f*) flowing through tissue volume (*v*). Dimension becomes volume per time per volume, or per time.

defined perfusion flow. This is defined as a bulk flow (Fig. 1A). If we use the microscopically measured flow data without understanding the difference between two definitions for the bulk flow and the perfusion flow, it might cause a slight confusion when comparing the two blood flow values. We therefore need to revisit the definitions for the microscopic bulk flow and the macroscopic perfusion flow.

As defined earlier, blood stream flowing in the vessel is called as bulk flow. The bulk flow in the vessel is determined as blood volume flowing per unit time. Here, the blood volume is given by a product of flow velocity and vascular cross section [$=\pi$ (radius)^2]. Therefore, a dimension of the bulk flow is defined as L^3T^{-1}.

On the other hand, the perfusion flow is determined as integrated bulk flow over unit volume of the brain tissue. Since the volume of the brain tissue is defined as the mass, a dimension of the perfusion flow is $L^3T^{-1}M^{-1}$. When we use a diffusible tracer, blood volume (L^3) is assumed to be equivalent to brain tissue mass (M) and then $L^3T^{-1}M^{-1}$ becomes T^{-1}. The perfusion flow was first determined quantitatively based on an indicator dilution principle as a reciprocal of MTT. The perfusion flow was also determined based on a particle fraction principle. The details of these principles will be described in Section 2.

1.3 DETERMINANTS OF CBF PHYSIOLOGY: VASCULAR NETWORKS AND STRUCTURES

The vessel in the brain composes complex networks. The vascular network is defined as a single connected vascular system, consisting of the artery, arteriole, capillary, venule, and vein. Starting from the aorta, the artery branches to smaller arteries and finally reaches arterioles and capillaries. In microscopic views, depending on the size of the vessel, it can be further subclassified as arteriole (50–100 μm in diameter), terminal arteriole (15–50 μm), arterial capillary (10–15 μm), (pure) capillary (<7 μm), venous capillary (5–10 μm), postcapillary vein (10–25 μm), collecting venule (25–40 μm), muscular venule (40–60 μm), and small collecting veins (<300 μm) (Rhodin, 1967, 1968). After these microcirculatory territories, microvessels again join and become thicker veins, and finally go back to the vena cava.

The artery branches and runs to cover the locations of the cortical surface and penetrate perpendicularly into cortical tissues. The surface artery and penetrating arteriole have different roles in supplying blood flow to the cortex (Mchedlishvili, 1986). The surface artery delivers blood flow over anatomy-based locations in response to sustained tissue demand, though the penetrating arteriole timely delivers blood flow based on variable neural activities. Following the capillaries, blood is drained into small venules and veins where blood can be pooled and buffered as a reservoir. Thus, veins are called as capacitance vessels.

The vascular tone is set by tension of the vascular smooth muscle cells and is regulated by various global factors (ie, chemical mediators and innervation system), as well as local factors (ie, autoregulation system and neurovascular regulation). The chemical mediators involve systemic endogenous (eg, catecholamine) or exogenous (eg, $PaCO_2$) mediators. The innervation system is sympathetic and parasympathetic nervous system (see the chapter "Neurogenic control of parenchymal arterioles in the cerebral cortex" by H. Hotta). The autoregulation system is a so-called myogenic regulation. This is defined as an intrinsic reaction of the smooth muscle cells; that is, their tension inversely reacts to the intraluminal pressure. Thus, the arterial diameter inversely responds to the perfusion pressure (for reviews, Longden et al., 2015). The neurovascular regulation is controlled at neural activation focus, but its mechanism is still to be solved (see also the chapters "Neurogenic control of parenchymal arterioles in the cerebral cortex" by H. Hotta, "Involvement of astrocytes in neurovascular communication" by M. Nuriya and H. Hirase, and "Vascular potassium channels in neurovascular coupling" by K. Yamada). In addition, local changes in the vascular diameter often induce the secondary imbalance of the perfusion pressure in the surrounding vessels around the foci of neural activation (see also chapter "Submillimeter-resolution fMRI: toward understanding local neural processing" by M. Fukuda et al.).

The penetrating arterioles branch to finer microvessels and reach capillaries. Capillary density is determined dependent on oxygen diffusion distance, which is estimated from Krough's cylinder. Since energy demand of the gray matter is four times of the white matter (Reivich et al., 1969), capillary density is proportional to this ratio. Capillary density in the literature was 300–500/mm^2 in the gray matter and 200–300/mm^2 in the white matter (Purves, 1972). It is a useful concept that development of vasculature is dependent on metabolic demand; hence the vascular density reflects a set point of regional activity.

1.4 DETERMINANTS OF CBF PHYSIOLOGY: BLOOD VISCOSITY AND RESISTANCES

The blood flow is determined by a balance of perfusion pressure (or pressure gradient) and vascular (blood) resistance. The perfusion pressure is an intraluminal pulse wave pressure, pushing blood from the upstream to the downstream, and is the highest at the aorta 120–140 mmHg and gradually decreases branch by branch and downs to 4–10 mmHg at the capillary end and the venous vessels. The perfusion pressure is dependent on systemic blood pressure, and the vascular resistance locally varies depending on a vascular diameter and blood viscosity. The viscosity is mainly divided into two components: plasma viscosity (determined by a content of blood serum proteins: albumin, globulin, and fibrinogen) and blood cell viscosity (represented by hematocrit, ie, erythrocyte density). The macroscopic methods measured cerebral hematocrit which represents a relative ratio to the systemic hematocrit, as 0.69 and 0.72 in PET (Lammertsma et al., 1984) and SPECT (Sakai et al., 1985), respectively. In accordance with these measurements, peripheral hematocrit in the microscopic territories is shown to be lower (about 10–30%) than the systemic circulation (Lipowsky et al., 1980). Those changes in peripheral hematocrit critically affect blood flow velocity in the brain microcirculation. A lower hematocrit increases in capillary flow velocity independent of systemic pressure (Hudetz et al., 1999). In contrast, the higher plasma viscosity reduces the changes in CBF responses to physiological perturbation, but no differences in the baseline CBF irrespective of the variable plasma viscosities (Tomiyama et al., 2000).

2 MACROSCOPIC METHOD

Macroscopic method measures the perfusion flow that is essential definition of blood flow to supply nutrition to the brain tissue. In a clinical stage CBF study most commonly uses this approach. The macroscopic method determines an average blood flow over certain tissue of region of interest, using RI tracers. In determining the perfusion flow, we have three types of RI tracers, diffusible, nondiffusible, and (chemical)-microsphere tracers (Fig. 2). The kinetics of the first two tracers is analyzed by an indicator dilution principle, and the last one is by a particle fraction principle.

2.1 DIFFUSIBLE TRACER

The diffusible tracer freely diffuses into the brain tissue at the first pass through the capillary. Since the capillary distributes the whole brain, when the tracer is fed into the brain, it diffuses into the whole brain (Fig. 2A). The diffusible tracer is then

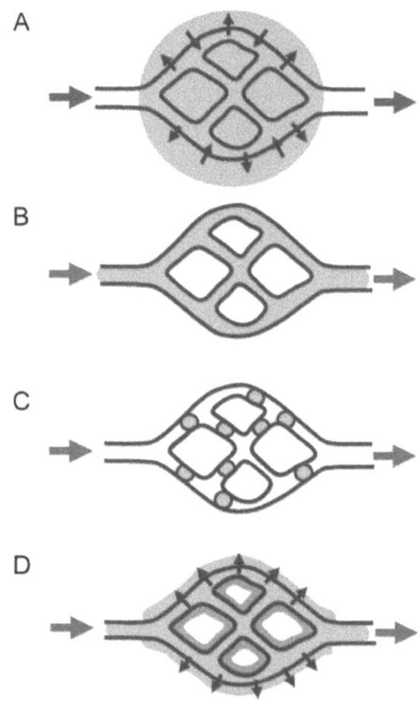

FIG. 2

Schematic draws of four types of tracers. There are four types of tracers for measuring cerebral blood flow. (A) Diffusible tracer: freely diffusible into the brain tissue through BBB. (B) Nondiffusible tracer: confined within blood does not cross BBB. (C) Microsphere tracer; small particle occluding microvessels, therefore forbidden to clinical use. (D) Chemical microsphere tracer. Once lipophilic compound of freely diffused into brain tissue, it converted to hydrophilic and then it has to stay there. (A) and (B) are analyzed based on the indicator dilution principle, and (C) and (D) are based on the particle fraction principle.

washed out and diluted by fresh blood. MTT is calculated based on an indicator dilution principle. Practically when the diffusible tracer is intravenously administered, it reaches the brain in a few 10 s, the tracer concentration is peaked in a couple of min, and it is washed out in a next 5–10 min. The kinetics of the tracer is dependent on a speed of the blood flow (Fig. 3). Therefore, following these processes, we can determine regional MTT, hence regional CBF.

An alternative analysis method is based on a compartment model (Fig. 4). Assuming two compartments of the blood and the brain, differential equation of the tracer concentration is described using an infusion rate constant (K_1) and an effusion rate constant (k_2). The differential equation is solved based on a dynamic measurement of the tracer concentration in the brain and blood. CBF and partition coefficient are given by determining K_1 and K_1/k_2, respectively. The partition coefficient is solubility ratio of the tracer between the brain and blood, and different between the gray matter and white matter, though it is usually assumed to be constant over the whole brain, and is approximated to be unity for H_2O. It is worthwhile to remind that in this tracer kinetics the spatial smoothing of adjacent heterogeneous tissues within the spatial resolution limit usually results in systematic underestimation of CBF. In a clinical stage, the method is simplified so as that we will calculate a table of the relationship between CBF and integral of tracer kinetics in the tissue using the measured arterial input function of the tracer concentration. Then, CBF is determined by looking-up the table of this relationship (Kanno and Lassen, 1979). The most PET-CBF methods employ this approach (Kanno et al., 1987).

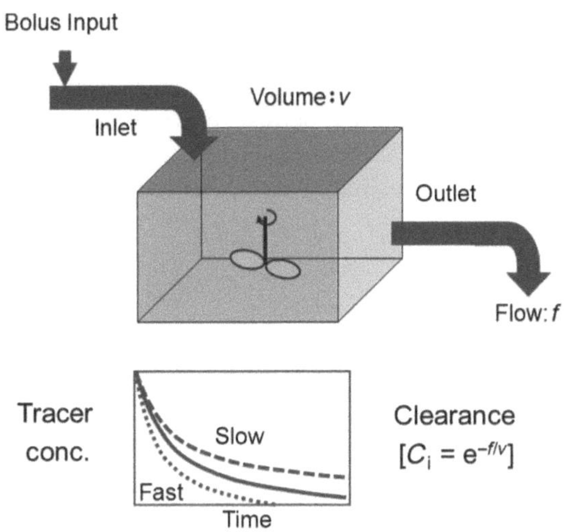

FIG. 3

Chamber model for clearance method. When a bolus input of diffusible tracer is given into closed chamber with volume (v), where is constant flow (f) and mixed immediately, tracer concentration is peaked and then washed out exponentially. The exponential slope is proportional to flow and inversely to volume.

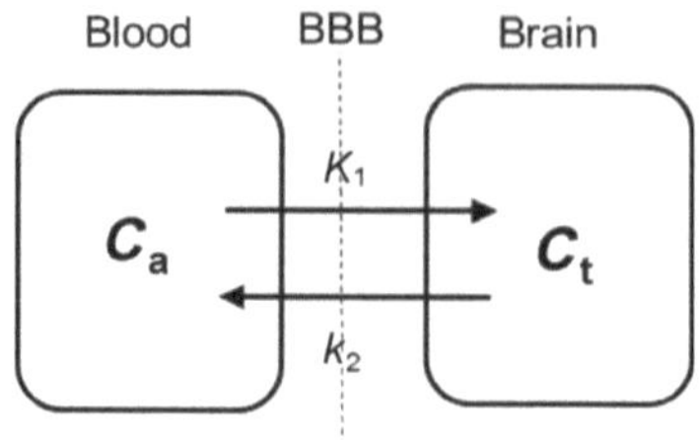

FIG. 4

Two-compartment model for diffusible tracer. Tracer kinetics for blood flow measurement use a two-compartment model. Tracer concentrations in the blood (C_a) and that in the brain (C_t) are changed in time. The tracer is freely transferred across BBB between two compartments with infusion rate constant (K_1) and effusion rate constant (k_2).

2.2 NONDIFFUSIBLE TRACER

The nondiffusible tracer does not diffuse into the brain tissue and stay inside of the vascular bed (Fig. 2B). Calculation of this tracer kinetics is again based on the indicator dilution principle. However, because the vascular space is 3–5% of the brain space, MTT of the nondiffusible tracer is also 3–5% of that of the diffusible tracer, ie, a few 10 s. In a clinical stage, contrast media in CT or MRI are used as the nondiffusible tracer. Because these modalities do not require heavy facilities for RI, they are often used to evaluate CBF as a CT or MRI "perfusion image." However, because of short MTT in nondiffusible tracer, even a few seconds transit delay of the collateral flow seen in cerebrovascular diseases causes serious errors in the CBF estimation, often resulting in right–left hemispheric misestimation. Therefore, interpretation of the CT or MRI "perfusion image" data obtained from stroke patients has to be carefully inspected in a qualitative and quantitative point of view.

2.3 MICROSPHERE TRACER

The microsphere tracer occludes the microvessels proportionally to the blood flow (Fig. 2C). Therefore, we can estimate CBF based on a particle fraction principle that calculates CBF from the measurements of tracer distribution. Practically, in clinical situation, because such occluding microsphere is not used in a human study, a sophisticated microsphere has been developed, that is, chemical microsphere (CMS). CMS has lipophilic character in the blood and easily diffuses into the brain tissue across BBB. However, once CMS diffuses into the brain tissue, CMS is metabolized by enzymes and converted to hydrophilic character. Then, the metabolized CMS cannot go back to the blood and remain there in the brain tissue (Nowotnik et al., 1985). Thus, the CMS is apparently trapped in proportion to CBF and distributed there in a flow-dependent manner (Fig. 2D). Actually, several CMSs labeled by ^{99m}Tc and ^{123}I are commercially available. Because the measurement of the arterial concentration of CMS is usually invasive, absolute quantification of CBF is difficult. Instead, relative quantification of CBF, eg, left–right ratio and/or diseased–healthy ratio, is available for clinical use. Because of widespread distribution of SEPCT devices, CMS CBF is now widely used in clinical circumstances.

2.4 ASSUMPTION FOR SPATIAL AND TEMPORAL HETEROGENEITY

Macroscopic methods require important assumption as for spatial and temporal heterogeneity. As far as we use RI tracers, the spatial resolution of this technique will not get high such as to detect circulation information of microvessel like a single capillary. The RI tracer provides an average CBF across over a unit of a few mm^3 of volume. The scale is still containing a few 10^2 of microvessels, a few 10^3 of neurons, and a few 10^4 of glias. This resolution is not small enough to be represented by single function and the distribution of blood flow is heterogeneous. Thus, in a macroscopic method the spatial heterogeneity within this unit volume is allowed to assume as being a single homogeneous function.

In general, such assumption will impose significant methodological underestimation of CBF in the indicator dilution principle. This comes from spatial limitation of the macroscopic methods. Regarding the temporal heterogeneity, the indicator dilution principle requires the steady-state condition of the vascular system during the measurement to assess MTT. It takes a few to 10 min for diffusible tracers and a few 10s of seconds for nondiffusible tracers. If the steady-state condition is not guaranteed during the measurement, the mathematic equation for the tracer kinetics is no more valid, and the averaging temporal heterogeneity usually results in slight underestimation.

3 MICROSCOPIC METHOD

Microscopic method measures the vessel dimensions and flow velocity of blood plasma/cells which are the only measureable parameters that determine CBF. By multiplying those measured mean velocity and cross-sectional area of the vessel, microscopic method provides the bulk flow (ie, volumetric flow rate) in the capillary and/or microvessels. Other parameters (ie, vascular resistance and hematocrit) cannot be directly measured with this technique.

3.1 FLUORESCENT TRACERS AND INSTRUMENTS

It was attempted to determine apparent viscosity by simultaneously measuring pressure drops and flow in the individual microvessels of mesentery (Lipowsky et al., 1978), but not in the brain. The reason why the vascular resistance in the brain microcirculation is undetermined is due to a complex 3D network structure of the brain microcirculation (Fig. 5). To overcome this limitation, pressure-sensitive probe could be desirable. So far, probes available for measurements of CBF in the microscopic tissue are a plasma marker with fluorescently conjugated biocompatible macromolecules and endogenous, fluorescently labeled blood cells.

The instruments to detect these fluorescent tracers are widely available, such as intravital fluorescence microscopy with high-speed video camera and confocal/two-photon laser scanning fluorescence microscopy. In addition to these classical equipments, a new system that enables us to measure volumetric flow rate in the 3D brain microcirculation with microscopic to mesoscopic (several millimeters) fields of view is being developed, such as Doppler optical coherence tomography, Doppler photoacoustic tomography, and high-resolution ultrasound apparatus. In this section, the details on the fluorescent microscopic techniques for the study of neurovascular coupling are explained. Other methods for measurements of microscopic/mesoscopic

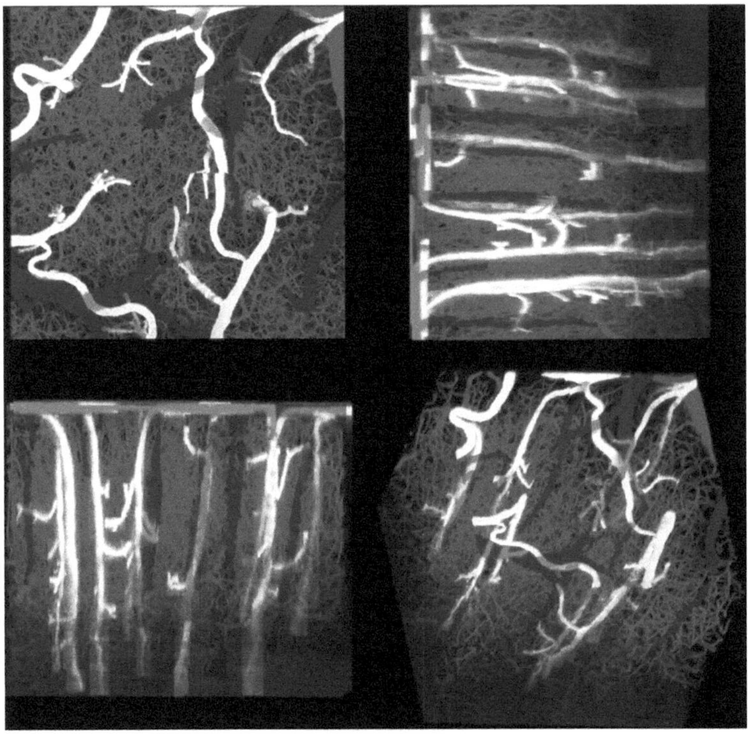

FIG. 5

Microvessels in cortical area of mouse. Image is connected from multiple pictures taken in the area of $0.9 \times 0.9 \times 0.8\ mm^3$ measured by two-photon microscopy (Sugashi et al., unpublished data). The vessels are colored as the arteries (*yellow; light gray* in the print version), capillaries (*orange; gray* in the print version), and veins (*blue; dark gray* in the print version).

blood flow can be referred to recent reviews (Baran and Wang, 2016; Purkayastha and Sorond, 2012; van den Berg et al., 2015). Although laser Doppler flowmetry and laser speckle flowgraphy are widely used for the basic studies of neurovacular coupling in the experimental animals, these instruments provide only an index of flow-related values, but never the absolute scales of the flow or velocity. It is good to compare relative changes in the CBF-related flow patterns under the controlled conditions with these techniques, but limited to quantitative comparisons of those flow indices between the subject-to-subject or location-to-location measurements. The details of the methods and principles of this technique can be referred in the excellent reviews (Briers, 2001; Draijer et al., 2009).

3.2 DIRECT ASSESSMENT OF SPATIOTEMPORAL HETEROGENEITY IN MICROVASCULAR FLOW

The important advantage of the microscopic method for flow measurements is to capture real-time parameters of individual microvessels. This feature would provide us the critical opportunity for clarifying the relation of cause and effect between the flow changes and functional activities of the brain cells by simultaneously measuring those flows, motion of vascular cells, and cellular activities in the neighborhood. Therefore, the live imaging of spatially and temporally varying hemodynamics in their natural conditions (without task or stimulation) itself will provide a hint to find the determinants in maintaining the homeostasis of cerebral circulation and metabolism.

3.3 MEASUREMENTS OF RBC SPEED

The fluorescent RBC technique was first developed by Sarelius and Duling (1982). They incubated purified RBCs collected from donor animals with biocompatible fluorescent dye (tetramethylrhodamine isothiocyanate) and applied them to the systemic circulation of the host animals. The labeled cells serve as a surrogate of the resident blood cells in perfusion and enable us to measure directly flow parameters, such as RBC velocity, flux, hematocrit, and transit time across the vascular networks, in any organs under fluorescence microscopy. The technique becomes widely spread, and modified for use with a variety of fluorescent dyes, including genetic fluorescent RBCs (Masamoto et al., 2013b).

With this labeled RBC technique, it was demonstrated that the mean capillary velocity in the brain microcirculation was equivalent between the mice (2.2 ± 1.6 mm s^{-1}) and rats (2.2 ± 1.3 mm s^{-1}) (Unekawa et al., 2010), although their body size was 10-fold different. It should be interested whether this principle extends to the mean value of the RBC velocity in the brain capillaries of the humans (ie, 2000-fold body size differences).

It was shown that to accurately measure the flow velocity of the labeled RBCs in the brain microcirculation, the instrument for imaging is necessary to meet a high-scanning frame acquisition (eg, >500 frame per second) (Unekawa et al., 2007). This is because the image acquisition with a low-frame rate scanner tends to miss the measurement of high-speed capillaries (eg, thoroughfare channels) or traffics of RBCs whose speed was occasionally high in the measured single capillaries. In addition, the methods relay on a tracking algorism for quantification of the cell displacement over the interval of the frame acquisition to calculate the cell velocity (ie, "particle tracking method"). Schiszler et al. (2005) developed the original software (KEIO-IS2) for automatically tracking the labeled RBCs and imaging two-dimensional distribution of RBC velocities in the rat brains. Their pioneer works shed light on the brain microcirculation properties during normal as well as pathological states, such as ischemic brains and cortical spreading depression (Tomita et al., 2011b; Toriumi et al., 2009; Unekawa et al., 2012, 2013).

Alternatively, without use of the high-speed imaging instruments, a single laser scanning microscopy has become widely used to determine the nonlabeled RBC flow. This technique, called as line-scanning velocimetry, uses fluorescent macromolecule probe as a contrast medium of plasma marker in the animal's circulation, and the nonlabeled blood cells which make a contrast as a dark shade (ie, nonfluorescence) with respect to the fluorescently labeled plasma. A single-line scanning in a longitudinal direction of the vessel enables us to produce a space–time image of the cell displacement over scanning time intervals. This measurement principle was first described by Ellis et al. (1992) and subsequently applied for the cell flow measurements in the brains with confocal laser scanning microscopy (Villringr et al., 1994) and two-photon laser scanning microscopy (Chaigneau et al., 2003; Kleinfeld et al., 1998). Furthermore, several algorisms to effectively convert the flow velocity from the space–time images of the cell motions have been developed, such as Hough, Radon, and Fourier transforms and their combinations (Autio et al., 2011; Dobbe et al., 2008; Drew et al., 2010; Duncan et al., 2010; Kim et al., 2012).

3.4 MEASUREMENTS OF BLOOD PLASMA SPEED

With injection of small volume of carbon black dye as a marker of the plasma flow into the animals, it was demonstrated that spatial distribution of the inverse of MTT in each pixel location represents a spatiotemporally varying CBF under normal, hypotension, and hypercapnia conditions (Schiszler et al., 2000). This technique was further applied for a long-term evaluation of flow dynamic changes in the ischemic brain mice up to 30 days after the occlusion surgery (Tomita et al., 2011a). This bulk flow measurements allowed quantitative comparisons across the regions and subjects, but not produced absolute values of CBF. To quantify the plasma flow speed, we recently developed a method for automatically measuring the transit time and traveling distance of the injected dye through the cortical surface vessels (Hoshikawa et al., under revision). The method directly uses traveling distance-dependent variations in the time–intensity curve (ie, MTT) over single or multiple vessels. Thus, in any situations with visualization of the vessels as well as measurements of the dye transits, flow velocity of the blood plasma can be mapped, which could be beneficial in clinical situations of surgical revascularization and clipping operations of cerebral aneurysm.

3.5 PLASMA VS RBC SPEED

Besides its clinical importance, there is a debatable argument that which components of the blood flow (plasma or blood cells) flow faster through the cerebral microcirculation. Theoretically, blood cells flow faster than the plasma in the small sizes of the vessels, which cause a decrease in peripheral hematocrit in these vessel regions. However, no consistent experimental data have been produced (Kawaguchi et al., 2012; Rovainen et al., 1993). This indicates spatial heterogeneity of plasma and blood cell flow in the parenchymal microcirculation. There are two kinds of the spatial heterogeneity potentially causing those inconsistent flow dynamics: one is a variable speed of the plasma and cell flow through the vessels and the other is a variable traveling distance over the multiple vascular networks. In particular, the latter depends on a depth-dependent difference in the traveling pathways from pial artery to veins. Future works should determine and integrate the complete flow structures of the both plasma and blood cells within the 3D networks of the parenchymal microcirculation. In particular, attention should be paid whether plasma or cell flow redistribution across the capillary networks plays a role in association with global increase in CBF during functional activation.

4 BRIDGING MACROSCOPIC AND MICROSCOPIC MEASUREMENTS OF CBF

4.1 COMPARISON OF PARAMETER MEASURED: HOW WE CAN INTEGRATE THOSE MEASUREMENTS?

The macroscopic method measures the perfusion flow in a closed system of the vascular network. Because the closed system assumes tissue volume of the vascular network having a single inlet and outlet, the method allows for the measurement of absolutely quantitative CBF. This means that it is possible to be compared two measurements of CBF obtained from two different conditions quantitatively. In contrast, the microscopic method measures the bulk flow in an open system which is not confined in the defined tissue volume. To link these macroscopic and microscopic flow data, it is therefore needed to introduce a common parameter which is measurable for the both methods. MTT measured in defined tissue volume could serve as a common parameter for this comparison, because MTT can be measured by a single vessel-basis dynamic scan of fluorescent tracers in the microscopic method and a voxel-basis in the macroscopic method. Those values should be compatible within the defined tissue volume.

Alternatively, a mathematical approach for integration of the microscopically determined flow parameters over geometrically defined tissue volume would allow for a voxel-basis simulation of the representative CBF. The calculated CBF can be compared to the actual CBF measured with the macroscopic method, such as for the changes in response to a variety of physiological perturbations. The question then arises, what is a representative value measured in the macroscopic method? Is this averaged over the multiple vessel flows in a measured tissue volume? or more weighted to the value in a specific type of the vessel? If we find a mismatch between the data in the simulation and macroscopic measurements, we can explore what the macroscopic method really measured.

4.2 DIFFERENCE IN HETEROGENEITY SCALES

One of the important differences between the macroscopic and microscopic methods is their own handling scale for heterogeneity of the flow parameters. The parameter variations within the defined tissue volume are smoothed out in the both temporal and spatial domains in the macroscopic CBF. Smoothing out the heterogeneity leads to possible underestimation/overestimation of the CBF values depending on an analysis method of the tracer kinetics.

On the other hand, the microscopic method directly measures heterogeneity of the flow parameters in the both fine spatial and temporal scales. Thus, a degree of the spatial and temporal heterogeneity caused by certain physiologic conditions can be used to characterize the flow states under the given conditions in the microscopic methods, unlike macroscopic methods. In turn, integration of these microscopic heterogeneities will help to estimate the above mismatches caused in the macroscopic CBF.

4.3 COMPARISON OF PARAMETER CHANGED IN RESPONSE TO PHYSIOLOGICAL AND BIOCHEMICAL PERTURBATIONS

There are several physiological methods to perturb CBF, and we can learn how these perturbations lead to changes in the microscopic flow parameters and result in these microscopic effects on the macroscopic CBF. For example, a $PaCO_2$ change is one of the most widely used methods for physiological perturbations. The macroscopic quantification has shown that the increase in $PaCO_2$ elevates CBF uniformly over the whole brains and increases CBF by 4–6% per 1 mmHg in $PaCO_2$ in humans (Ito et al., 2008; Tominaga et al., 1976). However, it remains unanswered what flow parameters in the microscopic scales produce these defined changes in the macroscopic CBF. Specifically, which part of the vessel changes in the flow velocity or volume is dominant to generate the absolute changes in CBF? Does this response homogeneously elicit the changes across the type of the vessels (ie, pial arteries to veins) irrespective of the brain regions?

Another perturbation method is an induction of hypertension/hypotension, ie, a change in perfusion pressure. It is well known that CBF sustains constant against perfusion pressure changes as far as the perfusion pressure is within the range of autoregulation. In this condition, a pressure gradient is microscopically changed depending on the type of the vessel, which leads to a compensative response of the vascular tones to drive or suppress the flow velocities, resulting in a macroscopically maintained constant CBF. The question how quickly adapt the vessel tones evoked by the pressure change is also unanswered.

Additionally, application of endothelium-specific and/or smooth muscle cell-specific vasoactive agents (eg, acetylcholine, adenosine, sodium nitroprusside, and other chemical/pharmacological agents) could be interested to use as a perturbation method to address how these vascular type-specific perturbations interfere the quantitative relationships between the macroscopic and microscopic CBF. Answering these questions with the microscopic method may give a further indication how the absolute CBF is regulated with respect to the vessel tones or capillary flow velocities to maintain the CBF homeostasis.

4.4 COMPARISON OF PARAMETER CHANGED IN RESPONSE TO FUNCTIONAL FOCAL PERTURBATION

Although most of the physiological perturbations explained in the earlier sections induce global changes in CBF over the whole brain regions, activation of neurovascular coupling (NVC) confines the flow perturbation at a local brain. So far, many studies have been carried out to address the physiological questions, such as a role of the microvascular response in generating the functional hyperemia, by measuring the stimulation-evoked changes in the flow parameters from the resting baseline states. Considering the variable baseline states, such as vascular morphology and density, energy metabolism, and cell density and activities of neurons and glias, across the brain regions, it would be worthwhile to carefully examine the regional differences in the macroscopic and microscopic CBF changes across the brain regions, such as visual, auditory, and somatosensory cortices. In addition, a variable stimulation paradigm (eg, duration, frequency, and intensity) and a type of the stimulation (natural or artificial) may produce a variable contribution of the flow parameter changes. It has been shown that short duration (10 s) of sensory stimuli mainly evoke vasodilation of the pial artery and penetrating arterioles in the mouse cortex, but longer duration (30 s) of the stimuli also involve the changes in the venous vessels in addition to the arterial volume changes (Drew et al., 2011). It remains undetermined how these evolutional changes in the vascular tones and flow velocities affect the absolute scale of the macroscopic CBF. Here, caution should be also paid that these changes in CBF must be normalized to the changes in neural activities such as measured by evoked field potentials, to quantitatively compare the hemodynamic responses.

4.5 FUTURE WORKS

One way to understand the microvessel behavior is employing a single physiological perturbation, observing carefully the responding microvessel responses, and comparing quantitatively the microscopic bulk flow with the macroscopic perfusion flow. The other sophisticated approach is combining multiple perturbations. Combining quantitatively magnitude of the response to a single perturbation with that of multiple perturbations will become a powerful tool to test superimposition/synergy effects in microvessel level. For example, in NVC with photic stimulation, CBF increase was 20–30% in the human visual cortex. When this photic stimulation is combined with hypercapnia/hypocapnia perturbation, the CBF increase was enhanced/reduced using PET-CBF (Shimosegawa et al., 1995). This combined effect was also confirmed in combining somatosensory NVC stimulation and hypercapnia/hypocapnia perturbation in the rat cortex (Matsuura et al., 2000). Even with these macroscopic methods, quantitative CBF measurements revealed such superimposition/synergy effect of the hemodynamic conjugate relationship between local NVC regulation and global hypercapnia/hypocapnia regulation.

Another next challenge is that one could pursue the long-term adaptive changes in the flow parameters in the both macroscopic and microscopic states by using specific animal models. The chronic hypoperfusion model in mouse treated with unilateral common carotid occlusion in which the perfusion pressure is slightly below the autoregulation limit mimics one of the chronic perturbation states provoked in patients with persistent vasodilation and exhausted vasodilatory capacities (Tajima et al., 2014). The animal model that is chronically exposed to hypoxia is also considered as a stable method for induction of chronic perturbation, which evoke capillary remodeling, vasodilation, and reduced responses of the activity-induced dilation of penetrating arterioles, but not the pial arteries, resulting in the decrease in flow response to NVC (Masamoto et al., 2013a, 2014; Sekiguchi et al., 2014; Takuwa et al., 2013; Yoshihara et al., 2013).

A long-term restructuring of the microvascular networks is apparently provoked during a process of aging. A reduction of the number density of the penetrating arterioles and parenchymal capillaries is well characterized in the aged animal brains (Lam et al., 2010; Sonntag et al., 1997). These anatomical features indicate an extension of the noncapillary space in the aged brain tissue, which cause a more severe damage to the hypoxic insults in the aged than in the younger brains (Lam et al., 2010). However, it remains unclear what level of macroscopically measured CBF causes an increased sensitivity to the cerebral ischemic insults. Combining these chronic animal models with physiological and NVC perturbations will provide direct insights into the underlying mechanisms of the plastic/adaptive changes in the microscopic vascular networks, neurovascular coupling, and their link to global CBF measured in clinical situations.

5 CONCLUDING REMARKS

Before closing this chapter for the measurement of CBF, we would like to point out a couple of comments why we eager to measure quantitative CBF. First, CBF is a fundamental parameter to represent brain function. However, CBF measurements always suffer from various physiological/biological perturbing factors and contain spatial/temporal heterogeneity which itself may represent the realities of brain functions. Therefore, establishment of the absolute quantitative method for CBF measurement is a starting point to understand how the brain is working. Second, PET studies have shown that activity-induced changes in oxygen metabolism are almost silent in response to neuronal activation compared to the changes in CBF. The regulation of CBF therefore seems more sensitive to the brain function than changes in oxygen metabolism as like CBF does actively adjust to the functional activity of the brain. Or alternatively, does CBF sustain the brain function high?

REFERENCES

Autio, J., Kawaguchi, H., Saito, S., Aoki, I., Obata, T., Masamoto, K., Kanno, I., 2011. Spatial frequency-based analysis of mean red blood cell speed in single microvessels: investigation of microvascular perfusion in rat cerebral cortex. PLoS One 6. e24056.

Baran, U., Wang, R.K., 2016. Review of optical coherence tomography based angiography in neuroscience. Neurophotonics 3 (1), 010902.

Briers, J.D., 2001. Laser Doppler, speckle and related techniques for blood perfusion mapping and imaging. Physiol. Meas. 22 (4), R35–R66.

Chaigneau, E., Oheim, M., Audinat, E., Charpak, S., 2003. Two-photon imaging of capillary blood flow in olfactory bulb glomeruli. Proc. Natl. Acad. Sci. U.S.A. 100 (22), 13081–13086.

Detre, J.A., Leigh, J.S., Williams, D.S., Koretsky, A.P., 1992. Perfusion imaging. Magn. Reson. Med. 23 (1), 37–45.

Dobbe, J.G., Streekstra, G.J., Atasever, B., van Zijderveld, R., Ince, C., 2008. Measurement of functional microcirculatory geometry and velocity distributions using automated image analysis. Med. Biol. Eng. Comput. 46, 659–670.

Draijer, M., Hondebrink, E., van Leeuwen, T., Steenbergen, W., 2009. Review of laser speckle contrast techniques for visualizing tissue perfusion. Lasers Med. Sci. 24 (4), 639–651.

Drew, P.J., Blinder, P., Cauwenberghs, G., Shih, A.Y., Kleinfeld, D., 2010. Rapid determination of particle velocity from space-time images using the Radon transform. J. Comput. Neurosci. 29, 5–11.

Drew, P.J., Shih, A.Y., Kleinfeld, D., 2011. Fluctuating and sensory-induced vasodynamics in rodent cortex extend arteriole capacity. Proc. Natl. Acad. Sci. U.S.A. 108 (20), 8473–8478.

Duncan, D.D., Lemaillet, P., Ibrahim, M., Nguyen, Q.D., Hiller, M., Ramella-Roman, J., 2010. Absolute blood velocity measured with a modified fundus camera. J. Biomed. Opt. 15, 056014.

Ellis, C.G., Ellsworth, M.L., Pittman, R.N., Burgess, W.L., 1992. Application of image analysis for evaluation of red blood cell dynamics in capillaries. Microvasc. Res. 44 (2), 214–225.

Frackowiak, R.S., Lenzi, G.L., Jones, T., Heather, J.D., 1980. Quantitative measurement of regional cerebral blood flow and oxygen metabolism in man using 15O and positron emission tomography: theory, procedure, and normal values. J. Comput. Assist. Tomogr. 4 (6), 727–736.

Hudetz, A.G., Wood, J.D., Biswal, B.B., Krolo, I., Kampine, J.P., 1999. Effect of hemodilution on RBC velocity, supply rate, and hematocrit in the cerebral capillary network. J. Appl. Physiol. (1985) 87 (2), 505–509.

Ito, H., Kanno, I., Ibaraki, M., Suhara, T., Miura, S., 2008. Relationship between baseline cerebral blood flow and vascular responses to changes in $PaCO_2$ measured by positron emission tomography in humans: implication of inter-individual variations of cerebral vascular tone. Acta Physiol (Oxf.) 193 (4), 325–330.

Kanno, I., Lassen, N.A., 1979. Two methods for calculating regional cerebral blood flow from emission computed tomography of inert gas concentrations. J. Comput. Assist. Tomogr. 3 (1), 71–76.

Kanno, I., Iida, H., Miura, S., Murakami, M., Takahashi, K., Sasaki, H., Inugami, A., Shishido, F., Uemura, K., 1987. A system for cerebral blood flow measurement using an H215O autoradiographic method and positron emission tomography. J. Cereb. Blood Flow Metab. 7 (2), 143–153.

Kawaguchi, H., Masamoto, K., Ito, H., Kanno, I., 2012. Image-based vessel-by-vessel analysis for red blood cell and plasma dynamics with automatic segmentation. Microvasc. Res. 84 (2), 178–187.

Kety, S.S., Schmidt, C.F., 1948. The nitrous oxide method for the quantitative determination of cerebral blood flow in man: theory, procedure and normal values. J. Clin. Invest. 27 (4), 476–483.

Kim, T.N., Goodwill, P.W., Chen, Y., Conolly, S.M., Schaffer, C.B., Liepmann, D., Wang, R.A., 2012. Line-scanning particle image velocimetry: an optical approach for quantifying a wide range of blood flow speeds in live animals. PLoS One 7. e38590.

Kleinfeld, D., Mitra, P.P., Helmchen, F., Denk, W., 1998. Fluctuations and stimulus-induced changes in blood flow observed in individual capillaries in layers 2 through 4 of rat neocortex. Proc. Natl. Acad. Sci. U.S.A. 95 (26), 15741–15746. Erratum in Proc. Natl. Acad. Sci. U.S.A. 1999, 96(14), 8307.

Lam, C.K., Yoo, T., Hiner, B., Liu, Z., Grutzendler, J., 2010. Embolus extravasation is an alternative mechanism for cerebral microvascular recanalization. Nature 465 (7297), 478–482.

Lammertsma, A.A., Brooks, D.J., Beaney, R.P., Turton, D.R., Kensett, M.J., Heather, J.D., Marshall, J., Jones, T., 1984. In vivo measurement of regional cerebral haematocrit using positron emission tomography. J. Cereb. Blood Flow Metab. 4 (3), 317–322.

Lassen, N.A., Hoedt-Rasmussen, K., 1966. Human cerebral blood flow measured by two inert gas techniques. Comparison of the Kety-Schmidt method and the intra-arterial injection method. Circ. Res. 19 (4), 681–694.

Lassen, N.A., Ingvar, D.H., 1961. The blood flow of the cerebral cortex determined by radioactive krypton. Experientia 17, 42–43.

Lassen, N.A., Ingvar, D.H., Skinhøj, E., 1978. Brain function and blood flow. Sci. Am. 239 (4), 62–71.

Lipowsky, H.H., Kovalcheck, S., Zweifach, B.W., 1978. The distribution of blood rheological parameters in the microvasculature of cat mesentery. Circ. Res. 43 (5), 738–749.

Lipowsky, H.H., Usami, S., Chien, S., 1980. In vivo measurements of "apparent viscosity" and microvessel hematocrit in the mesentery of the cat. Microvasc. Res. 19 (3), 297–319.

Longden, T.A., Hill-Eubanks, D.C., Nelson, M.T., 2016. Ion channel networks in the control of cerebral blood flow. J. Cereb. Blood Flow Metab. 36 (3), 492–512.

Masamoto, K., Kawaguchi, H., Ito, H., Kanno, I., 2013a. Dynamic two-photon imaging of cerebral microcirculation using fluorescently labeled red blood cells and plasma. Adv. Exp. Med. Biol. 765, 163–168.

Masamoto, K., Takuwa, H., Tomita, Y., Toriumi, H., Unekawa, M., Taniguchi, J., Kawaguchi, H., Itoh, Y., Suzuki, N., Ito, H., Kanno, I., 2013b. Hypoxia-induced cerebral angiogenesis in mouse cortex with two-photon microscopy. Adv. Exp. Med. Biol. 789, 15–20.

Masamoto, K., Takuwa, H., Seki, C., Taniguchi, J., Itoh, Y., Tomita, Y., Toriumi, H., Unekawa, M., Kawaguchi, H., Ito, H., Suzuki, N., Kanno, I., 2014. Microvascular sprouting, extension, and creation of new capillary connections with adaptation of the neighboring astrocytes in adult mouse cortex under chronic hypoxia. J. Cereb. Blood Flow Metab. 34 (2), 325–331.

Matsuura, T., Fujita, H., Kashikura, K., Kanno, I., 2000. Evoked local cerebral blood flow induced by somatosensory stimulation is proportional to the baseline flow. Neurosci. Res. 38 (4), 341–348.

Mchedlishvili, G.I., 1986. Arterial Behavior and Blood Circulation in the Brain. Plenum Press, New York.

Nowotnik, D.P., Canning, L.R., Cumming, S.A., Harrison, R.C., Higley, B., Nechvatal, G., Pickett, R.D., Piper, M., Bayne, V.J., Foster, A.N., Weisner, P.S., Neirinckx, R.D., Volkert, W.A., Troutner, D.E., Holmes, R.A., 1985. Development of a 99mTc-labelled radiopharmaceutical for cerebral blood flow imaging. Nucl. Med. Commun. 6, 499–506.

Ogawa, S., Tank, D.W., Menon, R., Ellermann, J.M., Kim, S.G., Merkle, H., Ugurbil, K., 1992. Intrinsic signal changes accompanying sensory stimulation: functional brain mapping with magnetic resonance imaging. Proc. Natl. Acad. Sci. U.S.A. 89 (13), 5951–5955.

Purkayastha, S., Sorond, F., 2012. Transcranial Doppler ultrasound: technique and application. Semin. Neurol. 32 (4), 411–420.

Purves, M.J., 1972. The physiology of the cerebral circulation. Monogr. Physiol. Soc. 28, 1–414.

Raichle, M.E., Martin, W.R., Herscovitch, P., Mintun, M.A., Markham, J., 1983. Brain blood flow measured with intravenous H2(15)O. II. Implementation and validation. J. Nucl. Med. 24 (9), 790–798.

Reivich, M., Slater, R., Sano, K., 1969. Cerebral Blood Flow. Springer, Berlin. In: The Physiology of the Cerebral Circulation by Purves, M.J., Cambridge University Press, London, p. 142, Fig. 5.15.

Rhodin, J.A., 1967. The ultrastructure of mammalian arterioles and precapillary sphincters. J. Ultrastruct. Res. 18 (1), 181–223.

Rhodin, J.A., 1968. Ultrastructure of mammalian venous capillaries, venules, and small collecting veins. J. Ultrastruct. Res. 25 (5), 452–500.

Rovainen, C.M., Woolsey, T.A., Blocher, N.C., Wang, D.B., Robinson, O.F., 1993. Blood flow in single surface arterioles and venules on the mouse somatosensory cortex measured with videomicroscopy, fluorescent dextrans, nonoccluding fluorescent beads, and computer-assisted image analysis. J. Cereb. Blood Flow Metab. 13 (3), 359–371.

Sakai, F., Nakazawa, K., Tazaki, Y., Ishii, K., Hino, H., Igarashi, H., Kanda, T., 1985. Regional cerebral blood volume and hematocrit measured in normal human volunteers by single-photon emission computed tomography. J. Cereb. Blood Flow Metab. 5 (2), 207–213.

Sarelius, I.H., Duling, B.R., 1982. Direct measurement of microvessel hematocrit, red cell flux, velocity, and transit time. Am. J. Physiol. 243, H1018–H1026.

Schiszler, I., Tomita, M., Fukuuchi, Y., Tanahashi, N., Inoue, K., 2000. New optical method for analyzing cortical blood flow heterogeneity in small animals: validation of the method. Am. J. Physiol. Heart Circ. Physiol. 279 (3), H1291–H1298.

Schiszler, I., Takeda, H., Tomita, M., Tomita, Y., Osada, T., Unekawa, M., Tanahashi, N., Suzuki, N., 2005. Software (KEIO-IS2) for automatically tracking red blood cells

(RBCs) with calculation of individual RBC velocities in single capillaries of rat brain. J. Cereb. Blood Flow Metab. 25, S541.

Sekiguchi, Y., Takuwa, H., Kawaguchi, H., Kikuchi, T., Okada, E., Kanno, I., Ito, H., Tomita, Y., Itoh, Y., Suzuki, N., Sudo, R., Tanishita, K., Masamoto, K., 2014. Pial arteries respond earlier than penetrating arterioles to neural activation in the somatosensory cortex in awake mice exposed to chronic hypoxia: an additional mechanism to proximal integration signaling? J. Cereb. Blood Flow Metab. 34 (11), 1761–1770.

Shimosegawa, E., Kanno, I., Hatazawa, J., Fujita, H., Iida, H., Miura, S., Murakami, M., Inugami, A., Ogawa, T., Itoh, H., et al., 1995. Photic stimulation study of changing the arterial partial pressure level of carbon dioxide. J. Cereb. Blood Flow Metab. 15 (1), 111–114.

Sonntag, W.E., Lynch, C.D., Cooney, P.T., Hutchins, P.M., 1997. Decreases in cerebral microvasculature with age are associated with the decline in growth hormone and insulin-like growth factor 1. Endocrinology 138 (8), 3515–3520.

Tajima, Y., Takuwa, H., Kokuryo, D., Kawaguchi, H., Seki, C., Masamoto, K., Ikoma, Y., Taniguchi, J., Aoki, I., Tomita, Y., Suzuki, N., Kanno, I., Saeki, N., Ito, H., 2014. Changes in cortical microvasculature during misery perfusion measured by two-photon laser scanning microscopy. J. Cereb. Blood Flow Metab. 34 (8), 1363–1372.

Takuwa, H., Masamoto, K., Yamazaki, K., Kawaguchi, H., Ikoma, Y., Tajima, Y., Obata, T., Tomita, Y., Suzuki, N., Kanno, I., Ito, H., 2013. Long-term adaptation of cerebral hemodynamic response to somatosensory stimulation during chronic hypoxia in awake mice. J. Cereb. Blood Flow Metab. 33 (5), 774–779.

Tominaga, S., Strandgaard, S., Uemura, K., Ito, K., Kutsuzawa, T., 1976. Cerebrovascular CO2 reactivity in normotensive and hypertensive man. Stroke 7 (5), 507–510.

Tomita, Y., Pinard, E., Tran-Dinh, A., Schiszler, I., Kubis, N., Tomita, M., Suzuki, N., Seylaz, J., 2011a. Long-term, repeated measurements of mouse cortical microflow at the same region of interest with high spatial resolution. Brain Res. 1372, 59–69.

Tomita, M., Tomita, Y., Unekawa, M., Toriumi, H., Suzuki, N., 2011b. Oscillating neurocapillary coupling during cortical spreading depression as observed by tracking of FITC-labeled RBCs in single capillaries. Neuroimage 56 (3), 1001–1010.

Tomiyama, Y., Brian Jr., J.E., Todd, M.M., 2000. Plasma viscosity and cerebral blood flow. Am. J. Physiol. Heart Circ. Physiol. 279 (4), H1949–H1954.

Toriumi, H., Tatarishvili, J., Tomita, M., Tomita, Y., Unekawa, M., Suzuki, N., 2009. Dually supplied T-junctions in arteriolo-arteriolar anastomosis in mice: key to local hemodynamic homeostasis in normal and ischemic states? Stroke 40 (10), 3378–3383.

Unekawa, M., Tomita, M., Osada, T., Tomita, Y., Toriumi, H., Suzuki, N., 2007. Sampling rate-dependent RBC velocity in intraparenchymal single capilaries of rat cerebral cortex. Microvasc. Rev. Commun. 1 (1), 12–15.

Unekawa, M., Tomita, M., Tomita, Y., Toriumi, H., Miyaki, K., Suzuki, N., 2010. RBC velocities in single capillaries of mouse and rat brains are the same, despite 10-fold difference in body size. Brain Res. 1320, 69–73.

Unekawa, M., Tomita, M., Tomita, Y., Toriumi, H., Suzuki, N., 2012. Sustained decrease and remarkable increase in red blood cell velocity in intraparenchymal capillaries associated with potassium-induced cortical spreading depression. Microcirculation 19 (2), 166–174.

Unekawa, M., Tomita, Y., Toriumi, H., Masamoto, K., Kanno, I., Suzuki, N., 2013. Potassium-induced cortical spreading depression bilaterally suppresses the electroencephalogram but only ipsilaterally affects red blood cell velocity in intraparenchymal capillaries. J. Neurosci. Res. 91 (4), 578–584.

van den Berg, P.J., Daoudi, K., Steenbergen, W., 2015. Review of photoacoustic flow imaging: its current state and its promises. Photoacoustics 3 (3), 89–99.

Villringer, A., Them, A., Lindauer, U., Einhäupl, K., Dirnagl, U., 1994. Capillary perfusion of the rat brain cortex. An in vivo confocal microscopy study. Circ. Res. 75 (1), 55–62.

Williams, D.S., Detre, J.A., Leigh, J.S., Koretsky, A.P., 1992. Magnetic resonance imaging of perfusion using spin inversion of arterial water. Proc. Natl. Acad. Sci. U.S.A. 89 (1), 212–216.

Yoshihara, K., Takuwa, H., Kanno, I., Okawa, S., Yamada, Y., Masamoto, K., 2013. 3D analysis of intracortical microvasculature during chronic hypoxia in mouse brains. Adv. Exp. Med. Biol. 765, 357–363.

CHAPTER

5

New horizons in neurometabolic and neurovascular coupling from calibrated fMRI

C.Y. Shu*, B.G. Sanganahalli*,†, D. Coman*,†, P. Herman*,†, F. Hyder*,†,1

**Yale University, New Haven, CT, United States*

†Magnetic Resonance Research Center (MRRC), Yale University, New Haven, CT, United States

[1]Corresponding author: Tel.: +1-203-785-6205; Fax: +1-203-785-6643, e-mail address: fahmeed.hyder@yale.edu

Abstract

Neurovascular coupling relates changes in neuronal activity to constriction/dilation of microvessels. However neurometabolic coupling, which is less well known, relates alterations in neuronal activity with metabolic demands. The link between the blood oxygenation level dependent (BOLD) signal and neural activity opened doors for functional MRI (fMRI) to be a powerful neuroimaging tool in the neurosciences. But due to the complex makeup of BOLD contrast, researchers began to investigate the relationship between BOLD signal and blood flow and/or volume changes during functional brain activation, which together provided the tools to measure oxygen consumption on the basis of the biophysical model of BOLD. This field is called calibrated fMRI, thereby allowed probing of both neurometabolic and neurovascular couplings for a variety of health conditions in animals and humans. Calibrated fMRI may provide brain disorder biomarkers that could be used for monitoring effective therapies.

Keywords

GABA, Glutamate, Field potentials, Multiunit activity, Neuroimaging

The association between the blood oxygenation level dependent (BOLD) signal and neural activity positioned function MRI (fMRI) as a powerful imaging modality for studying brain function (Ugurbil, 2012; see also chapter "Submillimeter-resolution fMRI: Toward understanding local neural processing" by M. Fukuda et al.), despite the complex makeup of BOLD contrast (Hyder and Rothman, 2012). To further understand the BOLD signal quantitatively, researchers sought to characterize the relationship between BOLD signal and blood flow (CBF) and/or volume (CBV) changes during functional brain activation, which together provided the tools to measure

Progress in Brain Research, Volume 225, ISSN 0079-6123, http://dx.doi.org/10.1016/bs.pbr.2016.02.003

cerebral metabolic rate of oxygen consumption (CMR_{O_2}) on the basis of the biophysical model of BOLD (Ogawa et al., 1993). Hence the field of calibrated fMRI emerged, thereby allowing probing of both neurometabolic and neurovascular couplings in the brain using multimodal techniques that measure neural activities, hemodynamics, and oxygen metabolism for a variety of health conditions in animals and humans.

The idea of "neurovascular coupling," which relates changes in neuronal activity and constriction/dilation of microvessels to decrease/increase of CBF (or CBV), is well known (Chen et al., 2014; Devor and Boas, 2012; Kozberg et al., 2013; Radhakrishnan et al., 2011; see also chapters (1) "Neurogenic control of parenchymal arterioles in the cerebral cortex" by H. Hotta; (2) "Involvement of astrocytes in neurovascular communication" by M. Nuriya and H. Hirase). However "neurometabolic coupling," which relates alterations in neuronal activity with CMR_{O_2} by independent mechanisms from neurovascular coupling, is less known (Devor et al., 2012; Herman et al., 2013b; Hyder et al., 2010; Thompson et al., 2004; Viswanathan and Freeman, 2007). We studied neurovascular/neurometabolic couplings based on relations of LFP/MUA with BOLD, CBF, CBV, and CMR_{O_2} (Bailey et al., 2013; Herman et al., 2009, 2013a,b, 2015; Sanganahalli et al., 2009, 2011, 2013). Insights into differences in neurometabolic and neurovascular couplings across regions in disease and health will provide brain disorder biomarkers that could be used for monitoring effective therapies.

1 MEASURING CMR_{O_2} WITH CALIBRATED fMRI

It is well established that glucose consumption (CMR_{glc}), particularly its oxidation, efficiently produces energy to support brain function (Siesjo, 1978). However in the late 1980s some PET data revealed that CMR_{glc} and CBF changed significantly during brain activation, but CMR_{O_2} did not (Fox and Raichle, 1986; Fox et al., 1988). These PET results, along with early theoretical calculations of the energy demand of neuronal signaling by Creutzfeldt, supported a model of brain function in which neuronal signaling had minimal oxidative energy costs (Creutzfeldt, 1975). This conclusion was surprising given the disproportionately high energy costs of human brain function at rest (Aiello and Wheeler, 1995; Sokoloff, 1991).

To test the oxidative energy demand of functional brain activation without having to rely on fMRI, independent magnetic resonance spectroscopy (MRS) studies were performed in animal and human visual cortex. ^{13}C MRS is a very powerful tool to study metabolism when it is used in combination with a ^{13}C-labeled substrate, such as [1-^{13}C]glucose. Infusion of [1-^{13}C]glucose (into the blood stream) results in ^{13}C turnover into glutamate, glutamine, GABA, lactate, aspartate, etc., in brain tissue. Since ^{13}C turnover into these pools is time dependent, coupled differential equations can be used to estimate a range of fluxes (including neuronal/glial glucose oxidation, glutamate-GABA-glutamine cycling; Hyder et al., 2006). Furthermore proton-observed carbon-edited (POCE) MRS technique allows the detection of ^{13}C turnover with the higher sensitivity of ^{1}H.

At Yale, POCE MRS was used in an anesthetized rat model to demonstrate a large increase in neuronal glucose oxidation in the somatosensory cortex during forepaw activation (Hyder et al., 1996, 1997). Because the earliest interpretations of BOLD signal increase were interpreted to reflect minimal rise in CMR_{O_2} (Kim and Ugurbil, 1997; Ogawa et al., 1993), this Yale finding was especially controversial at the time since a large increase in CMR_{O_2} in the presence of a positive BOLD signal was considered an artifact based on the prior PET findings. POCE MRS studies of the human brain conducted independently at the Universities of Minnesota and Nottingham also showed a substantial increase in CMR_{O_2} with visual stimulation in the primary visual cortex (Chen et al., 2001; Chhina et al., 2001). Parallel to the aforementioned studies, optical imaging with its superior temporal resolution (on the order of 200 ms) was able to show that postactivation oxygen consumption increased ~2 s before a new wave of blood flow arrived (Malonek and Grinvald, 1996), and thus explained that a positive BOLD signal during neural activation was not due to negligible CMR_{O_2} change, but a significant CMR_{O_2} increase in the presence of an ample rise in blood flow. Their work combined with later findings supported the initial finding at Yale (Barinaga, 1997), but controversial nonetheless.

The term calibrated fMRI was first used in the literature in 1998, where Davis et al. exploited the CMR_{O_2} −CBF and CBV–CBF couplings embedded within the BOLD response ($\Delta S/S$) to extract $\Delta CMR_{O_2}/CMR_{O_2}$ (Davis et al., 1998),

$$\Delta S/S_0 = M\left(1 - \left(CMR_{O_2}/CMR_{O_{2,0}}\right)^{\beta} CBF/CBF_0\right)^{\alpha-\beta}, \quad (1)$$

where the subscript "0" represents the baseline state for each parameter. M represents the maximum baseline BOLD signal change, which is described in Section 2; α describes the CBV–CBF coupling and is assumed to be 0.4 based on PET primate studies of CBV–CBF relationship with hypercapnia challenge (Grubb et al., 1974), but as discussed in Section 3, β is measurable with MRI as well (Hyder et al., 2001) and it describes the CMR_{O_2} −CBF coupling and is assumed to be 1.5 in human studies based on simulations of BOLD effect at 1.5 T (Boxerman et al., 1995), but as discussed in Section 4, it is assumed to approach one at higher magnetic fields (Ogawa et al., 1993).

The early development of calibrated fMRI studies, both in animals and humans, greatly benefited from MRI methods for CBV and CBF imaging. In the 1990s several groups focused on novel MRI methods for CBV and CBF imaging, which proved to be seminal for development of calibrated fMRI (Detre et al., 1992; Kennan et al., 1998; Kim, 1995; Mandeville et al., 1998; Williams et al., 1992).

Just as hemoglobin is nature's endogenous MRI contrast agent to measure blood oxygenation changes, exogenous MRI contrast agents that are specifically blood-borne can reflect alterations in CBV with almost the same assumptions (ie, more paramagnetic agent, lower MRI signal due to increase in R_2^* or R_2). Since each red blood cell packs hemoglobin molecules in high density, these micrometer-sized particulates (ie, red blood cells) when traversing down capillaries create large paramagnetic fields (ie, BOLD effect). Similarly, nanometer-sized particulates containing high density of paramagnetic iron oxide moieties when made blood-borne create similar superparamagnetic fields. In both cases, the spatial distribution of magnetic susceptibility gradients generated beyond blood vessel walls affect distal tissue water proton transverse relaxation rates.

Principles of CBF measured by MRI are similar to that in PET (see chapter "Vascular Potassium Channels in NVC" by K. Yamada), but require magnetic labeling of arterial blood water entering the brain and detecting subsequent MRI data from the brain, which reflects the degree to which the label has decayed by mixing with unlabeled brain tissue water. This CBF technique is based on arterial spin labeling (ASL) using the sensitivity of the longitudinal relaxation rate of tissue water to reflect blood perfusion. Because magnetic labeling of arterial blood with ASL proves to be a sensitive enough perfusion tracer in brain tissue, the CBF method does not require exogenous contrast agents, and thus can be imaged reliably in both humans and animals.

In addition to measuring BOLD, CBF, and/or CBV, accurate quantification of CMR_{O_2} with calibrated fMRI requires a number of scaling constants—M, α, and β—to be predetermined. These parameters will be discussed in the following Sections 2–4.

2 THE *M* PARAMETER

Since magnetic properties of changing blood oxygenation (Y) affect the tissue water MRI signal through intravoxel spin dephasing (Kennan et al., 1994), changes in Y can be differentially captured with the transverse relaxation rates as measured by gradient-echo $\left(R_2^*\right)$ and spin-echo (R_2). If the relaxation components in R_2^* and R_2 assigned to nonsusceptibility-based effects are ignored (Kida et al., 2000), then under well-shimmed conditions

$$R_2^* \cong R_2'(Y) + R_2(Y) \tag{2}$$

$$R_2 \cong R_2(Y), \tag{3}$$

where $R_2'(Y)$ and $R_2(Y)$ represent the blood oxygenation-dependent relaxation components that are reversible (eg, static magnetic fields inhomogeneity and slow diffusion regime) and irreversible (ie, intermediate to fast diffusion regime), respectively (Kida et al., 2000).

M represents the maximum baseline BOLD signal change which, under the approximation that the change in R_2 due to BOLD is small, is hypothesized to represent the product of $R_2'(Y)$ and echo time (TE).

$$M = R_2'(Y) \cdot \text{TE}. \tag{4}$$

$R_2'(Y)$ is directly measurable by the difference between R_2^* and R_2 under well-shimmed conditions (Hyder et al., 2000, 2001; Kida et al., 1999, 2000; Shu et al., 2016a)

$$R_2'(Y) = R_2^* - R_2, \tag{5}$$

where the first and second terms on the right-hand side are measured by gradient-echo and spin-echo MRI, respectively.

The main pitfall with the direct *M* measurement is the residual components, which are unaccounted for in the relaxation rate terms (Kida et al., 2000; Shu et al., 2016a). In human-calibrated studies the use of gas challenges—with O_2 (ie, hyperoxia), CO_2 (ie, hypercapnia), or combination of both (ie, carbogen)—is the preferred approach to derive *M* (for a recent review, see Hoge, 2012). If CMR_{O_2} remains unaffected (ie, no neural activity change) with the gas perturbations whereas other factors (eg, BOLD, CBF, and CBV) change, then parameter *M* can be derived with the gas perturbations and it can be applied in the functional activation study. These gas perturbations are assumed to be physiologically benign. Both the CBV–CBF and BOLD–CBF couplings in gas challenge are presumed to be the same as in functional activation.

To derive *M*, it should be noted that the experimental data obtained for BOLD and CBF changes during the gas exposures represent a very small pseudo-linear portion of a highly nonlinear (ie, asymptotic) relationship between large CBF changes in relation to smaller BOLD signal changes (Fig. 1). The measured CBF increases during gas exposure are less than 100% (Davis et al., 1998; Gauthier and Hoge, 2012), whereas the exponential fitting that defines the *M*-value from the asymptotic limit revealing the dynamic range of the BOLD signal is reached when CBF increases by several times higher (Gauthier and Hoge, 2013; Hoge et al., 1999a). Moreover, it is difficult to reach large CBF changes in humans without inducing adverse cognitive effects and changes in neural or metabolic activities (Rodgers et al., 2013; Sicard and Duong, 2005; Thesen et al., 2012; Xu et al., 2011; Zappe et al., 2008). Because the data fitting to derive *M*-value is susceptible to inaccuracies of both BOLD and CBF measurements, small experimental errors in BOLD or CBF measurements can substantially deviate the *M*-value derived from asymptotic fitting (Fig. 1).

As recently summarized (Shu et al., 2016a), a very wide range of *M*-values has been reported in human-calibrated fMRI experiments (Fig. 2; Chiarelli et al., 2007b; Davis et al., 1998; Gauthier et al., 2011; Hoge et al., 1999a; Kastrup et al., 1999; Stefanovic et al., 2006). Since it remains uncertain if the range of *M*-values reported in human-calibrated fMRI studies are due to regional and/or state differences, we measured *M* from direct R_2' measurements under well-shimmed conditions and examined regional and brain-state dependencies of *M*-values in the rat brain (Shu et al., 2016a). Most human-calibrated fMRI studies are conducted in the awake state but with animal models different anesthesia can be applied to achieve different levels of global brain activity (Hyder et al., 2013a). α-Chloralose, as a GABAa receptor agonist, inhibits principle cells during resynchronization of the neural network (Garrett and Gan, 1998). Medetomidine is an α_2 adrenoreceptor agonist which depresses mainly glutamate release and thereby causes inhibition of excitatory glutamatergic transmission (Angenstein et al., 2010; Sinclair, 2003). Since these different states can exhibit large differences in neural activity (Herman et al., 2011), MRI studies to measure *M* between α-chloralose and medetomidine were conducted (Shu et al., 2016a).

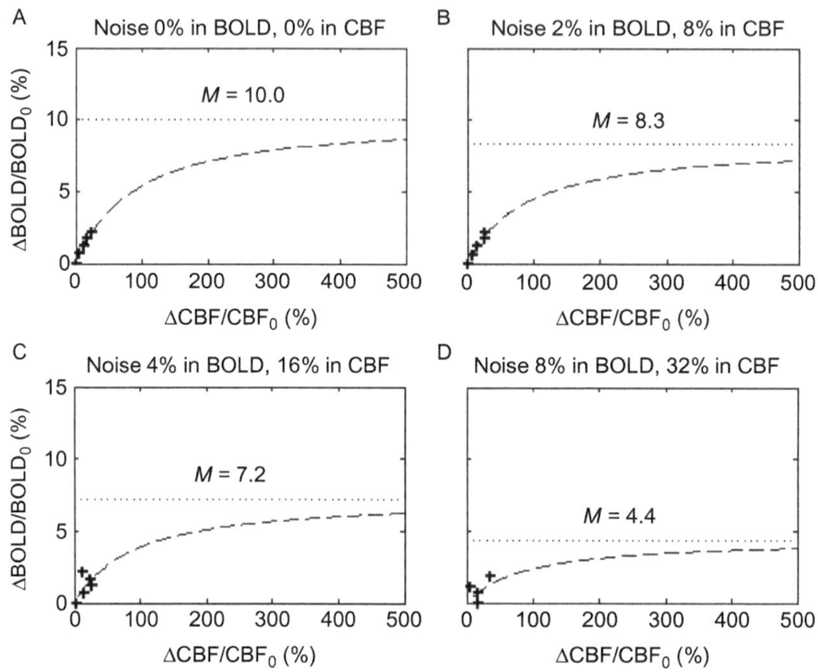

FIG. 1

Determination of *M* for calibrated fMRI from graded hypercapnia experiments in humans. The value of *M* is usually determined from an assumed asymptotic relationship between changes in BOLD and CBF (indicated by the *dotted horizontal lines*). Since typically CBF increases are below 50% (Hoge et al., 1999a), the asymptote is determined from a pseudo-linear portion of BOLD and CBF measurements. Simulations show that as noise level in the BOLD and CBF measurements increase (from 0% to 8% for BOLD and from 0% to 32% for CBF), the fitted *M*-value decreased by as much as a factor of 2 from the theoretical value. Shown above are one example out of 500 simulations with different noise levels, where *M*-values were (A) 10.0 with 0% noise, (B) 8.3 with 2% noise in BOLD and 8% noise in CBF, (C) 7.2 with 4% noise in BOLD and 16% noise in CBF, and (D) 4.4 with 8% noise in BOLD and 32% in CBF. Over 500 simulations, fitted *M*-values at these noise levels were 10.0 ± 0.0, 9.3 ± 1.3, 7.3 ± 1.7, and 5.5 ± 2.0, respectively.

Modified from Shu, C.Y., Herman, P., Coman, D., Sanganahalli, B.G., Wang, H., Juchem, C., Rothman, D.L., de Graaf, R.A., Hyder, F., 2016a. Brain region and activity-dependent properties of M for calibrated fMRI. Neuroimage 125, 848–856.

In this study (Fig. 3), a much smaller range of regional differences in M-values was observed across all regions-of-interest (ROIs) in the brain. Similarly, Blockley et al. measured M-values by two different methods in human cerebral cortex: hypercapnia challenge and R_2' method. They found small difference in M-values between visual cortex and the whole gray matter with the R_2' method, but regional variations in M-values increased significantly with the hypercapnia challenge

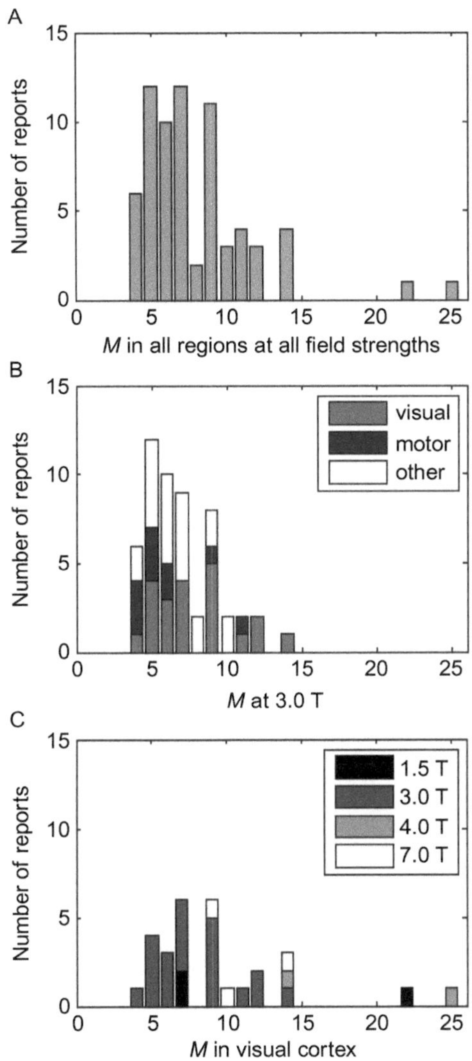

FIG. 2

Summary of *M*-values in human-calibrated fMRI studies published between 1998 and 2015 (Alderliesten et al., 2014; Ances et al., 2008, 2009; Blockley et al., 2015; Bulte et al., 2012; Chiarelli et al., 2007a,b; Davis et al., 1998; Driver et al., 2012; Gauthier and Hoge, 2012, 2013; Gauthier et al., 2011, 2012; Goodwin et al., 2009; Hoge et al., 1999a,b; Kastrup et al., 1999, 2002; Kim and Ugurbil, 1997; Kim et al., 1997; Krieger et al., 2014a,b; Leontiev et al., 2007; Lin et al., 2009, 2010; Mark and Pike, 2012; Mark et al., 2011; Mohtasib et al., 2012; Moradi et al., 2012; Perthen et al., 2008; Restom et al., 2008; St. Lawrence et al., 2003; Stefanovic et al., 2004, 2006; Uludag et al., 2004; Wise et al., 2013; Yücel et al., 2014). (A) Distribution of *M*-values across various magnetic field strengths. Each bin contains

(Blockley et al., 2015). Their study, in agreement with our observations, suggests that regional M variations may, in part, be dependent on the method used (ie, hypercapnia vs R_2' mapping), where R_2' mapping shows less variations in M-values across regions.

The higher M-values observed in white matter relative to gray matter could be partly attributed to the susceptibility of myelin, as a previous study found that at 9.4 T R_2' measured in the white matter of normal mouse was higher by 10 s^{-1} than R_2' measured in demyelinated white matter of mouse on a cuprizone diet (Lee et al., 2012). With different brain activity levels M (or R_2') does not change in white matter, but M (or R_2') changes in gray matter. Since the lack of change in white matter with activity could also be due to alterations in myelin susceptibility and iron content canceling out changes due to vascular factors, future studies will be needed to further assess this possibility. Given that calibrated fMRI is mainly focused toward gray matter changes, the white matter should be masked for quantification of CMR_{O_2} changes from calibrated fMRI data.

In terms of magnitude of M between α-chloralose and medetomidine, a two-way ANOVA test suggested M-values in medetomidine was significantly lower than M-values in α-chloralose in both cortical and subcortical gray matter, but not in white matter and diencephalon (Fig. 4). Particularly, in both states the neocortical M-values were invariant among themselves and the regional differences were not as large as variations in cortical M-values observed in the human-calibrated fMRI literature (Fig. 2). The regional pattern of M-value was conserved across brain states (Fig. 4). Moreover, both in this study and a previous study (Kida et al., 2000), M (or R_2') is decreased (ie, baseline BOLD signal increased) when the brain activity/metabolism is increased. It should be noted, however, that absolute differences of CBF and CMR_{O_2} between α-chloralose and medetomidine states are expected. In the future, characterizing dependencies of M-values over a range of neural activities will give insights as to how the dynamic range of the BOLD signal is related to baseline neuronal activity or metabolic rates.

FIG. 2—Cont'd

M-values rounded up to the nearest integer. Majority of *M*-values ranged from 5 to 15. (B) Distribution of *M*-values at 3.0 T with respect to different cortical regions: visual cortex (*gray bars*), motor cortex (*black bars*), and all other regions (*white bars*). There was no clear trend for *M*-values among these regions. (C) Distribution of *M*-values in the visual cortex with respect to different field strengths: 1.5 T (*black bars*), 3.0 T (*dark gray bars*), 4.0 T (*light gray bars*), and 7.0 T (*white bars*). No magnetic field dependency of *M*-values on field strength was observed.

Modified from Shu, C.Y., Herman, P., Coman, D., Sanganahalli, B.G., Wang, H., Juchem, C., Rothman, D.L., de Graaf, R.A., Hyder, F., 2016a. Brain region and activity-dependent properties of M for calibrated fMRI. Neuroimage 125, 848–856.

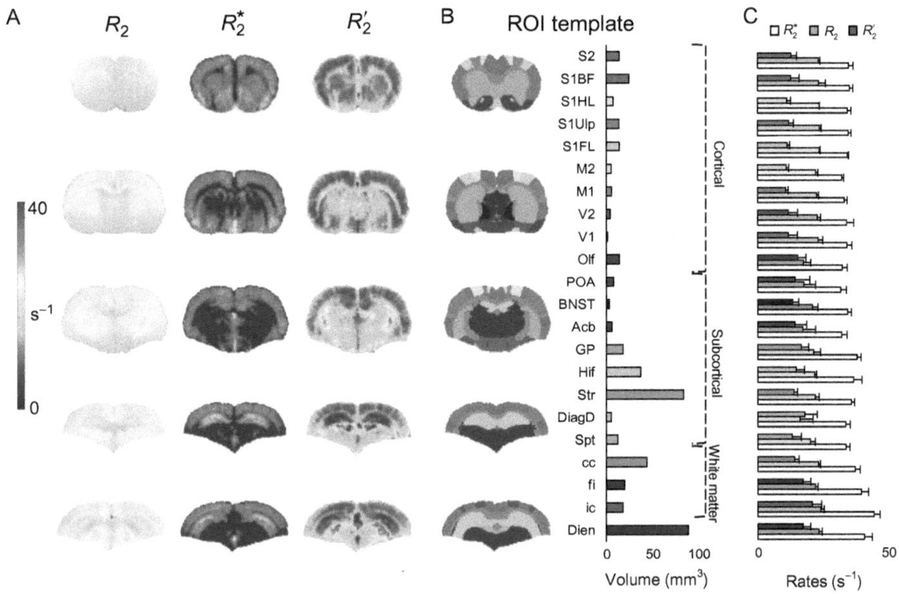

FIG. 3

Summary of transverse relaxation rates (R_2, R_2^*, and R_2') measured at 9.4 T in rats with α-chloralose. (A) Mean R_2, R_2^*, and R_2' maps for representative slices from frontal (*top*) to occipital (*bottom*). (B) ROI templates (onto which all the measured maps were registered) and their volumes are shown. Template includes 10 ROIs in the cerebral cortex, eight ROIs in the subcortical gray matter, three ROIs in white matter, and the diencephalon. The subcortical and cortical regions accounted for 39% and 23%, respectively, of the total volume, whereas white matter and diencephalon covered 18% and 20%, respectively. The abbreviations are as follow: internal capsule (*ic*), septum (*Spt*), diagonal domain (*DiagD*), striatum (*Str*), diencephalon (*Dien*), hippocampal formation (*Hif*), pallidum (*GP*), accumbens (*Acb*), fimbria fornix (*fi*), preoptic area (*POA*), bed nuclei stria terminalis (*BNST*), corpus callosum(*cc*), olfactory structure (*Olf*), visual cortex: primary (*V1*), visual cortex: secondary (*V2*), motor cortex: primary (*M1*), motor cortex: secondary (*M2*), primary somatosensory cortex: fore-limb (*S1FL*), primary somatosensory cortex: upper-lip (*S1ULP*), primary somatosensory cortex: hind-limb (*S1HL*), somatosensory cortex: barrel fields (*S1BF*), secondary somatosensory cortex (*S2*). (C) Mean R_2, R_2^*, and R_2' in each ROI, where the error bars indicate standard deviation (SD). In general, the regional variations were smaller in cortex than in the other regions.

Modified from Shu, C.Y., Herman, P., Coman, D., Sanganahalli, B.G., Wang, H., Juchem, C., Rothman, D.L., de Graaf, R.A., Hyder, F., 2016a. Brain region and activity-dependent properties of M for calibrated fMRI. Neuroimage 125, 848–856.

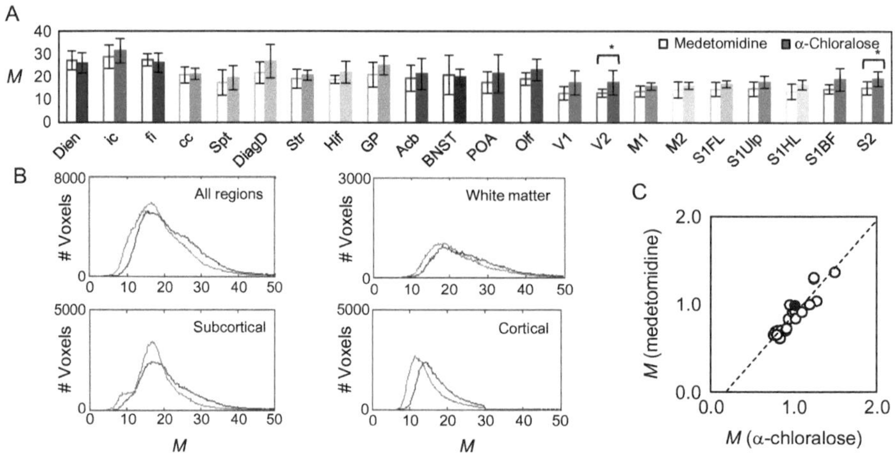

FIG. 4

Comparison of M across different brain states. (A) M for each ROI (mean ± SD) with α-chloralose (*filled bars*) vs medetomidine (*open bars*). The ROI definition was given in Fig. 3 legend. Regions of maximum and minimum M were very similar among α-chloralose and medetomidine. Student's t-test between M_α and M_{med} gave significance two small cortical regions (V2, S2; $p<0.05$). Two-way ANOVA test indicated that M_{med} were significantly lower than M_α in both cortical ($p<0.001$) and subcortical regions ($p<0.05$), but not in white matter ($p>0.05$) and diencephalon ($p>0.05$). (B) Distribution of M in all regions, for white matter, subcortical, and cortical regions. Measurements made in the α-chloralose group are traced in *black lines*, and those in the medetomidine group are traced in *red* (*light gray* in the print version) *lines*. (C) With M in cc (*center black circle*) as the normalization factor, magnitudes of normalized M_α in all ROIs were highly correlated to normalized M_{med} ($r^2=0.85$), which corresponded to the trend observed in (A).

Modified from Shu, C.Y., Herman, P., Coman, D., Sanganahalli, B.G., Wang, H., Juchem, C., Rothman, D.L., de Graaf, R.A., Hyder, F., 2016a. Brain region and activity-dependent properties of M for calibrated fMRI. Neuroimage 125, 848–856.

3 THE CONSTANT α

The earliest fMRI study of human brain function in 1991 was with CBV contrast using repeated bolus injections of a paramagnetic MRI contrast agent (ie, gadolinium based) that has a relatively short half-life in blood circulation (Belliveau et al., 1990). A few years later, however, the CBV method in animals was based on intravascular-borne exogenous superparamagnetic MRI contrast agents that maintained a steady blood concentration over several hours (Kennan et al., 1998; Mandeville et al., 1998). The susceptibility effect of the agent was so strong that it could drown out the diamagnetic BOLD effect during functional hyperemia (ie, due to more oxyhemoglobin at the venous end by increased blood flow). Recently, some new CBV methods have been developed for human studies that do not require exogenous

MRI contrast agents (Lu et al., 2003; Stefanovic and Pike, 2004). But early human-calibrated fMRI studies combined only BOLD and CBF measurements to estimate CMR_{O_2} changes during functional activation (Davis et al., 1998; Hoge et al., 1999b; Kastrup et al., 1999; Kim and Ugurbil, 1997; Kim et al., 1999), whereas animal-calibrated fMRI studies also included CBV measurements because the contrast agents were limited to animals (Hyder et al., 2001; Kida et al., 1999, 2000; Mandeville et al., 1998).

Most early human-calibrated fMRI studies assumed a specific value of α (ie, 0.4) because CBV could not be independently measured, and thus it had to be derived from changes in CBF as indicated from hypercapnia PET studies in primates (Grubb et al., 1974),

$$(\mathrm{CBV}/\mathrm{CBV}_0) = (\mathrm{CBF}/\mathrm{CBF}_0)^{\alpha}, \tag{6}$$

where the subscript "0" represents the baseline state for each parameter. Animal studies at Harvard and Yale suggested that the α-value could range from 0.1 to 0.4 because CBV and CBF dynamics were not fixed during the time course of the functional activation (Hyder et al., 2001; Mandeville et al., 1999). These suggestions have now been independently supported by subsequent findings from studies in rat, cat, and human brain where CBV and CBF have been measured in the same session (Chen and Pike, 2009; Jin and Kim, 2008; Kida et al., 2007; Shen et al., 2008), where during a functional challenge the α-value is found to range from 0.1 to 0.2. Closer inspection of dynamic changes in CBV and CBF over the course of the BOLD response during sensory stimulation in the rat, as shown in Fig. 5, suggest that α is initially quite small (ie, ~0.1) and gradually become larger (ie, >0.2) suggesting that calibrated fMRI should consider a dynamically varying α to accurately reflect ΔCMR_{O_2} (Kida et al., 2007).

More recently, van Raaij et al. (2012) measured α in the cortical and subcortical gray matter of rats during rest with functional microultrasound imaging. In the cortex, α was nearly half the value in the subcortex. Considering all findings regarding α, it is likely that the choice of α-value should depend on the presence of stimulation and the region studied. Particularly, α is reduced upon onset of stimulation, and α in the cortical gray matter is lower than in the subcortical gray matter. Given the regional-dependent (van Raaij et al., 2012) and functionally dynamic (Kida et al., 2007) nature of α, measuring CBV directly instead of deriving it from CBF could improve the accuracy of M and CMR_{O_2}.

4 THE CONSTANT β

In 1993, Ogawa et al. modeled the dependencies of R_2^* in the presence of water diffusion as, roughly:

$$R_2^* \propto v \quad \text{(for large vessels)} \tag{7a}$$

$$R_2^* \propto v^2 \quad \text{(for small vessels)} \tag{7b}$$

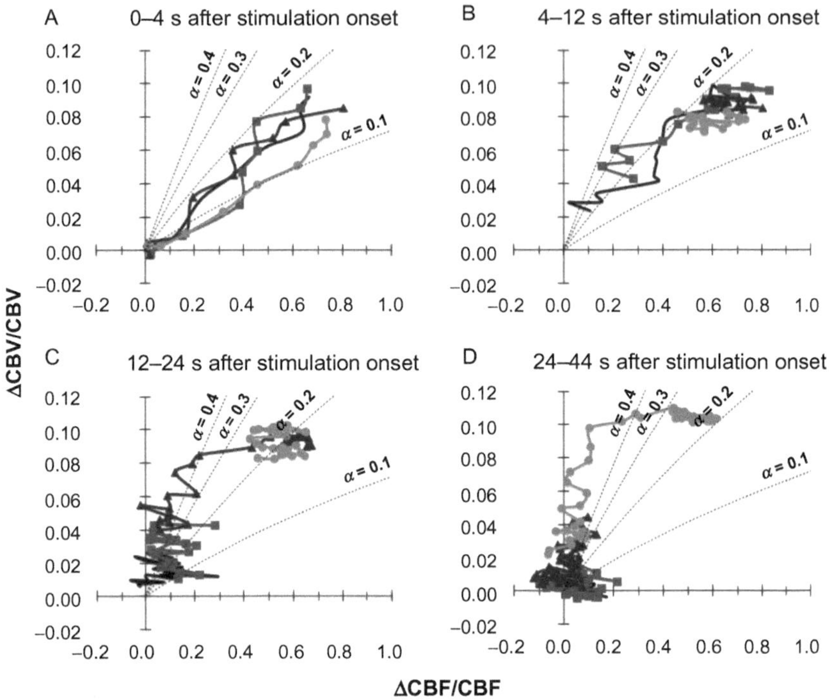

FIG. 5

Detailed comparison between ΔCBF and ΔCBV for all stimulus durations in terms of the α-value. The relationships are shown for (A) 0–4 s, (B) 4–12 s, (C) 12–24 s, and (D) 24–44 s after stimulation onset. Data were averaged over all animals ($n=12$ from eight animals). The *dotted contour lines* indicate the theoretical relationship between ΔCBF and ΔCBV for different α-values (ie, 0.1, 0.2, 0.3, and 0.4). The different symbols indicate the different stimulus durations (*black line*=4 s, *red* (*gray* in the print version) line with *square*=8 s, *blue* (*dark gray* in the print version) *line* with *triangle*=16 s, and *orange* (*light gray* in the print version) *line* with *circle*=32 s).

Modified from Kida, I., Rothman, D.L., Hyder, F., 2007. Dynamics of changes in blood flow, volume, and oxygenation: implications for dynamic functional magnetic resonance imaging calibration. J. Cereb. Blood Flow Metab. 27, 690–696.

where v accounts for the susceptibility difference between blood and tissue and is proportional to blood deoxyhemoglobin. In other words, large vessels affect R_2^* linearly, while small vessels affect R_2^* quadratically. This dependence of R_2^* on the susceptibility produced by blood within vessels was termed "β" by Davis et al. (Davis et al., 1998). In simulating the relationship for a more realistic physiological environment, where both small and large vessels exist, Davis et al. chose a value of 1.5 for β, in part due to simulations results (Boxerman et al., 1995; Kennan et al., 1994).

Not only does β varies with vascular size, β is maximal if the blood vessels' axes are perpendicular to B_0. But if the blood vessels are positioned parallel to the magnetic field, β is equal to zero (Yablonskiy and Haacke, 1994). In a blood vessel network, geometry of individual vessels is a less serious factor, but diffusion and vessel size still dominate. In terms of diffusion, β is 2 in fast diffusion regime. In slow and intermediate diffusion regime, however, β is between 1 and 2, depending on the choice of spin-echo or gradient-echo sequence for measuring R_2 (Kennan et al., 1994). But in large vessels, such as arterioles and venules, where diffusion is slow, diffusion does not play any remarkable role in the signal formation. However it is in the smaller capillaries, and at low fields, where field inhomogeneities are much reduced, that diffusion effect on relaxation rates become significant (Yablonskiy and Haacke, 1994). Hence in most high field (eg, $\geq$7 T) animal studies have usually assumed a β-value of 1 (Kida et al., 2000). While it is difficult to separately measure all the factors that contribute to β in vivo, it is feasible to measure R_2' in vivo (Blockley et al., 2015; Fujita et al., 2003; Kida et al., 2000; Shu et al., 2016a). In fact, β could be determined experimentally through analyzing the relationship between R_2' and susceptibility levels in blood. In another study, we mapped β in vivo across multiple regions in the white and gray matter and across different brain activity states by imaging R_2' (the difference between R_2^* and R_2) as a function of contrast agent (ie, Feraheme) dosage (Shu et al., 2016b; Fig. 6), where β was found to be uniform across the cortical swathe, with an average value of 0.8.

Besides separately measuring R_2^* and R_2 to find R_2' (Shu et al., 2016a), other methods have been developed in the past to measure R_2' in a single scan (Hyder et al., 2000; Ma and Wehrli, 1996; Stables et al., 1998; Yablonskiy and Haacke, 1997). A recent study which compared these methods for human brain at 3 T concluded that values of R_2' measured vary with the choice of methods, mainly because in each voxel signal decayed multiexponentially rather than monoexponentially. In such cases β will indeed depend on the technique used for measuring R_2' (Ni et al., 2015; Ulrich and Yablonskiy, 2015). At lower fields there is a significant relaxation component from the intravascular compartment, which could induce some multiexponential behaviors to relaxations measured, specifically at lower spatial resolution. On the contrary, the study by Shu et al. (2016b) at 9.4 T, benefited from high-magnetic field strength with its inherent higher spatial resolution and reduced intravascular weighting to achieve monoexponential R_2' decay. Under these experimental conditions, R_2' and β should not vary significantly between R_2' methods, but further measurements are needed to confirm this observation.

At 9.4 T, ΔR_2 was generally very small compared to ΔR_2^*, additionally, ΔR_2 was uniform throughout the brain, whereas ΔR_2^* (and therefore $\Delta R_2'$) was higher in superficial layers of the cortex (Fig. 6). Previous imaging studies showed that cortical microvessels are thinner in deeper layers (Lapi et al., 2008; Zhao et al., 2006), but despite the gradients observed in cortical $\Delta R_2'$ due to vessel size or blood volume fraction difference, β-values among all neocortical regions appeared homogeneous (Fig. 6), in contrast to former simulations and analyses (Boxerman et al., 1995; Ogawa et al., 1993; Yablonskiy and Haacke, 1994).

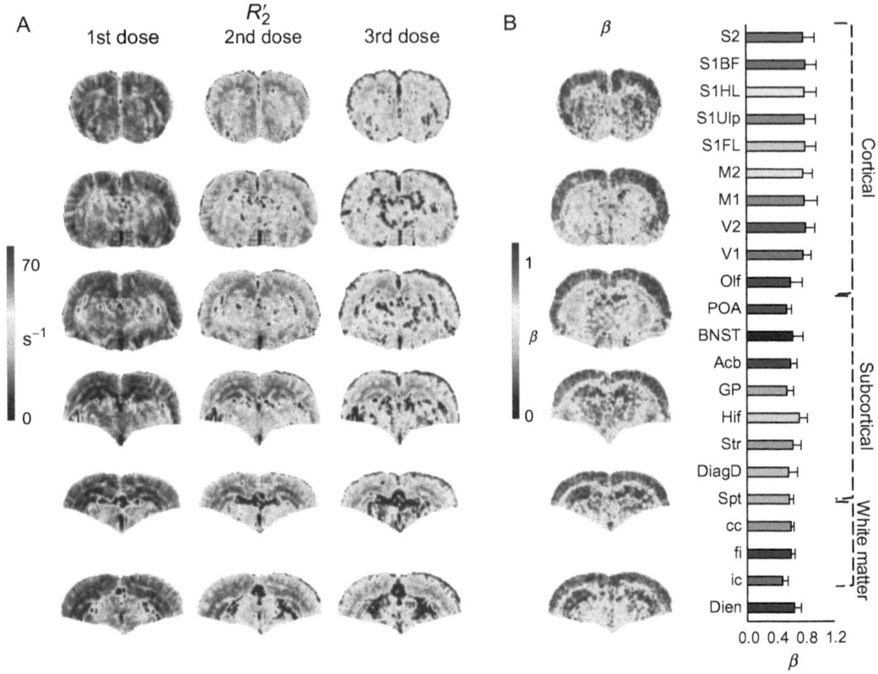

FIG. 6

A summary of dose-dependent R_2' and the mean β in each ROI (under α-chloralose). (A) R_2' images for the same subject after one, two, and three sequential doses of Feraheme. (B) β was fitted to the dose-dependent R_2' values. Cortical *gray matter* showed uniformly higher β compared to other regions. ROI areas was defined according to the template shown and defined in Fig. 3. The selected ROI's used to calculate mean and SDs of β across 22 ROIs. Consistent with the images, the mean β was 0.78 among the cortical *gray matter* regions, 0.56 in the *white matter*, and 0.61 in the subcortex.

Modified from Shu, C.Y., Sanganahalli, B.G., Coman, D., Herman, P., Rothman, D.L., Hyder, F., 2016b. Quantitative β mapping for calibrated fMRI. Neuroimage 126, 219–228.

Given that the MRI contrast was obtained at 9.4 T, there is minimal contributions from intravascular components. While the contrast induced by the iron oxide particles represent all vessels, the major contributions are likely to be from capillaries and venules as they together would represent nearly 80% of total vascular volume (Kim et al., 2006). It is possible, however, that the enhanced contrast in R_2' maps for higher dose of iron oxide particles could represent arterioles, given their superficial locations in the cortex. A difference between contrast generated by iron oxide and deoxyhemoglobin (ie, intrinsic BOLD contrast) is that there will be relatively greater contrast at the venous end for BOLD due to higher susceptibility as a consequence of lower fractional oxygen binding to hemoglobin. However due to the extensive branching and convolutions of the vascular bed along with the high diffusivity of oxygen across walls of small blood vessels, there is an averaging of oxygen partial pressure across the venules and capillary beds which will considerably reduce heterogeneity. Thus the smoothness of β-values in the cortex suggests relatively homogeneous coupling between metabolism and hemodynamics. However, further studies with vascular staining results are needed to confirm β's physiological dependence on vessel size across regions.

β's dependency on brain states was also examined with the use of α-chloralose and medetomidine anesthesia (Fig. 7). The medetomidine group exhibited about a 100% increase in neural activity from α-chloralose according to electrophysiological recordings (Fig. 7A), but both the magnitude and pattern of β among the ROIs were very similar under both states, which suggests that β is not dependent on brain states. Therefore, the same β-value should be applicable for subjects under different baseline brain activities, from being deeply anesthetized to being awake. However, β might vary with regional disease effects.

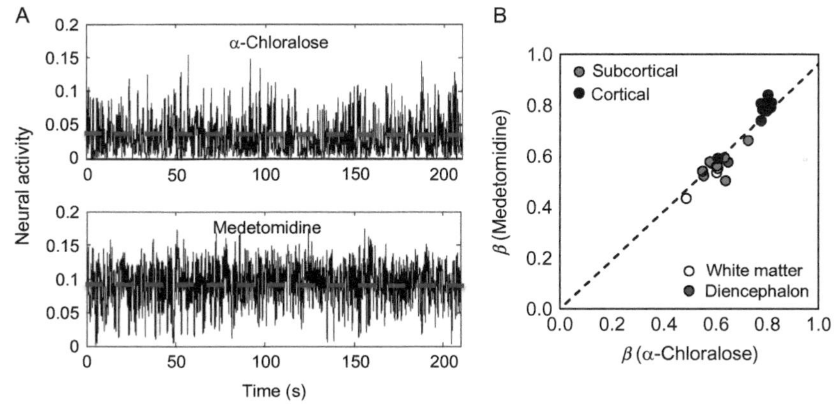

FIG. 7

Comparison between β measured at different brain activity levels (α-chloralose vs medetomidine). (A) The neural activity measurements (based on spike density function) showed a roughly two times higher activity level with medetomidine anesthesia (*red* (*dark gray* in the print version) *dashed lines* are the average activity). (B) Relationship between regional β with α-chloralose (β_α) and medetomidine (β_{med}) showed strong correlation ($r^2=0.91$), indicating a very similar pattern for β in all ROIs across brain states. Regions of both maximum and minimum for β were nearly identical among α-chloralose and medetomidine. Among all ROIs, maximum β was among the primary somatosensory regions. ANOVA test suggested that β_{med} were not significantly different from β_α ($p>0.05$). Likewise, Student's *t*-test did not give significant difference in any ROI between α-chloralose and medetomidine.

Modified from Shu, C.Y., Sanganahalli, B.G., Coman, D., Herman, P., Rothman, D.L., Hyder, F., 2016b. Quantitative β mapping for calibrated fMRI. Neuroimage 126, 219–228.

5 TRANSLATIONAL APPLICATIONS OF CALIBRATED fMRI

Most methods for measuring M rely on assumptions of isometabolism during hypercapnia (or hyperoxia) challenge, and they require additional gas challenge set up which may be difficult to implement for many laboratories (Hoge, 2012; Hyder and Rothman, 2012; Pike, 2012). On the other hand, due to the complex nature of β, there had been no existing methods to measure α and β in vivo until recently (Kida et al., 2007; Shen et al., 2008; Shu et al., 2016a,b). Simulating evoked changes in CMR_{O_2} measured with calibrated fMRI over the range of β found in literature (0.9–2), a variation in CMR_{O_2} as high as 40% is observed for fixed levels of increase in BOLD signal and CBF. Moreover, since M by hypercapnia is estimated using a simplified form of Eq. (1), inaccuracy in β will propagate into M. To minimize error in CMR_{O_2}, it is best to measure M, α, and β independently per subject (Kida et al., 2007; Shen et al., 2008; Shu et al., 2016a,b).

An advantage of the R_2'-based approach is that M and β can be obtained in the same session by repeated scanning of R_2 and R_2^*, with zero to three doses of Feraheme (an FDA approved contrast agent) which can be delivered through an intravenous line. Throughout the calibration procedure, subjects are not required to wear a gas-delivering mask as would be for hypercapnia (or hyperoxia) challenges. Furthermore a protocol can be designed, prior to injecting Feraheme, to evaluate α on a voxel-wise basis in the same session by measuring CBF change during a hypercapnia challenge. After Feraheme is injected, CBV change can be measured during an additional hypercapnia challenge. By utilizing Eq. (6), a global α map can be obtained with the measured changes in CBF and CBV. The CBV map with enhanced contrast and spatial specificity with Feraheme (Qiu et al., 2012) is expected to improve the accuracy of α measurements.

Imaging of relative CMR_{O_2} between two brain conditions can be a meaningful physiological biomarker in clinical studies of the brain (Hyder et al., 2013a,b, 2016). In disease, baseline metabolic activity may be altered, which can introduce potential confounding factors for interpreting relative CMR_{O_2} during activation. Thus there is a need to further expand the clinical applications of calibrated fMRI. It has been proposed that by using both hypercapnia and hyperoxia challenges, in an interleaved or combined manner, one can measure CMR_{O_2} at rest (Bulte et al., 2012; Gauthier and Hoge, 2012; Wise et al., 2013). This is because the hypercapnia portion of the experiment measures M, while with a known value of M the hyperoxia challenge allows one to derive resting oxygen extraction fraction (OEF). Then, absolute CMR_{O_2} can be calculated using two separate gas challenges

$$CMR_{O_2} = CaO_2 \cdot OEF \cdot CBF, \tag{8}$$

where CaO_2 is the arterial oxygen content and OEF is related to concentration of deoxyhemoglobin in the blood (ie, [dHb]). While absolute CMR_{O_2} for a given state using Eq. (8) still requires some assumptions (ie, gas challenges do not affect brain activity or metabolism), relative CMR_{O_2} between states using Eq. (1) may still be a more meaningful biomarker in clinical studies given that independent mapping of R_2', α, β, and CBF can be used to determine basal CMR_{O_2} change between states on an individual subject basis.

Overall findings from this chapter could improve precision of future calibrated fMRI studies by focusing on regional and global measurements of CBF, M, α, and β in gray matter across different resting activity levels as well as during sensory stimuli. Functional brain imaging with calibrated fMRI can examine neurometabolic and neurovascular activities in the brain both in health and disease.

ACKNOWLEDGMENT

Supported by National Institutes of Health Grants (R01 MH-067528 to FH, P30 NS-052519 to FH, R01 AG-034953 to D.L.R.).

REFERENCES

Aiello, L.C., Wheeler, P., 1995. The expensive-tissue hypothesis: the brain and the digestive-system in human and primate evolution. Curr. Anthropol. 36, 199–221.

Alderliesten, T., De Vis, J.B., Lemmers, P.M., van Bel, F., Benders, M.J., Hendrikse, J., Petersen, E.T., 2014. Simultaneous quantitative assessment of cerebral physiology using respiratory-calibrated MRI and near-infrared spectroscopy in healthy adults. Neuroimage 85 (Pt. 1), 255–263.

Ances, B.M., Leontiev, O., Perthen, J.E., Liang, C., Lansing, A.E., Buxton, R.B., 2008. Regional differences in the coupling of cerebral blood flow and oxygen metabolism changes in response to activation: implications for BOLD-fMRI. Neuroimage 39, 1510–1521.

Ances, B.M., Liang, C.L., Leontiev, O., Perthen, J.E., Fleisher, A.S., Lansing, A.E., Buxton, R.B., 2009. Effects of aging on cerebral blood flow, oxygen metabolism, and blood oxygenation level dependent responses to visual stimulation. Hum. Brain Mapp. 30, 1120–1132.

Angenstein, F., Krautwald, K., Scheich, H., 2010. The current functional state of local neuronal circuits controls the magnitude of a BOLD response to incoming stimuli. Neuroimage 50, 1364–1375.

Bailey, C.J., Sanganahalli, B.G., Herman, P., Blumenfeld, H., Gjedde, A., Hyder, F., 2013. Analysis of time and space invariance of BOLD responses in the rat visual system. Cereb. Cortex 23, 210–222.

Barinaga, M., 1997. What makes brain neurons run? Science 276, 196–198.

Belliveau, J.W., Rosen, B.R., Kantor, H.L., Rzedzian, R.R., Kennedy, D.N., Mckinstry, R.C., Vevea, J.M., Cohen, M.S., Pykett, I.L., Brady, T.J., 1990. Functional cerebral imaging by susceptibility-contrast NMR. Magn. Reson. Med. 14, 538–546.

Blockley, N.P., Griffeth, V.E., Simon, A.B., Dubowitz, D.J., Buxton, R.B., 2015. Calibrating the BOLD response without administering gases: comparison of hypercapnia calibration with calibration using an asymmetric spin echo. Neuroimage 104, 423–429.

Boxerman, J.L., Hamberg, L.M., Rosen, B.R., Weisskoff, R.M., 1995. MR contrast due to intravascular magnetic susceptibility perturbations. Magn. Reson. Med. 34, 555–566.

Bulte, D.P., Kelly, M., Germuska, M., Xie, J., Chappell, M.A., Okell, T.W., Bright, M.G., Jezzard, P., 2012. Quantitative measurement of cerebral physiology using respiratory-calibrated MRI. Neuroimage 60, 582–591.

Chen, J.J., Pike, G.B., 2009. BOLD-specific cerebral blood volume and blood flow changes during neuronal activation in humans. NMR Biomed. 22, 1054–1062.

Chen, W., Zhu, X.H., Gruetter, R., Seaquist, E.R., Adriany, G., Ugurbil, K., 2001. Study of tricarboxylic acid cycle flux changes in human visual cortex during hemifield visual stimulation using (1)H-[(13)C] MRS and fMRI. Magn. Reson. Med. 45, 349–355.

Chen, B.R., Kozberg, M.G., Bouchard, M.B., Shaik, M.A., Hillman, E.M., 2014. A critical role for the vascular endothelium in functional neurovascular coupling in the brain. J. Am. Heart Assoc. 3, e000787.

Chhina, N., Kuestermann, E., Halliday, J., Simpson, L.J., MacDonald, I.A., Bachelard, H.S., Morris, P.G., 2001. Measurement of human tricarboxylic acid cycle rates during visual activation by (13)C magnetic resonance spectroscopy. J. Neurosci. Res. 66, 737–746.

Chiarelli, P.A., Bulte, D.P., Gallichan, D., Piechnik, S.K., Wise, R., Jezzard, P., 2007a. Flow-metabolism coupling in human visual, motor, and supplementary motor areas assessed by magnetic resonance imaging. Magn. Reson. Med. 57, 538–547.

Chiarelli, P.A., Bulte, D.P., Wise, R., Gallichan, D., Jezzard, P., 2007b. A calibration method for quantitative BOLD fMRI based on hyperoxia. Neuroimage 37, 808–820.

Creutzfeldt, O.D., 1975. Neurophysiological correlates of different functional states of the brain. In: Ingvar, D.H., Lassen, N.A. (Eds.), Brain Work. The Coupling of Function, Metabolism and Blood Flow in the Brain. Alfred Benzon Symposium VIII. Academic Press, New York.

Davis, T.L., Kwong, K.K., Weisskoff, R.M., Rosen, B.R., 1998. Calibrated functional MRI: mapping the dynamics of oxidative metabolism. Proc. Natl. Acad. Sci. U.S.A. 95, 1834–1839.

Detre, J.A., Leigh, J.S., Williams, D.S., Koretsky, A.P., 1992. Perfusion imaging. Magn. Reson. Med. 23, 37–45.

Devor, A., Boas, D., 2012. Neurovascular imaging. Front. Neuroenergetics 4, 1.

Devor, A., Sakadzic, S., Srinivasan, V.J., Yaseen, M.A., Nizar, K., Saisan, P.A., Tian, P., Dale, A.M., Vinogradov, S.A., Franceschini, M.A., Boas, D.A., 2012. Frontiers in optical imaging of cerebral blood flow and metabolism. J. Cereb. Blood Flow Metab. 32, 1259–1276.

Driver, I.D., Hall, E.L., Wharton, S.J., Pritchard, S.E., Francis, S.T., Gowland, P.A., 2012. Calibrated BOLD using direct measurement of changes in venous oxygenation. Neuroimage 63, 1178–1187.

Fox, P.T., Raichle, M.E., 1986. Focal physiological uncoupling of cerebral blood flow and oxidative metabolism during somatosensory stimulation in human subjects. Proc. Natl. Acad. Sci. U.S.A. 83, 1140–1144.

Fox, P.T., Raichle, M.E., Mintun, M.A., Dence, C., 1988. Nonoxidative glucose consumption during focal physiologic neural activity. Science 241, 462–464.

Fujita, N., Shinohara, M., Tanaka, H., Yutani, K., Nakamura, H., Murase, K., 2003. Quantitative mapping of cerebral deoxyhemoglobin content using MR imaging. Neuroimage 20, 2071–2083.

Garrett, K.M., Gan, J., 1998. Enhancement of gamma-aminobutyric acid A receptor activity by alpha-chloralose. J. Pharmacol. Exp. Ther. 285, 680–686.

Gauthier, C.J., Hoge, R.D., 2012. Magnetic resonance imaging of resting OEF and CMRO using a generalized calibration model for hypercapnia and hyperoxia. Neuroimage 60, 1212–1225.

Gauthier, C.J., Hoge, R.D., 2013. A generalized procedure for calibrated MRI incorporating hyperoxia and hypercapnia. Hum. Brain Mapp. 34, 1053–1069.

Gauthier, C.J., Madjar, C., Tancredi, F.B., Stefanovic, B., Hoge, R.D., 2011. Elimination of visually evoked BOLD responses during carbogen inhalation: implications for calibrated MRI. Neuroimage 54, 1001–1011.

Gauthier, C.J., Desjardins-Crepeau, L., Madjar, C., Bherer, L., Hoge, R.D., 2012. Absolute quantification of resting oxygen metabolism and metabolic reactivity during functional activation using QUO2 MRI. Neuroimage 63, 1353–1363.

Goodwin, J.A., Vidyasagar, R., Balanos, G.M., Bulte, D., Parkes, L.M., 2009. Quantitative fMRI using hyperoxia calibration: reproducibility during a cognitive Stroop task. Neuroimage 47, 573–580.

Grubb Jr., R.L., Raichle, M.E., Eichling, J.O., Ter-Pogossian, M.M., 1974. The effects of changes in PaCO2 on cerebral blood volume, blood flow, and vascular mean transit time. Stroke 5, 630–639.

Herman, P., Sanganahalli, B.G., Blumenfeld, H., Hyder, F., 2009. Cerebral oxygen demand for short-lived and steady-state events. J. Neurochem. 109 (Suppl. 1), 73–79.

Herman, P., Sanganahalli, B.G., Hyder, F., Eke, A., 2011. Fractal analysis of spontaneous fluctuations of the BOLD signal in rat brain. Neuroimage 58, 1060–1069.

Herman, P., Sachdev, R.N.S., Yu, Y., Sanganahalli, B.G., Blumenfeld, H., Mccormick, D.A., Hyder, F., 2013a. State-dependent resting-state coherence between neural activity and blood flow. In: International Society for Cerebral Blood Flow and Metabolism, pp. 264–265.

Herman, P., Sanganahalli, B.G., Blumenfeld, H., Rothman, D.L., Hyder, F., 2013b. Quantitative basis for neuroimaging of cortical laminae with calibrated functional MRI. Proc. Natl. Acad. Sci. U.S.A. 110, 15115–15120.

Herman, P., Sanganahalli, B.G., Coman, D., Blumenfeld, H., Jiang, L., Rothman, D.L., Hyder, F., 2015. Neurophysiological and neuroenergetic basis of spontaneous BOLD signal fluctuations in resting-state fMRI connectivity maps. In: International Society for Magnetic Resonance in Medicine, p. 5103.

Hoge, R.D., 2012. Calibrated fMRI. Neuroimage 62, 930–937.

Hoge, R.D., Atkinson, J., Gill, B., Crelier, G.R., Marrett, S., Pike, G.B., 1999a. Investigation of BOLD signal dependence on cerebral blood flow and oxygen consumption: the deoxyhemoglobin dilution model. Magn. Reson. Med. 42, 849–863.

Hoge, R.D., Atkinson, J., Gill, B., Crelier, G.R., Marrett, S., Pike, G.B., 1999b. Linear coupling between cerebral blood flow and oxygen consumption in activated human cortex. Proc. Natl. Acad. Sci. U.S.A. 96, 9403–9408.

Hyder, F., Rothman, D.L., 2012. Quantitative fMRI and oxidative neuroenergetics. Neuroimage 62, 985–994.

Hyder, F., Chase, J.R., Behar, K.L., Mason, G.F., Siddeek, M., Rothman, D.L., Shulman, R.G., 1996. Increased tricarboxylic acid cycle flux in rat brain during forepaw stimulation detected with $^{1}H[^{13}C]$NMR. Proc. Natl. Acad. Sci. U.S.A. 93, 7612–7617.

Hyder, F., Rothman, D.L., Mason, G.F., Rangarajan, A., Behar, K.L., Shulman, R.G., 1997. Oxidative glucose metabolism in rat brain during single forepaw stimulation: a spatially localized $^{1}H[^{13}C]$ nuclear magnetic resonance study. J. Cereb. Blood Flow Metab. 17, 1040–1047.

Hyder, F., Kennan, R.P., Kida, I., Mason, G.F., Behar, K.L., Rothman, D., 2000. Dependence of oxygen delivery on blood flow in rat brain: a 7 tesla nuclear magnetic resonance study. J. Cereb. Blood Flow Metab. 20, 485–498.

Hyder, F., Kida, I., Behar, K.L., Kennan, R.P., Maciejewski, P.K., Rothman, D.L., 2001. Quantitative functional imaging of the brain: towards mapping neuronal activity by BOLD fMRI. NMR Biomed. 14, 413–431.

Hyder, F., Patel, A.B., Gjedde, A., Rothman, D.L., Behar, K.L., Shulman, R.G., 2006. Neuronal-glial glucose oxidation and glutamatergic-GABAergic function. J. Cereb. Blood Flow Metab. 26, 865–877.

Hyder, F., Sanganahalli, B.G., Herman, P., Coman, D., Maandag, N.J., Behar, K.L., Blumenfeld, H., Rothman, D.L., 2010. Neurovascular and neurometabolic couplings in dynamic calibrated fMRI: transient oxidative neuroenergetics for block-design and event-related paradigms. Front. Neuroenergetics 2, 18. http://dx.doi.org/10.3389/fnene.2010.00018.

Hyder, F., Fulbright, R.K., Shulman, R.G., Rothman, D.L., 2013a. Glutamatergic function in the resting awake human brain is supported by uniformly high oxidative energy. J. Cereb. Blood Flow Metab. 26, 865–877.

Hyder, F., Rothman, D.L., Bennett, M.R., 2013b. Cortical energy demands of signaling and nonsignaling components in brain are conserved across mammalian species and activity levels. Proc. Natl. Acad. Sci. U.S.A. 110, 3549–3554.

Hyder, F., Herman, P., Bailey, C.J., Moller, A., Globinsky, R., Fulbright, R.K., Rothman, D.L., Gjedde, A., 2016. Uniform distributions of glucose oxidation and oxygen extraction in gray matter of normal human brain: no evidence of regional differences of aerobic glycolysis. J. Cereb. Blood Flow Metab. Published online before print January 11, 2016, http://dx.doi.org/10.1177/0271678X15625349.

Jin, T., Kim, S.G., 2008. Cortical layer-dependent dynamic blood oxygenation, cerebral blood flow and cerebral blood volume responses during visual stimulation. Neuroimage 43, 1–9.

Kastrup, A., Kruger, G., Glover, G.H., Moseley, M.E., 1999. Assessment of cerebral oxidative metabolism with breath holding and fMRI. Magn. Reson. Med. 42, 608–611.

Kastrup, A., Kruger, G., Neumann-Haefelin, T., Glover, G.H., Moseley, M.E., 2002. Changes of cerebral blood flow, oxygenation, and oxidative metabolism during graded motor activation. Neuroimage 15, 74–82.

Kennan, R.P., Zhong, J., Gore, J.C., 1994. Intravascular susceptibility contrast mechanisms in tissues. Magn. Reson. Med. 31, 9–21.

Kennan, R.P., Scanley, B.E., Innis, R.B., Gore, J.C., 1998. Physiological basis for BOLD MR signal changes due to neuronal stimulation: separation of blood volume and magnetic susceptibility effects. Magn. Reson. Med. 40, 840–846.

Kida, I., Hyder, F., Kennan, R.P., Behar, K.L., 1999. Toward absolute quantitation of bold functional MRI. Adv. Exp. Med. Biol. 471, 681–689.

Kida, I., Kennan, R.P., Rothman, D.L., Behar, K.L., Hyder, F., 2000. High-resolution CMR(O2) mapping in rat cortex: a multiparametric approach to calibration of BOLD image contrast at 7 Tesla. J. Cereb. Blood Flow Metab. 20, 847–860.

Kida, I., Rothman, D.L., Hyder, F., 2007. Dynamics of changes in blood flow, volume, and oxygenation: implications for dynamic functional magnetic resonance imaging calibration. J. Cereb. Blood Flow Metab. 27, 690–696.

Kim, S.G., 1995. Quantification of relative cerebral blood flow change by flow-sensitive alternating inversion recovery (FAIR) technique: application to functional mapping. Magn. Reson. Med. 34, 293–301.

Kim, S.G., Ugurbil, K., 1997. Comparison of blood oxygenation and cerebral blood flow effects in fMRI: estimation of relative oxygen consumption change. Magn. Reson. Med. 38, 59–65.

Kim, S.G., Tsekos, N.V., Ashe, J., 1997. Multi-slice perfusion-based functional MRI using the FAIR technique: comparison of CBF and BOLD effects. NMR Biomed. 10, 191–196.

Kim, S.G., Rostrup, E., Larsson, H.B., Ogawa, S., Paulson, O.B., 1999. Determination of relative CMRO2 from CBF and BOLD changes: significant increase of oxygen consumption rate during visual stimulation. Magn. Reson. Med. 41, 1152–1161.

Kim, T., Hendrich, K.S., Masamoto, K., Kim, S.-G., 2006. Arterial versus total blood volume changes during neural activity-induced cerebral blood flow change: implication for BOLD fMRI. J. Cereb. Blood Flow Metab. 27, 1235–1247.

Kozberg, M.G., Chen, B.R., Deleo, S.E., Bouchard, M.B., Hillman, E.M., 2013. Resolving the transition from negative to positive blood oxygen level-dependent responses in the developing brain. Proc. Natl. Acad. Sci. U.S.A. 110, 4380–4385.

Krieger, S.N., Gauthier, C.J., Ivanov, D., Huber, L., Roggenhofer, E., Sehm, B., Turner, R., Egan, G.F., 2014a. Regional reproducibility of calibrated BOLD functional MRI: implications for the study of cognition and plasticity. Neuroimage 101, 8–20.

Krieger, S.N., Ivanov, D., Huber, L., Roggenhofer, E., Sehm, B., Turner, R., Egan, G.F., Gauthier, C.J., 2014b. Using carbogen for calibrated fMRI at 7Tesla: comparison of direct and modelled estimation of the M parameter. Neuroimage 84, 605–614.

Lapi, D., Marchiafava, P.L., Colantuoni, A., 2008. Geometric characteristics of arterial network of rat pial microcirculation. J. Vasc. Res. 45, 69–77.

Lee, J., Shmueli, K., Kang, B.T., Yao, B., Fukunaga, M., Van Gelderen, P., Palumbo, S., Bosetti, F., Silva, A.C., Duyn, J.H., 2012. The contribution of myelin to magnetic susceptibility-weighted contrasts in high-field MRI of the brain. Neuroimage 59, 3967–3975.

Leontiev, O., Dubowitz, D.J., Buxton, R.B., 2007. CBF/CMRO2 coupling measured with calibrated BOLD fMRI: sources of bias. Neuroimage 36, 1110–1122.

Lin, A.L., Fox, P.T., Yang, Y., Lu, H., Tan, L.H., Gao, J.H., 2009. Time-dependent correlation of cerebral blood flow with oxygen metabolism in activated human visual cortex as measured by fMRI. Neuroimage 44, 16–22.

Lin, A.L., Fox, P.T., Hardies, J., Duong, T.Q., Gao, J.H., 2010. Nonlinear coupling between cerebral blood flow, oxygen consumption, and ATP production in human visual cortex. Proc. Natl. Acad. Sci. U.S.A. 107, 8446–8451.

Lu, H., Golay, X., Pekar, J.J., Van Zijl, P.C., 2003. Functional magnetic resonance imaging based on changes in vascular space occupancy. Magn. Reson. Med. 50, 263–274.

Ma, J., Wehrli, F.W., 1996. Method for image-based measurement of the reversible and irreversible contribution to the transverse-relaxation rate. J. Magn. Reson. B 111, 61–69.

Malonek, D., Grinvald, A., 1996. Interactions between electrical activity and cortical microcirculation revealed by imaging spectroscopy: implications for functional brain mapping. Science 272, 551–554.

Mandeville, J.B., Marota, J.J., Kosofsky, B.E., Keltner, J.R., Weissleder, R., Rosen, B.R., Weisskoff, R.M., 1998. Dynamic functional imaging of relative cerebral blood volume during rat forepaw stimulation. Magn. Reson. Med. 39, 615–624.

Mandeville, J.B., Marota, J.J., Ayata, C., Zaharchuk, G., Moskowitz, M.A., Rosen, B.R., Weisskoff, R.M., 1999. Evidence of a cerebrovascular postarteriole windkessel with delayed compliance. J. Cereb. Blood Flow Metab. 19, 679–689.

Mark, C.I., Pike, G.B., 2012. Indication of BOLD-specific venous flow-volume changes from precisely controlled hyperoxic vs. hypercapnic calibration. J. Cereb. Blood Flow Metab. 32, 709–719.

Mark, C.I., Fisher, J.A., Pike, G.B., 2011. Improved fMRI calibration: precisely controlled hyperoxic versus hypercapnic stimuli. Neuroimage 54, 1102–1111.

Mohtasib, R.S., Lumley, G., Goodwin, J.A., Emsley, H.C., Sluming, V., Parkes, L.M., 2012. Calibrated fMRI during a cognitive Stroop task reveals reduced metabolic response with increasing age. Neuroimage 59, 1143–1151.

Moradi, F., Buracas, G.T., Buxton, R.B., 2012. Attention strongly increases oxygen metabolic response to stimulus in primary visual cortex. Neuroimage 59, 601–607.

Ni, W., Christen, T., Zun, Z., Zaharchuk, G., 2015. Comparison of R2′ measurement methods in the normal brain at 3 Tesla. Magn. Reson. Med. 73, 1228–1236.

Ogawa, S., Menon, R.S., Tank, D.W., Kim, S.G., Merkle, H., Ellermann, J.M., Ugurbil, K., 1993. Functional brain mapping by blood oxygenation level-dependent contrast magnetic resonance imaging. A comparison of signal characteristics with a biophysical model. Biophys. J. 64, 803–812.

Perthen, J.E., Lansing, A.E., Liau, J., Liu, T.T., Buxton, R.B., 2008. Caffeine-induced uncoupling of cerebral blood flow and oxygen metabolism: a calibrated BOLD fMRI study. Neuroimage 40, 237–247.

Pike, G.B., 2012. Quantitative functional MRI: concepts, issues and future challenges. Neuroimage 62, 1234–1240.

Qiu, D., Zaharchuk, G., Christen, T., Ni, W.W., Moseley, M.E., 2012. Contrast-enhanced functional blood volume imaging (CE-fBVI): enhanced sensitivity for brain activation in humans using the ultrasmall superparamagnetic iron oxide agent ferumoxytol. Neuroimage 62, 1726–1731.

Radhakrishnan, H., Wu, W., Boas, D., Franceschini, M.A., 2011. Study of neurovascular coupling by modulating neuronal activity with GABA. Brain Res. 1372, 1–12.

Restom, K., Perthen, J.E., Liu, T.T., 2008. Calibrated fMRI in the medial temporal lobe during a memory-encoding task. Neuroimage 40, 1495–1502.

Rodgers, Z.B., Jain, V., Englund, E.K., Langham, M.C., Wehrli, F.W., 2013. High temporal resolution MRI quantification of global cerebral metabolic rate of oxygen consumption in response to apneic challenge. J. Cereb. Blood Flow Metab. 33, 1514–1522.

Sanganahalli, B.G., Herman, P., Blumenfeld, H., Hyder, F., 2009. Oxidative neuroenergetics in event-related paradigms. J. Neurosci. 29, 1707–1718.

Sanganahalli, B.G., Herman, P., Rothman, D.L., Blumenfeld, H., Hyder, F., 2011. Comparative oxidative demands in cortex and subcortex revealed by high field calibrated fMRI. In: International Society for Magnetic Resonance in Medicine, p. 5764.

Sanganahalli, B.G., Herman, P., Behar, K.L., Blumenfeld, H., Rothman, D.L., Hyder, F., 2013. Functional MRI and neural responses in a rat model of Alzheimer's disease. Neuroimage 79, 404–411.

Shen, Q., Ren, H., Duong, T.Q., 2008. CBF, BOLD, CBV, and $CMRO_2$ fMRI signal temporal dynamics at 500-msec resolution. J. Magn. Reson. Imaging 27, 599–606.

Shu, C.Y., Herman, P., Coman, D., Sanganahalli, B.G., Wang, H., Juchem, C., Rothman, D.L., De Graaf, R.A., Hyder, F., 2016a. Brain region and activity-dependent properties of M for calibrated fMRI. Neuroimage 125, 848–856.

Shu, C.Y., Sanganahalli, B.G., Coman, D., Herman, P., Rothman, D.L., Hyder, F., 2016b. Quantitative β mapping for calibrated fMRI. Neuroimage 126, 219–228.

Sicard, K.M., Duong, T.Q., 2005. Effects of hypoxia, hyperoxia, and hypercapnia on baseline and stimulus-evoked BOLD, CBF, and CMRO2 in spontaneously breathing animals. Neuroimage 25, 850–858.

Siesjo, B.K., 1978. Brain Energy Metabolism. Wiley and Sons, Ltd., New York.

Sinclair, M.D., 2003. A review of the physiological effects of alpha2-agonists related to the clinical use of medetomidine in small animal practice. Can. Vet. J. 44, 885–897.

Sokoloff, L., 1991. Relationship between functional activity and energy metabolism in the nervous system: whether, where and why? In: Lassen, N.A., Ingvar, D.H., Raichle, M.E., Friberg, L. (Eds.), Brain Work and Mental Activity. Munksgaard, Copenhagen.

St. Lawrence, K.S., Ye, F.Q., Lewis, B.K., Frank, J.A., McLaughlin, A.C., 2003. Measuring the effects of indomethacin on changes in cerebral oxidative metabolism and cerebral blood flow during sensorimotor activation. Magn. Reson. Med. 50, 99–106.

Stables, L.A., Kennan, R.P., Gore, J.C., 1998. Asymmetric spin-echo imaging of magnetically inhomogeneous systems: theory, experiment, and numerical studies. Magn. Reson. Med. 40, 432–442.

Stefanovic, B., Pike, G.B., 2004. Human whole-blood relaxometry at 1.5 T: assessment of diffusion and exchange models. Magn. Reson. Med. 52, 716–723.

Stefanovic, B., Warnking, J.M., Pike, G.B., 2004. Hemodynamic and metabolic responses to neuronal inhibition. Neuroimage 22, 771–778.

Stefanovic, B., Warnking, J.M., Rylander, K.M., Pike, G.B., 2006. The effect of global cerebral vasodilation on focal activation hemodynamics. Neuroimage 30, 726–734.

Thesen, T., Leontiev, O., Song, T., Dehghani, N., Hagler Jr., D.J., Huang, M., Buxton, R., Halgren, E., 2012. Depression of cortical activity in humans by mild hypercapnia. Hum. Brain Mapp. 33, 715–726.

Thompson, J.K., Peterson, M.R., Freeman, R.D., 2004. High-resolution neurometabolic coupling revealed by focal activation of visual neurons. Nat. Neurosci. 7, 919–920.

Ugurbil, K., 2012. The road to functional imaging and ultrahigh fields. Neuroimage 62, 726–735.

Ulrich, X., Yablonskiy, D.A., 2015. Separation of cellular and BOLD contributions to T2* signal relaxation. Magn. Reson. Med. 75, 606–615.

Uludag, K., Dubowitz, D.J., Yoder, E.J., Restom, K., Liu, T.T., Buxton, R.B., 2004. Coupling of cerebral blood flow and oxygen consumption during physiological activation and deactivation measured with fMRI. Neuroimage 23, 148–155.

van Raaij, M.E., Lindvere, L., Dorr, A., He, J., Sahota, B., Foster, F.S., Stefanovic, B., 2012. Quantification of blood flow and volume in arterioles and venules of the rat cerebral cortex using functional micro-ultrasound. Neuroimage 63, 1030–1037.

Viswanathan, A., Freeman, R.D., 2007. Neurometabolic coupling in cerebral cortex reflects synaptic more than spiking activity. Nat. Neurosci. 10, 1308–1312.

Williams, D.S., Detre, J.A., Leigh, J.S., Koretsky, A.P., 1992. Magnetic resonance imaging of perfusion using spin inversion of arterial water. Proc. Natl. Acad. Sci. U.S.A. 89, 212–216.

Wise, R.G., Harris, A.D., Stone, A.J., Murphy, K., 2013. Measurement of OEF and absolute CMRO2: MRI-based methods using interleaved and combined hypercapnia and hyperoxia. Neuroimage 83, 135–147.

Xu, F., Uh, J., Brier, M.R., Hart Jr., J., Yezhuvath, U.S., Gu, H., Yang, Y., Lu, H., 2011. The influence of carbon dioxide on brain activity and metabolism in conscious humans. J. Cereb. Blood Flow Metab. 31, 58–67.

Yablonskiy, D.A., Haacke, E.M., 1994. Theory of NMR signal behavior in magnetically inhomogeneous tissues: the static dephasing regime. Magn. Reson. Med. 32, 749–763.

Yablonskiy, D.A., Haacke, E.M., 1997. An MRI method for measuring T2 in the presence of static and RF magnetic field inhomogeneities. Magn. Reson. Med. 37, 872–876.

Yücel, M.A., Evans, K.C., Selb, J., Huppert, T.J., Boas, D.A., Gagnon, L., 2014. Validation of the hypercapnic calibrated fMRI method using DOT–fMRI fusion imaging. Neuroimage 102 (Pt. 2), 729–735.

Zappe, A.C., Uludag, K., Oeltermann, A., Ugurbil, K., Logothetis, N.K., 2008. The influence of moderate hypercapnia on neural activity in the anesthetized nonhuman primate. Cereb. Cortex 18, 2666–2673.

Zhao, F., Wang, P., Hendrich, K., Ugurbil, K., Kim, S.G., 2006. Cortical layer-dependent BOLD and CBV responses measured by spin-echo and gradient-echo fMRI: insights into hemodynamic regulation. Neuroimage 30, 1149–1160.

CHAPTER

Submillimeter-resolution fMRI: Toward understanding local neural processing

6

M. Fukuda*,[1], A.J. Poplawsky*, S.-G. Kim[†]

**University of Pittsburgh, Pittsburgh, PA, United States*

[†]Center for Neuroscience Imaging Research, Institute for Basic Science, Sungkyunkwan University, Suwon, South Korea

[1]Corresponding author: Tel.: +1-412-383-6695; Fax: +1-412-383-6799, e-mail address: mif5@pitt.edu

Abstract

Functional magnetic resonance imaging (fMRI) measures the hemodynamic response to active neurons. The most prevailing fMRI contrast, blood oxygenation level-dependent (BOLD) contrast, specifically results from a complex interplay between oxygen metabolism, blood flow, and blood volume reactions. Therefore, because fMRI indirectly measures brain function, the spatial accuracy of these blood-borne signal changes to the true neuronal activity comes into question, especially at high resolutions. To better interpret the neural basis of fMRI, these properties have been studied in localized neural circuits of the brain using fMRI with blood flow, blood volume, and BOLD contrasts, as well as optical intrinsic signal imaging (OISI), which shares similar signal sources as fMRI. Here, we review how submillimeter-scale high-resolution fMRI and OISI in the visual cortex columnar and olfactory bulb laminar models have advanced our basic knowledge of the spatial localization of the individual hemodynamic signals and neurovascular coupling mechanisms.

Keywords

BOLD, CBV, CBF, Blood flow, Blood volume, Neurovascular coupling, Column, Layer

1 INTRODUCTION

Neuronal activity in the brain and the associated increase in metabolic demand are accompanied by hemodynamic responses, such as changes in blood oxygenation, blood flow, and blood volume. Functional magnetic resonance imaging (fMRI) measures these hemodynamic responses and allows us to noninvasively map the spatial patterns of neuronal activity in the brain; with the goal of accurately inferring localized neuronal activity from this secondary vascular response. As MRI technology

Progress in Brain Research, Volume 225, ISSN 0079-6123, http://dx.doi.org/10.1016/bs.pbr.2016.03.003

advances and the physical limitations of this technique continue to improve, submillimeter spatial-resolution fMRI is becoming increasingly possible. However, it is currently unclear how well the hemodynamic changes are localized to the neurally active sites as the spatial resolution increases. In other words, does the site of the fMRI changes accurately mark the location of the increased neuronal activity? Since the hemodynamic response originates from the blood, the spatial specificity of the fMRI change relies on how precisely the blood supply is regulated in response to the neuronal activity and which vessels the changes originate from. The change in blood oxygenation mainly occurs in the capillaries and venous vessels. While the changes in blood flow can occur in any vessels, the changes in blood volume dominantly occurs in arterial vessels (Kim et al., 2007), although changes in the venous blood volume seems nonnegligible for certain conditions (Drew et al., 2011; Fukuda et al., 2013; Kim and Kim, 2011).

In addition to the spatial localization of the fMRI response, the neural origin of fMRI must be identified. What neuronal events are responsible for evoking the fMRI response? Neuronal events can be divided into subthreshold synaptic and suprathreshold spiking activities. Are subthreshold synaptic activities sufficient or are action potentials required for evoking an fMRI signal change? Which neuronal events does the fMRI response reflect, pre- or postsynaptic activity? Similarly, at the level of neural circuits, which can be divided into input to and output from the circuit plus processing within the circuit, the question will be which part is reflected in the fMRI response: input, output, or intracircuit processing? Furthermore, excitatory and inhibitory neurons, which are often located within a close proximity in the brain and are active at the same time, can both contribute to the hemodynamic response. Does the activation of inhibitory neurons produce the negative fMRI response as they hyperpolarize the membrane potentials of the excitatory neurons, even though they are metabolically active?

Since fMRI has become an increasingly important tool for studying human brain functions, neurovascular coupling and the physiological basis of fMRI have become fields of great interest in neuroscience. Although our knowledge of neurovascular coupling has been significantly improved, as is found in several insightful review papers (Attwell et al., 2010; Hillman, 2014; Logothetis, 2008; Petzold and Murthy, 2011), many outstanding issues still remain unresolved. In this review, we will mainly focus on how optical intrinsic signal imaging (OISI), which shares a similar signal source as fMRI, and submillimeter-scale high-resolution fMRI in animal models of visual cortical columns and olfactory bulb layers have advanced our basic knowledge of the spatial localization of the individual hemodynamic signals, as well as neurovascular coupling mechanisms.

2 VASCULAR CONSTRAINTS

Increased neuronal activity is primarily maintained by aerobic respiration, where glucose is metabolized with oxygen to produce energy. Since the brain lacks stores of glucose and oxygen, it requires a continuous supply of these fuels in order to meet its high energy demands. Oxygen and glucose are delivered from the blood to active neurons through capillaries that cover the whole brain. Thus, capillary density correlates well with the demand for oxidative metabolism, as is identified by 2-deoxyglucose (2-DG) and cytochrome oxidase (Borowsky and Collins, 1989; Keller et al., 2011; Weber et al., 2008; Zheng et al., 1991). The average distance from a neuron to the nearest microvessel (defined as having a 3.5–4.0 μm diameter) is only ~13 μm in the cerebral cortex (Tsai et al., 2009), which is, theoretically, the finest resolution fMRI can achieve to accurately represent neuronal activity. Here we briefly overview cortical vascular structures to consider vascular constraints for the spatial resolution of fMRI. Those who are interested in a more comprehensive description of the cortical vascular anatomy, please consult the work of Duvernoy et al. (1981) for the human cortex and a review paper by Hirsch et al. (2012) for the primate cortex (see also Chapter "Bridging macroscopic and microscopic methods for the measurements of cerebral blood flow: Toward finding the determinants in maintaining the CBF homeostasis" by Kanno and Masamoto).

In the cerebral cortex, the vasculature can be divided into the large pial vessels on the cortical surface and the microvascular network in the parenchyma below the brain surface (Fig. 1). The pial artery (red) and vein (blue) are responsible for grossly

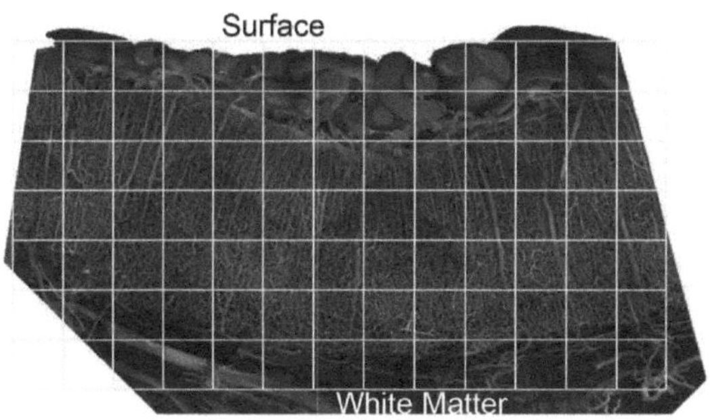

FIG. 1

Cortical vasculature. Scanning electron micrograph of a vascular corrosion cast of the primate primary visual cortex around the calcarine sulcus. Large vessels are localized to the surface of the brain. Vascular density in the gray matter is obviously higher than that in the white matter. Oxygenated blood is delivered into the parenchyma via penetrating arterioles that branch radially from the large pial arteries (*red, gray* in the print version). The penetration depths of the arterioles vary; some penetrate into the deep cortical layers, while others only penetrate into superficial and middle layers. The penetrating arterioles further branch into microvessels (*gray*), including capillaries that deliver blood to the active neurons in individual cortical layers. Then, the blood is drained by emerging venules that radially connect back to the large pial veins (*blue, dark gray* in the print version). Each *square* in the *yellow* (*white* in the print version) *grid* is 0.5×0.5 mm^2. A typical in-plane voxel resolution for human fMRI is 3×3 mm^2 (ie, 6×6 cells of the *yellow, white* in the print version, *grid*), while an in-plane resolution for high-resolution human fMRI is usually less than 1×1 mm^2 (2×2 cells).

Coronal section, adapted with permission from fig. 2 in Hirsch, S., Reichold, J., Schneider, M., Szekely, G., Weber, B., 2012. Topology and hemodynamics of the cortical cerebrovascular system. J. Cereb. Blood Flow Metab. 32, 952–967.

supplying oxygenated blood and glucose to individual cortical areas and removing deoxygenated blood and waste from the brain, respectively. They are parental vessels for numerous penetrating arterioles and emerging venules that radially plunge into the parenchyma, and, thus, hemodynamic responses from the pial vessels can only provide a supramillimeter resolution due to their size and distal location from the activated neurons. Particularly, hemodynamic-based imaging techniques that are sensitive to venous drainage, such as blood oxygenation level-dependent (BOLD) contrast, can give rise to spurious activation at sites remote from the neuronal activity (Turner, 2002). The penetrating arterioles (red) further connect to intricate microvessel networks (gray) to supply blood to the parenchyma. Then, the blood drains from the microvessel network to the emerging venules (blue) and finally returns to the pial veins. The number of branches from penetrating arterioles peaks around layer 4 in the cortex, whereas that from emerging venules becomes larger toward the brain surface (Blinder et al., 2013). The fractional vascular volume for all vessels is largest on the cortical surface, due to the large pial vessels, then decreases within the first 100 μm below the pial, and peaks again around the middle layer (Tsai et al., 2009).

The density of penetrating arterioles varies across species and becomes greater in higher order animals (Mchedlishvili and Kuridze, 1984). For instance, in rats, the average distance between neighboring penetrating arterioles is ~130 μm and the territory of a single penetrating arteriole extends up to 350 μm (Nishimura et al., 2007). Thus, if individual penetrating arterioles are independently regulated, then hemodynamic-based imaging techniques sensitive to the penetrating arteriole can achieve a 100–200 μm spatial resolution. Emerging venules are more numerous than the penetrating arterioles by a factor of ~3 in mice (Blinder et al., 2013), and 1.8 (Shih et al., 2013) to 2.6 (Nguyen et al., 2011) in rats. However, the inverse ratio was reported in primates, whereby the venule to arteriole ratios are reported as 0.61 (Weber et al., 2008) and 0.33–0.66 (Guibert et al., 2010). Thus, penetrating arterioles are more numerous than the emerging venules in primates. The higher density of emerging veins in mice and rats may be necessary for the more efficient removal of waste resulting from the higher metabolic rates typical of small animals. Further, a single penetrating arteriole typically drains into several emerging veins, while several penetrating arterioles drain into a single emerging vein as reported in rodents (Woolsey et al., 1996), suggesting that the territory for the venous drainage is less specific than the arterial supply.

Relationships between the neural and vascular organization have also been studied. One question is whether the microvascular density across layers follows the cortical neuronal density. While the neuronal density sharply increases at layers 4 and 6 for the mouse primary sensory cortex, the density of microvessels (<6 μm) varies relatively little across layers with a blunt peak around layer 4 (Tsai et al., 2009). Thus, the laminar variations in neuronal density are not perfectly correlated with the laminar microvascular density. Since synaptic activity is the most energetically demanding process in the cortex (Howarth et al., 2012) and synaptic density is not perfectly correlated with neuronal density (O'Kusky and Colonnier, 1982), the microvascular density may correlate with synaptic density instead. Indeed, the primary visual cortex of the primate also shows the highest capillary density in layer 4 and the metabolic profile as measured by cytochrome oxidase activity perfectly matches the capillary density (Weber et al., 2008). Layer 4 is also known to have the highest spontaneous firing rate (Snodderly and Gur, 1995) and the highest local field potential amplitudes during rest (Maier et al., 2010) in V1. Another question is whether the vasculature follows the same functional organization as cortical columns (Gardner, 2010). Neurons belonging to the same local microcircuit that has a specific functional property are often clustered together in the brain. Such clusters are referred to as functional structures. Some functional structures are only apparent in a specific layer, such as the whisker barrels in layer 4 of the rodent somatosensory cortex (Woolsey and Van der Loos, 1970), while others form columns across the cortical layers, such as the ocular dominance and orientation columns (Blasdel and Salama, 1986; Hubel and Wiesel, 1962, 1963). The dimensions of these functional structures are generally a few hundred micrometers across and are similar to the size of the domain of a single penetrating arteriole to emerging venule module (Nguyen et al., 2011; Nishimura et al., 2007). Woolsey et al. (1996) first examined the relationship between the arrangement of penetrating arterioles and emerging venules to the underlying whisker barrels in layer 4, but they found no correlation between them. More recently, Blinder et al. (2013) have compared topographical relationships of barrels mapped with OISI (see Section 3) to the penetrating arterioles and emerging venules. They found no topological relationship between the microvascular networks and the position and boundary of the neuronal columns, despite a slight, yet significant, bias for the centroid of the barrel to be located near the penetrating arteriole. However, no relation between the barrel centroid and the emerging venules was found. Adams et al. (2015) also reported that penetrating arterioles did not discriminate between patches and interpatches of cytochrome oxidase blobs in the primary visual cortex of primates. Therefore, microvasculatures downstream of the penetrating arterioles but before the emerging venules should be a major source for mapping functional structures.

3 FUNCTIONAL COLUMN-SPECIFIC HEMODYNAMIC RESPONSES

3.1 SIGNAL SOURCE OF OIS

OISI is a technique to record changes in the reflected light intensity from the brain surface that accompanies neuronal activity (Grinvald et al., 1986). It is widely used to map functional structures in the neocortex (Bonhoeffer and Grinvald, 1993; Ts'o et al., 1990), such as orientation columns in the cat primary visual cortex (Fig. 2). For example, to map orientation columns, while the exposed visual cortical surface is constantly illuminated by a certain wavelength of visible light (500 nm (green)–650 nm (red)), images of the cortical surface are captured with a high-pixel resolution ($\sim 10 \times 10$ μm) video camera during visual presentation of a moving full-field grating to the animal (A). The OIS is believed to stem from concentration changes in

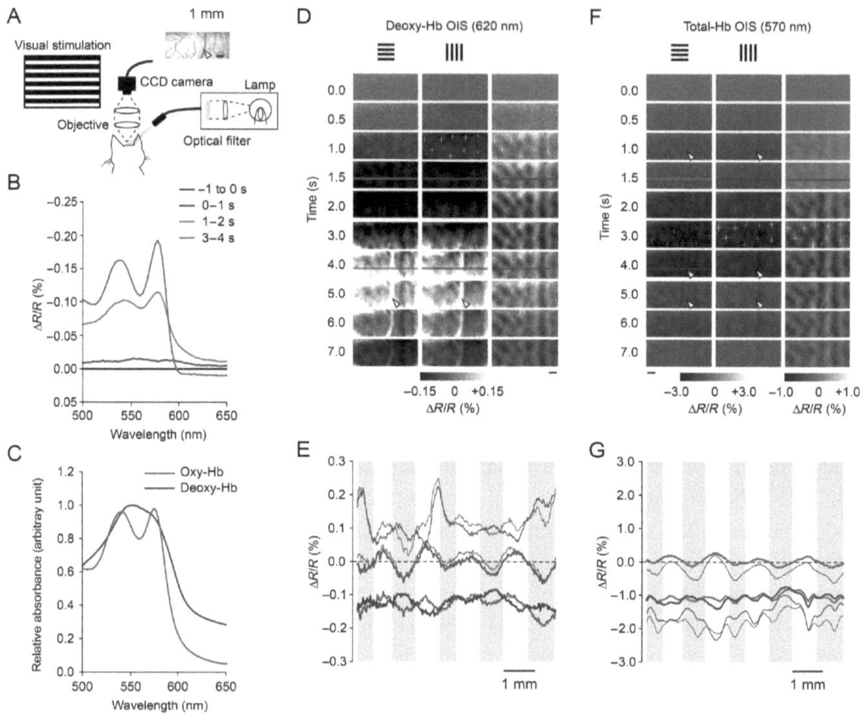

FIG. 2

Optical intrinsic signal imaging. (A) Experimental schematic for mapping iso-orientation domains in the cat visual cortex with OISI. The exposed cortical surface of the anesthetized cat is illuminated by filtered light (typically red or green) and reflected light from the cortex is captured by a CCD camera via a microscope objective during the presentation of a full-field moving grating. Top right panel shows a cortical surface image taken with *green* (*light gray* in the print version) *light* illumination (540 nm wavelength). The *yellow* (*white* in the print version) arrowhead points to a large pial draining vein. Scale bar = 1 mm. (B) Action spectra of OIS evoked by a 2-s long visual stimulation in the visual cortex of a rat (for cat, see fig. 1 in Vanzetta and Grinvald (2001)). (C) In vitro hemoglobin absorption spectra obtained from blood containing 1% intralipid used as a scattering medium (average of five blood samples from rats). The oxygen saturation of hemoglobin was 100% to obtain the oxy-Hb absorption spectrum and changed to 0% by adding 0.1% yeast to obtain the deoxy-Hb spectrum (see Sato et al. (2002)). (D) Spatiotemporal development of deoxy-Hb-weighted OIS evoked by the 2-s presentation of full-field view moving gratings. The wavelength used is 620 nm. Images in the left, middle, and right columns show the single-condition maps for a horizontal grating presentation, single-condition maps for a vertical grating, and the differential maps (horizontal–vertical), respectively. The *y*-axis represents the time from the stimulation onset (0 s). *Yellow* (*white* in the print version) *arrowheads* on the maps at 5.0 s indicate the location of a pial draining vein. Iso-orientation domains for horizontal (*red, light gray* in the print version, "+" symbol) and vertical (*yellow, white* in the print version, "+") gratings are marked on the maps at 1.0 s. Scale bar = 1 mm.
(E) Profiles of deoxy-Hb-weighted OIS along the lines on the images at 1.5 s and 4.0 s in (D).

oxy- and deoxyhemoglobin (Hb) because the hemoglobin is the most abundant intrinsic chromophore and the major light absorber in the brain at these visible wavelengths (Frostig et al., 1990). Indeed, when changes in the reflected light intensity evoked by visual stimulation are plotted against the visible wavelengths of illuminated light, these action spectra of OIS resemble the hemoglobin light absorption spectra. Initially the action spectrum of OIS has one peak around 550 nm (blue trace in B) like the absorption spectrum of deoxyhemoglobin (deoxy-Hb) (blue in C), and then the spectra appear with two peaks around 540 and 580 nm (green and red in B) like the absorption spectrum of oxyhemoglobin (oxy-Hb) (red in C). Another noticeable feature of the OIS action spectra is the polarity of the signal change at the wavelengths above ~600 nm. While initially reflected light intensity decreases after stimulation at the all wavelengths (blue and green traces in B), later reflected light intensity above 600 nm starts to increase from baseline (red trace in B). Since light absorption by deoxy-Hb is dominant above 600 nm, OIS taken at these wavelengths mainly reflects changes in deoxy-Hb content (referred to as deoxy-Hb-weighted OIS). On the other hand, OIS taken at the wavelengths below 600 nm mainly reflect changes in the sum of the oxy- and deoxy-Hb, or total hemoglobin (total-Hb) content since light absorbed by oxy- and deoxy-Hb are almost equal at these wavelengths (referred to as total-Hb-weighted OIS).

To better separate oxy- and deoxy-Hb components from OIS, Malonek and Grinvald (1996) fitted OIS action spectra evoked by a 2-s long visual stimulation on the cat visual cortex to the linear sum of oxy- and deoxy-Hb absorption spectra, and decomposed the OIS into oxy- and deoxy-Hb concentration changes. They found that the deoxy-Hb concentration increased immediately after the onset of stimulation reached maximum at ~2 s and went below the baseline toward a minimum at ~6 s. The increase in the oxy-Hb concentration always lagged behind the increase in deoxy-Hb concentration by 0.5–1.0 s and reached a maximum at ~5 s. They further reported that an increase in the total-Hb concentration occurs earlier than an increase in blood flow measured by laser-Doppler flowmetry (Malonek et al., 1997). Thus, for the deoxy-Hb-weighed OIS, the initial decrease in reflected light intensity (referred

---

Red (*light gray* in the print version) and *blue* (*dark gray* in the print version) *traces* are OIS evoked by horizontal and vertical gratings, respectively. *Green (gray* in the print version) *traces* are obtained by subtracting *blue* (*dark gray* in the print version) from *red* (*light gray* in the print version) *traces*. The portions of the *green* (*gray* in the print version) *trace* with negative reflection changes at 1.5 s are marked with *gray bars* to highlight the iso-orientation domains assigned to the horizontal grating bars. (F) Spatiotemporal development of the total-Hb-weighted OIS from the same cat. The wavelength used is 570 nm. Locations of iso-orientation domains at 3.0 s for horizontal (*red, light gray* in the print version, "+") and vertical (*yellow, white* in the print version, "+") gratings mark the same location as the deoxy-Hb-weighted OIS maps in (D). *Yellow* (*white* in the print version) *arrowheads* on the maps at 1.0, 4.0, and 5.0 s indicate the location of a pial draining vein. Other conventions are the same as (D). Scale bar = 1 mm. (G) Profiles of total-Hb-weighted OIS. The same *gray bars* as (E) are used for ease of comparison. Other conventions are the same as (E). See also, Fukuda et al. (2005).

to as the early negative deoxy-Hb OIS) can be caused by the initial increase in deoxy-Hb content due to an increased oxygen consumption by active neurons plus the early increase in total-Hb content, whereas the later increase in reflected light intensity from baseline (referred to as the late positive deoxy-Hb OIS) can be caused by the decrease in deoxy-Hb content from its prestimulus baseline due to increased oxygenated blood flow. However, it should be noted that this late positive signal reflects the draining of blood, rather than blood supply (Fukuda et al., 2006b; Nagaoka et al., 2006). Since BOLD fMRI contrast also measures changes in the concentration of deoxy-Hb that leads to alterations in the MR signal intensity (for review, see Kim and Ogawa, 2012), the deoxy-Hb-weighted OIS, particularly the late positive signal, is often treated as an analog of the BOLD fMRI signal. Additionally, the total-Hb-weighted OIS is often treated as an analog of cerebral blood volume (CBV)-weighted fMRI signal with the assumption that hematocrit is constant during activation.

3.2 SPATIAL SPECIFICITY OF OISI

Since OISI shares similar signal sources as fMRI, studying the localization of OIS to functional structures, which have dimensions of a few hundred micrometers, has brought insights into the spatial specificity of fMRI. Take, for example, orientation column mapping with deoxy-Hb-weighted OIS (Fig. 2D). Sweeping full-field gratings in front of the anesthetized cat generally causes an initial signal darkening followed by lightening over the visual cortical surface regardless of the grating orientation. These single-condition activity maps are shown as percent signal changes from the prestimulus baseline (compare images in the left and middle columns). Large increases in reflected light intensity are also evoked in the large pial vein regardless of the grating orientation (indicated by the yellow arrowhead in images at 5.0 s). Iso-orientation domains are readily visible as black and white band-like patches (marked by red and yellow "+" signs, respectively) on their differential maps (right column in Fig. 2D), where the single-condition activity maps for the vertical grating presentation are subtracted from the horizontal grating maps to remove common signal changes, including the pial vessel response. These iso-orientation domains, however, are visible as discrete dark patches only during the initial 1 s from the stimulation onset in the single-condition activity maps (see the red and yellow "+" signs in the left and middle columns). The darkening spreads beyond the iso-orientation domains at later times (see 1.5 s). This spread can be caused by a less specific blood supply, subthreshold neuronal activity (Das and Gilbert, 1995), and light scattering (Orbach and Cohen, 1983). As is shown in the example profiles (E) along the thick red and blue lines on the images at 1.5 s (D), the local modulations to the orthogonal orientation stimuli show complementary patterns (ie, peaks in the red trace correspond to troughs in the blue trace). By subtracting the blue from the red trace, the common, nonspecific orientation modulations are removed in order to yield only the orientation-specific modulations (see the green thick trace). Thus, iso-orientation maps are conventionally represented by these differential maps rather than the single-condition activity maps. During the lightening phase of the OIS changes measured over the cortex (3–5 s), the iso-orientation domains are less recognizable in the single-condition activity maps (images in the left and middle columns). Although the spatial profiles of the OIS changes along the thin red and blue lines on the images at 4.0 s still show local modulations, the complementary patterns are not apparent and the signal is severely distorted around the pial veins (E). Nevertheless, the differential signal at 4.0 s (green thin trace) is very similar to that at 1.5 s (green thick trace), resulting in the same iso-orientation maps over the full 7 s (images in the right column in D). This means that if the positive deoxy-Hb-weighted OIS is comparable with the positive BOLD signal, then the BOLD signal evoked by horizontal grating should be smaller than that evoked by vertical grating in iso-orientation domains for the horizontal grating. This is counterintuitive because a larger neuronal activity generally produces a larger BOLD signal.

The results from the optical imaging spectroscopy studies by Malonek and Grinvald (1996) are essentially the same as those from deoxy-Hb-weighted OIS (Fig. 2D and E). While the early increase in deoxy-Hb concentration was highly localized to the iso-orientation domains, the late decrease in the deoxy-Hb concentration and the increase in oxy-Hb concentration (referred to as the late hyperoxygenation signal) were poorly localized to the iso-orientation domains. These results suggest that the early oxygen metabolic signal has a submillimeter resolution, while the late hyperoxygenation signal has a lower spatial resolution. Consistent with this notion, Kim et al. (2000) successfully demonstrated that the early negative BOLD signal clearly resolved the iso-orientation domains, while the later positive signal component (ie, the positive BOLD response) did not. However, since the presence of the early negative BOLD fMRI response is not consistent across studies, (Jezzard et al., 1997) unlike OIS, the source of the early negative deoxy-Hb-weighted OIS came into question. Although extensive studies have been performed to determine whether this early OIS negativity stems from deoxy-Hb or total-Hb content changes, the controversy has not been fully settled (Buxton, 2001; Devor et al., 2003; Frostig et al., 1990; Kohl et al., 2000; Lindauer et al., 2001; Mayhew et al., 1998, 1999, 2000; Nemoto et al., 1999; Sheth et al., 2004; Shtoyerman et al., 2000; Sirotin et al., 2009; Thompson et al., 2003; Vanzetta and Grinvald, 1999, 2001). However, since the early negative deoxy-Hb OIS was intact even after suppressing blood flow and volume responses (Fukuda et al., 2006b; Masamoto et al., 2008; Vazquez et al., 2010), we believe that its major source is an increase in deoxy-Hb content due to increased oxygen metabolism.

On the other hand, for the total-Hb-weighted OIS, grating visual stimulation causes just darkening of the cortex (images in the left and middle columns in Fig. 2F) and is an analog of CBV fMRI. No signal change is observed in the large pial vein (yellow arrowheads on images at 1.0, 4.0, and 5.0 s), suggesting that blood volume does not change in venous vessels. Unlike the deoxy-Hb-weighted OIS, iso-orientation domains are not recognizable in the single-condition maps at any time points (see images in the left and middle columns). Thus, the orientation-specific modulations that are complementary between orthogonal grating orientations are also not apparent in the example profiles (G) along the red and blue lines (F).

However, a difference between the two signals (red minus blue traces) clearly shows orientation-specific modulations (green traces), similar to those shown in E. Thus, the differential maps from the total-Hb-weighted OIS (images in the right column) are almost identical with those from the deoxy-Hb-weighted OIS. These results suggest that blood volume should be regulated at a submillimeter scale (Fukuda et al., 2005; Hess et al., 2000; Vanzetta et al., 2004, 2005). Despite the possible existence of a fine blood supply control mechanism, the spatial spread of the blood volume response is larger than that of the oxygen metabolic response. The ratio of the orientation-specific modulation amplitude (green trace) to the orientation-nonspecific modulation amplitude (mean of red and blue traces) for total-Hb-weighted OIS is ~0.25 at 1.5 s and 4 s, while that for the early negative deoxy-Hb-weighted OIS is ~0.8 at 1.5 s. To further investigate the spatial extent of the total-Hb-weighted OIS, Vazquez et al. (2014) photostimulated neurons expressing light-sensitive channelrhodopsin-2 with a single point of light in the forepaw area of the somatosensory cortex, a brain region that does not form functional columns. They found that the full-width at half-maximum of the evoked total-Hb-weighted OIS ranged from 103 to 175 μm. Thus, submillimeter-scale functional structures separated by a few hundred micrometers should be resolved by the blood volume response. In addition, a submillimeter-scale regulation of the blood flow was also suggested in the ocular dominance column maps of primates measured with laser speckle contrast imaging (Wang and Roe, 2012).

3.3 SPATIAL SPECIFICITY OF fMRI

With the differential method, ocular dominance column-like activation patterns have been demonstrated with high-resolution BOLD fMRI in the human visual cortex in response to monocular eye stimulation (Cheng et al., 2001; Dechent and Frahm, 2000; Goodyear and Menon, 2001; Menon and Goodyear, 1999; Menon et al., 1997; Yacoub et al., 2007). However, it was not clear whether voxels, where the positive BOLD response to one eye stimulation is larger than that of the other eye stimulation, should be assigned to either left or right eye columns because of the unknown BOLD point spread function (Duong et al., 2000) and the inference from the OISI results. To determine the stimulation identity of columns, we compared orientation column maps obtained from fMRI with those from OISI because the relationship of a given stimulus orientation and iso-orientation domain is known in the OIS maps (Fig. 3). To improve the specificity and sensitivity of the functional signals, we injected a blood plasma contrast agent intravenously into the animals. With this intravascular contrast agent, an increase in plasma volume evoked by stimulation will induce an increase in the amount of contrast agent within the voxel and will consequently decrease the MRI signal. Thus, fMRI with this contrast agent is predominantly weighted toward measuring CBV-weighted changes (Kennan et al., 1998; Mandeville et al., 1998; van Bruggen et al., 1998; Zhao et al., 2005). Iso-orientation maps obtained by CBV fMRI match well with those obtained by total-Hb-weighted OIS (A), indicating that regions with the highest orientation-selective blood volume

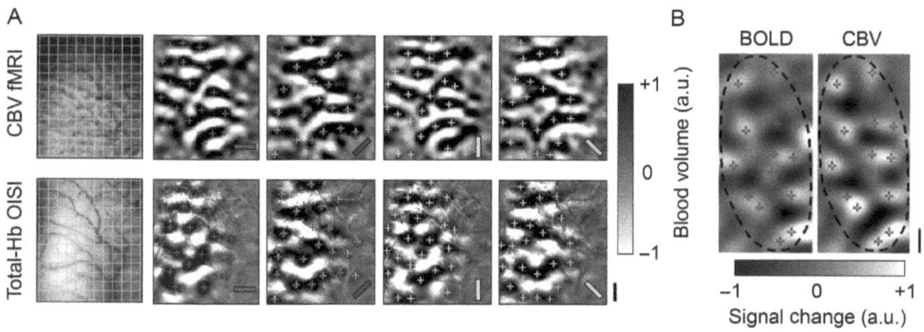

FIG. 3

Orientation column mapping with functional MRI. (A). Comparison between iso-orientation maps obtained by CBV fMRI (*top row*) and those by total-Hb-weighted OISI (*bottom row*) from the same cat after pial vessel coregistration (see the left-most images in each row, grids are $0.5 \times 0.5\ mm^2$.) Only results from the presentation of grating orientations at 0° (*horizontal bars*), 45°, 90°, and 135° are shown. To assist in comparisons, plus signs are placed in regions of increased blood volume in individual total-Hb-weighted OISI maps and then overlaid on the fMRI maps. Scale bar = 1 mm. (B). Iso-orientation domains responsible for horizontal bars obtained by BOLD (*left*) and CBV fMRI (*right*). To assist in comparisons, roughly the same cortical region is indicated by the *black dashed contours* and the *red* (*gray* in the print version) *plus signs* are placed in regions of increased blood volume to a 0° simulation and are overlaid on the BOLD 0° iso-orientation maps. In the bright color domains, the BOLD responses to the presentation of *horizontal bars* are larger than the *vertical bars*. Scale bar = 1 mm.

Panels A and B are adapted from fig. 8 in Fukuda, M., Moon, C.H., Wang, P., Kim, S.G., 2006a. Mapping iso-orientation columns by contrast agent-enhanced functional magnetic resonance imaging: reproducibility, specificity, and evaluation by optical imaging of intrinsic signal. J. Neurosci. 26, 11821–11832 and fig. 7 in Moon, C.H., Fukuda, M., Park, S.H., Kim, S.G., 2007. Neural interpretation of blood oxygenation level-dependent fMRI maps at submillimeter columnar resolution. J. Neurosci., 27, 6892–6902, respectively.

signal changes correspond to the sites of increased neuronal activity (Fukuda et al., 2006a). Further, iso-orientation maps obtained by CBV fMRI are congruent with those obtained by BOLD fMRI (B), indicating that regions of the highest orientation-selective BOLD signal correspond to the sites of increased neuronal activity (Moon et al., 2007). Thus, unlike the late positive deoxy-Hb-weighted OIS (Fig. 2) or the decomposed deoxy-Hb signal (Malonek and Grinvald, 1996), the BOLD fMRI signal evoked by horizontal grating is larger than that evoked by vertical grating in iso-orientation domains for the horizontal grating (Kim and Fukuda, 2008; Moon et al., 2007). Additionally, the increase in both plasma blood volume and total-Hb content in the same iso-orientation domains suggests that arterioles rather than capillaries should be responsible for the column-specific blood volume regulation because capillaries do not seem to dilate to sensory stimulation (Chaigneau et al., 2003; Kleinfeld et al., 1998; Takano et al., 2006).

4 LAYER-SPECIFIC fMRI RESPONSE IN THE NEOCORTEX

Mapping functional columns in the visual cortex with the positive BOLD (Cheng et al., 2001; Moon et al., 2007; Yacoub et al., 2008), blood volume (Fukuda et al., 2006a; Zhao et al., 2005), and blood flow fMRI (Duong et al., 2001) demonstrates that all of these hemodynamic signals have a submillimeter resolution. However, the spatial localization of each hemodynamic signal is not apparent because a differential method was used to map the functional columns. Therefore, the spatial specificity of the different hemodynamic signals is unclear for a single-stimulus condition, which is more typical of an fMRI study. Further, since the columnar structure spans across the cortical layers and each layer is recurrently connected, it was not clear whether these hemodynamic signals can resolve functional processing in individual layers (see also Chapter "Neurogenic control of parenchymal arteries in the cerebral cortex" by Hotta). For instance, is the vascular response to neuronal activity in a single layer reflected in arteriole activation that spans across all cortical layers or is it reflected in microvessel dilation only within the same layer? The answer to this question critically impacts the neuronal interpretation of high-resolution fMRI activation maps and our ability to accurately use fMRI to study laminar processing. To investigate whether the fMRI response in individual layers truly reflects the local neuronal activity, high-resolution fMRI studies have been performed in the cortical laminar model.

The neocortex is subdivided into six main layers (Fig. 4A). In the visual cortex, layer 4 contains spiny stellate cells that primarily receive input from the thalamus, layers 2 and 3 contain pyramidal output neurons that project to other cortical areas, and layers 5 and 6 contain pyramidal output neurons that project to subcortical areas. Layer 6 also receives limited input from the thalamus. Feedback connections from other areas project back to layers 2, 3, and 5. Therefore, individual layers have specialized synaptic connections and functions that could be explored using submillimeter resolution fMRI. For instance, Tootell et al. (1988) reported that sustained visual stimulation evoked the largest 2-DG uptake in layers 4 and 6. This result suggests that fMRI responses evoked by sensory stimulation should peak in layers 4 and 6 first if it reflects the local evoked neuronal activity.

Lu et al. (2004) compared fMRI responses evoked by whisker stimulation between BOLD and CBV signals across the primary somatosensory cortical layers of rats. As a marker of evoked neuronal activity, they also measured cFos expression evoked by whisker stimulation and found the greatest expression in layer 4. They found the BOLD response was greatest at the cortical surface and superficial layers. But, in some rats, the BOLD response was also evoked in deeper layers. In contrast to BOLD, the CBV response consistently peaked around the middle layer where cFos expression was highest. Consistent results have been obtained from other groups in other cortical areas and species (Chen et al., 2013; Harel et al., 2006; Jin and Kim, 2008; Ress et al., 2007; Zhao et al., 2006). Of these, Jin and Kim (2008) compared the spatiotemporal dynamics of BOLD, cerebral blood flow (CBF) and CBV fMRI responses in the cat visual cortical layers evoked by the presentation of full-field moving gratings (Fig. 4B and C). The BOLD signal first appeared in layer 4 but later

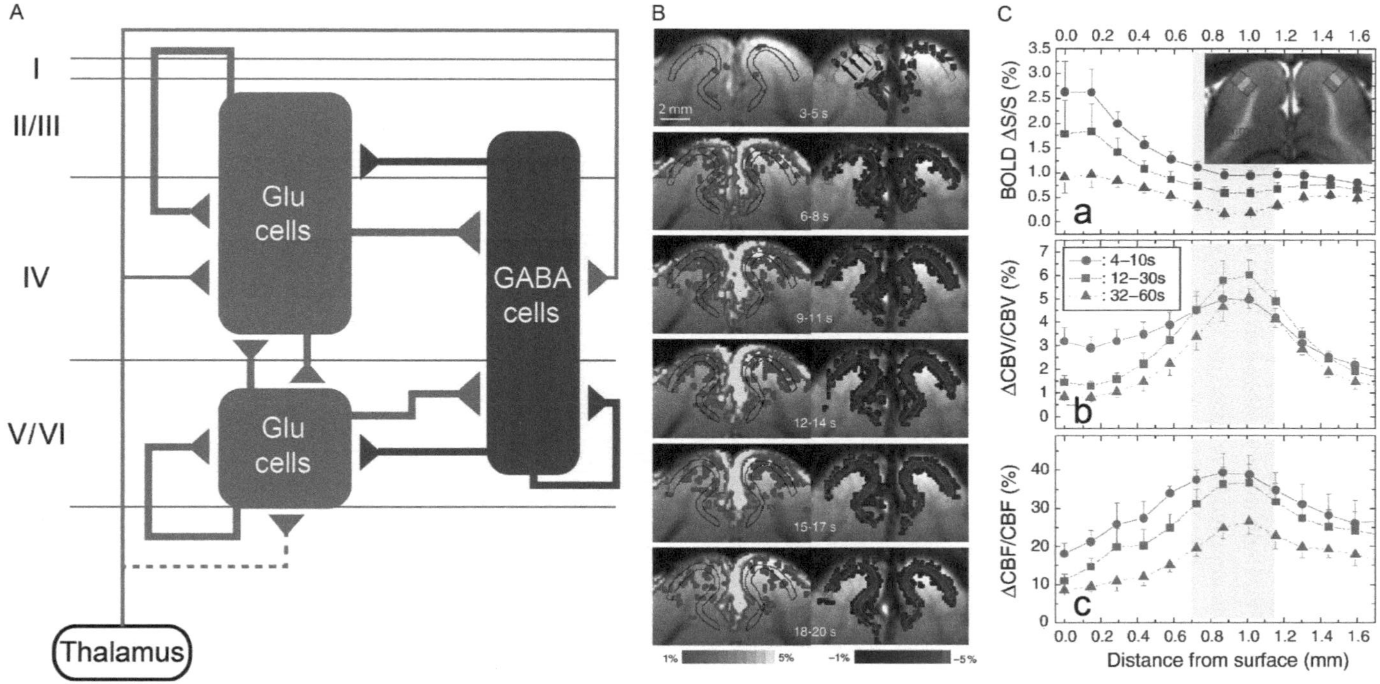

FIG. 4

Cortical layer-specific BOLD, CBV, and CBF fMRI responses. (A). A simple schematic of the visual cortical microcircuit with cortical layers. Excitatory glutamatergic (Glu) cells (*red*, *gray* in the print version) and inhibitory GABAergic cells (*blue*, *dark gray* in the print version) are distributed throughout all layers. Excitatory and inhibitory synapses are depicted as *red* and *blue* colors, respectively. Both Glu and GABA cells in layer IV receive monosynaptic excitatory thalamic input. Note that all cells have reciprocal and strong recurrent connections. (B), Time-dependent percent signal change maps of BOLD (*left*) and CBV-weighted (*right*) fMRI at 3 s time windows in the coronal section of the cat primary visual cortex. The *black* and *green* (*light gray* in the print version) *contours* indicate the approximate locations of layer 4. Decrease in the CBV-weighted signal represents an increase in CBV. During stimulation, the CBV signal first appears, presumably, in penetrating arterioles and then becomes increasingly more localized to layer 4 over time, whereas the BOLD signal first appears in layer 4 but becomes strongest at the cortical surface and cerebrospinal fluid area with time. (C). Time-dependent profiles of BOLD (a), CBV (b), and CBF (c) responses compared to the cortical depth during three time periods (4–10 s, *circle*; 12–32 s, *square*; and 32–60 s, *triangle*) of the 60-s stimulation data. Profiles were obtained from two rectangular ROIs in the primary visual cortex shown in the T_1-weighted image (inset). The *yellow* (*white* in the print version) *bars* show the cortical depth corresponding to layer 4. Error bar = standard error of the mean ($n=6$ cats for BOLD and CBV, and $n=5$ cats for CBF).

Panel A is adapted with permission from Douglas, R.J., Martin, K.A., 1991. A functional microcircuit for cat visual cortex. J. Physiol. 440, 735-769 and Logothetis, N.K., 2008. What we can do and what we cannot do with fMRI. Nature 453, 869–878, and B and C from figs. 3 and 5 in Jin, T., Kim, S.G., 2008. Cortical layer-dependent dynamic blood oxygenation, cerebral blood flow and cerebral blood volume responses during visual stimulation. NeuroImage 43, 1–9.

became strongest at the cortical surface. This is consistent to the notion that the spatial specificity of the positive BOLD signal is better at an early time point before the large draining vessels begin to respond (Goodyear and Menon, 2001; Sheth et al., 2004). On the other hand, the CBV signal first appeared in, presumably, the penetrating arterioles and then became increasingly more localized to layer 4 over time. Interestingly, localization to layer 4 was improved over time for the CBV signal, but not for CBF. They discussed that while the CBV fMRI response originated from vessels of all sizes and with different time characteristics, the CBF fMRI response mostly emerges from the microvessels and surrounding tissue and, thus, its spatial specificity is only weakly time dependent. Better spatial localization of the CBV signal at a later time period was also found in the whisker barrel map with total-Hb-weighted OIS (Berwick et al., 2008) and the orientation column maps with CBV fMRI (Moon et al., 2013). However, Sheth et al. (2004) reported that total-Hb-weighted OIS evoked by whisker stimulation was better localized to the active barrel at an earlier time point. The exact reason for this discrepancy is unknown, but one possibility is that the early OIS response might stem from penetrating arterioles that tend to be closely located to the centroid of the barrel (Blinder et al., 2013).

These results suggest that the CBV and CBF signals more accurately indicate the location of increased neuronal activity than BOLD. However, the baseline blood volume of the cortex has a similar laminar profile as the fMRI response; specifically with the largest microvessel volume in layer 4. In this way, it is not known if the measured fMRI response is indicative of the greatest neural activation in this layer or if it is a simple reflection of the baseline conditions. In addition, different patterns of fMRI responses were also reported between a full visual field stimulation and a surround visual field stimulation. With the full visual field stimulation, Smirnakis et al. (2007) obtained results consistent to the earlier layer-specific fMRI studies, with the largest CBV response in the middle layers and largest BOLD response at the brain surface and superficial layers. However, with the large surround stimulation, they observed an increased CBV fMRI response in an unstimulated V1 region despite the lack of a significant BOLD fMRI response in this same region. They concluded that this CBV response had a nonspecific vascular origin and questioned whether it truly had a superior spatial specificity compared to BOLD. Moreover, Goense et al. (2012) reported that the surround visual stimulation evoked a positive BOLD response in the stimulated V1 region and a negative BOLD response in the unstimulated V1 region. In the positive BOLD region, increased CBV and CBF responses were observed in the middle cortical layers. On the other hand, in the negative BOLD region, decreased CBV in the middle layers and decreased CBF in the superficial layers were observed. Despite the difficulty to explain this observation in the negative BOLD region, the decreased blood flow in the superficial layers seems to be consistent to earlier observations of surround suppression (Devor et al., 2007; Shmuel et al., 2002, 2006). Their finding suggests that different neurovascular coupling mechanisms exist between the positive and negative BOLD signal changes. More importantly, these results suggest that the layer-specific fMRI response can be modulated differently by different kinds of stimulation. For instance,

Wagner et al. (1981) reported that uptake of 2-DG across the visual cortical layers varied depending on the patterns of visual stimulation used. While a moving horizontal grating pattern caused the greatest 2-DG uptake in layer 4, Gaussian visual noise motion increased the uptake in layers 2 and 5 more than in layer 4. This can be tested using fMRI.

5 LAYER-SPECIFIC fMRI RESPONSE IN THE MAIN OLFACTORY BULB

To further explore layer-specific vascular regulation, the rat olfactory bulb model is advantageous because the laminar processing circuit is more simple and has fewer recurrent connections between layers compared to the cortex. These features allow for synapses in single discrete layers of the bulb to be preferentially evoked using different stimuli, which is advantageous to test whether laminar fMRI responses are independent of the constant baseline blood volume.

5.1 OLFACTORY BULB LAMINAR CIRCUIT AND VASO-ARCHITECTURE

The rat olfactory bulb contains six main layers that can be easily identified using endogenous anatomical MRI contrast that can be verified using a traditional histological method with Nissl staining (Poplawsky et al., 2015). In turn, these layers compartmentalize specific neuronal populations and synaptic events (Fig. 5A, for review, see Shepherd, 1972; Shipley and Ennis, 1996). During natural odor stimulation, odorant molecules bind to receptors on olfactory receptor neurons (ORNs) located in the olfactory epithelium (located anterior to the bulb) and evoke action potentials that propagate through unmyelinated axons that enter the bulb through the most superficial bulb layer, the olfactory nerve layer (ONL), and terminate within glomeruli of the glomerular layer (GL). ORNs express only one type of odorant receptor and a single glomerulus receives input from only one ORN receptor type. Mitral cells (MCs), with their somas located in the mitral cell layer (MCL), and tufted cells (TCs) are the main output neurons of the bulb that send their apical dendrites to the glomeruli and receive input from the ORNs. These output cells also receive inputs from inhibitory periglomerular cells (PGs) at their apical dendrites in glomeruli of GL and from inhibitory granule cells (GCs) at their lateral dendrites in the external plexiform layer (EPL). These contacts are often reciprocal dendrodendritic synapses (Fig. 5A insets, for review, see Schoppa and Urban, 2003). The PGs also make presynaptic contact with the ORN axon terminals and inhibit glutamate release. Evoked action potentials of the output cells are then sent out of the bulb en route to their cortical targets through the lateral olfactory tract (LOT). Finally, feedback projections from outside the bulb travel through the anterior commissure (AC) and terminate mainly at the proximal dendrites of GCs in the granule cell layer (GCL). More recent studies, however, show that the feedback projections to GCs are mostly disynaptic via local inhibitory deep short axon cells (dSA) in the GCL. Furthermore, the

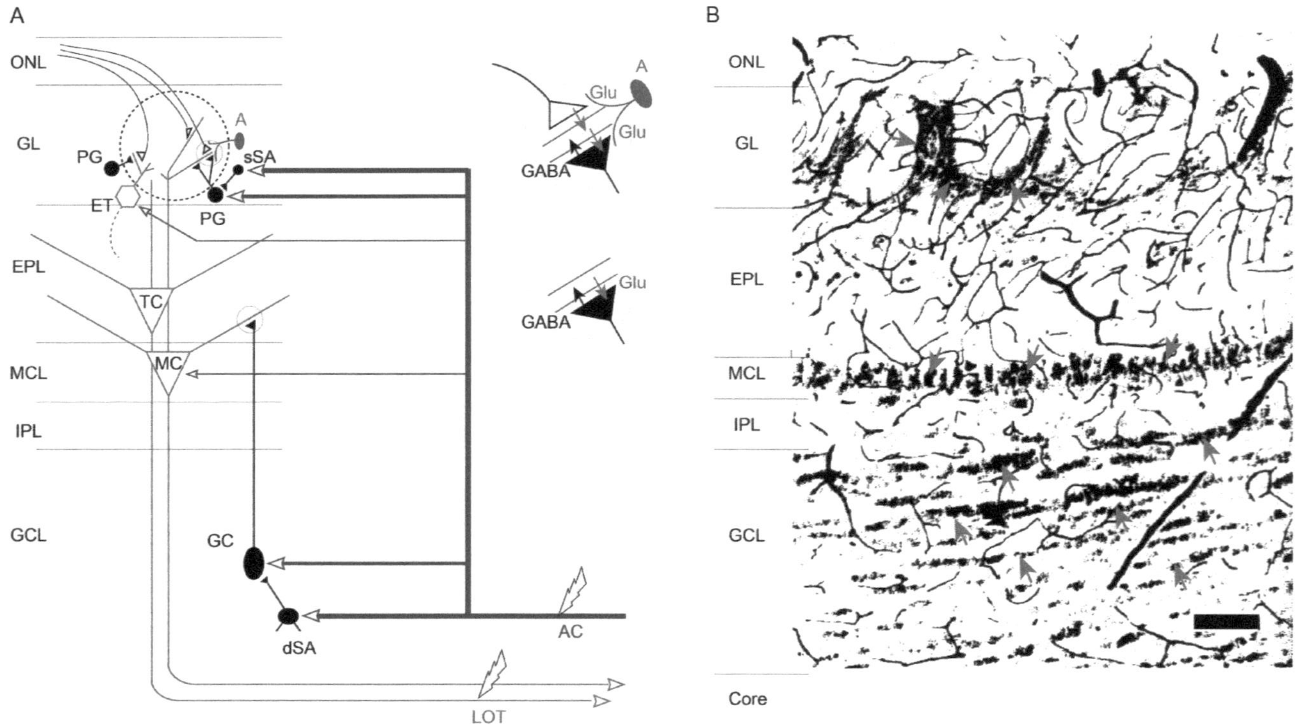

FIG. 5

Microcircuit and vascular structure of the olfactory bulb. (A) Schematic of the olfactory bulb circuit. The olfactory bulb consists of six concentric layers that include, from the most superficial to the deepest, the olfactory nerve layer (*ONL*), glomerular layer (*GL*), external plexiform layer (*EPL*), mitral cell layer (*MCL*), internal plexiform layer (*IPL*), and granule cell layer (*GCL*). Each layer contains distinct cell types and synapses. Axons from olfactory receptor neurons, which bind and detect the odorant molecules, travel from the olfactory epithelium to the bulb feedback projections also terminate on PGs and inhibitory superficial short axon cells (sSA) in GL, as well as some excitatory cells (Boyd et al., 2012; Markopoulos et al., 2012). Although the feedback projections have diverse connections to the olfactory bulb microcircuits, the net effect of the feedback on the output cells is to inhibit odor-evoked spiking (Boyd et al., 2012).

As is similar to the cortex (Fig. 1), the bulb vasculature can also be divided into large surface vessels and parenchymal vessels underneath the surface. The baseline total blood volume of the bulb is greatest at the surface (Surf) and decreases with each layer and is the least in the bulb core (Poplawsky and Kim, 2014), while the baseline blood volume of only microvessels, including capillaries, is greatest in GL and decreases with laminar depth toward GCL (Fig. 5B) that matches well with the laminar profile of the metabolic demand measured by 2-DG (Sharp et al., 1977). Capillaries are very sparse or nonexistent at the bulb surface and ONL (Chaigneau et al., 2003) as well as in the core. Similar to the cortex, the microvessel blood volumes are greatest in the sensory input layers: GL of the bulb and layer 4 of the cortex. Therefore, fMRI activity in GL during odor stimulation is the most comparable to cortical studies and we will briefly review glomerular mapping with fMRI in the next section.

---

through ONL, and terminate on the apical dendrites of excitatory mitral cells (*MCs*) and tufted cells (*TCs*) in glomeruli located in GL (ie, sensory input layer). MCs and TCs are the principal output cells of the bulb that project to the olfactory cortex via the lateral olfactory tract (*LOT*). The activity of MCs and TCs are modulated by two major classes of inhibitory neurons: granule cells (*GCs*) located in GCL and periglomerular cells (*PGs*) located in GL. PGs form reciprocal synapses (inset top) on the apical dendrites of MCs and TCs in a glomerulus, while GCs make reciprocal synapses (inset bottom) on the lateral dendrites of MCs and TCs in EPL. These inhibitory PGs and GCs receive excitatory feedback input from extrinsic higher order olfactory structures via the anterior commissure (*AC*) and local inhibitory short axon cells (superficial short axon cells (*sSA*) for PG and deep short axon cells (*dSA*) for GC) that also receive inputs from AC. Both AC and LOT stimulations will have a net suppression effect on the output of MCs and TCs. Other abbreviations: *IPL*, internal plexiform layer; *ET*, external tufted cell (excitatory); *A*, astrocyte. (B) Blood vessels and cytoarchitecture of the olfactory bulb layers. Vessels density is high in the superficial bulb layers, such as GL and upper EPL, decreases with laminar depth, and becomes sparse in GCL. Note the difference in vessel density between the upper and lower layers of EPL. Vessels in GCL tend to lie parallel to the layer. Characteristic distributions of cell bodies (*red, gray* in the print version, *arrows*) are noted: PGs in GL surround individual glomeruli, excitatory MCs are aligned in MCL, and inhibitory GCs form distinctive bands parallel to the layers in GCL. Scale bar = 0.1 mm.

Panels A and B are adapted with permission from Oswald, A.M., Urban, N.N., 2012. There and back again: the corticobulbar loop. Neuron 76, 1045–1047 and from fig. 1 in Korol, D.L., Brunjes, P.C., 1992. Unilateral naris closure and vascular development in the rat olfactory bulb. Neuroscience 46, 631–641, respectively.

5.2 ODOR STIMULATION EVOKES UNIQUE SPATIAL PATTERNS OF GLOMERULAR ACTIVITY IN THE GLOMERULAR LAYER

A single odorant binds with certain affinities to ORN receptors in the olfactory epithelium. These binding affinities are different depending on the odorant molecule and airborne concentration, which is reflected in the amplitude of the glomerulus activation. In turn, one specific glomerulus in the bulb receives input from only one type of ORN and each glomerulus is spatially segregated from the others. Thus, different odor stimulations elicit a different spatial pattern of activations. These whole bulb activation maps were traditionally examined by measuring the glomeruli with increased metabolic activity, such as 2-DG (Johnson and Leon, 2000; Lancet et al., 1982; Stewart et al., 1979) and c-fos (Guthrie et al., 1993; Onoda, 1992; Sallaz and Jourdan, 1993) histological methods. However, these images were acquired postmortem and were limited to a single odor stimulation condition. Directly measuring neuronal activity of the glomeruli through voltage-gated dyes (Spors and Grinvald, 2002) and calcium imaging (Bozza et al., 2004; Wachowiak and Cohen, 2003), and indirectly by the hemodynamic-based OISI (Meister and Bonhoeffer, 2001; Rubin and Katz, 1999; Uchida et al., 2000) revealed similar spatial-pattern activation maps. These methods are performed in vivo, so multiple odor conditions can be measured in the same animal, but are limited to measuring activations in the dorsal bulb only. Therefore, fMRI is an advantageous tool to investigate glomerular activation maps because it is noninvasive and can measure activations in the whole bulb and entire brain.

Functional MRI using BOLD contrast was first used to map activation of glomeruli to odorants in coronal sections of the rat olfactory bulb (Xu et al., 2000; Yang et al., 1998). During odor stimulation, the largest BOLD activations were observed in ONL and GL, with little to no BOLD responses in deeper bulb layers. To further examine odor-specific activation maps, BOLD responses were evaluated in GL only to form flat maps of glomerular activation, similar to previous 2-DG analyses (Xu et al., 2003). Odorants with related chemical structures (eg, with slightly different carbon chain lengths) showed unique BOLD activation patterns, supporting the use of fMRI for mapping active glomeruli. However, BOLD contrast is complex and is known to be contaminated by vascular changes nonspecific to the site of neural activity, such as large velocity changes in draining veins. To examine this, Poplawsky and Kim (2014) compared BOLD fMRI maps to CBV fMRI maps in the same rats. This study found that BOLD signal changes were primarily located in the bulb surface vessels, ONL and GL, while CBV changes peaked in GL and EPL. Activation in ONL may be due to the layer-specific neurovascular coupling of the ORN unmyelinated axons, but few to no capillaries are located in this layer and these axons may employ anaerobic respiration for energy (Lecoq et al., 2009). Therefore, the neural basis of the BOLD responses in ONL is debatable. Significant activations for both fMRI contrasts were observed in GL, the synaptic input layer, but the location of significant BOLD activations were poorly correlated with the sites of CBV activations. This severely limits whether either fMRI contrast accurately reflects the underlying neural activity. In visual cortical column mapping, the spatial mismatch between BOLD and CBV fMRI maps was negligibly small because the differential method was used to generate the column maps (see Figs. 2 and 3). Thus, we further acquired CBV and BOLD fMRI while preferentially activating other bulb layers.

5.3 THE CBV fMRI RESPONSE COINCIDES WITH THE LAYER-SPECIFIC NEURONAL ACTIVATION, WHILE THE BOLD fMRI RESPONSE POORLY CORRESPONDS

Previous BOLD and CBV fMRI studies with sensory stimulation in the cortex and olfactory bulb show the greatest fMRI response primarily in the sensory input layer (ie, layer 4 of V1 for visual stimulation, GL of the bulb for odor stimulation). However, these input layers also contain the largest baseline total and microvessel blood volumes, which confounds the interpretation of these evoked hemodynamic responses. To test whether the fMRI response accurately reflects the neuronal activity in each layer, different layers must be independently evoked, which is difficult in the cortex because the individual layers are often interconnected (Fig. 4A). On the contrary, synaptic activity in single layers of the olfactory bulb can be evoked with different stimulations. Thus, we measured BOLD and CBV fMRI responses across the bulb layers while we preferentially evoked synaptic activity in GL during natural odor stimulation (5% amyl acetate in mineral oil), EPL during electrical stimulation of LOT, and GCL during electrical stimulation of AC (Fig. 6). The layer-specific neuronal activity is, therefore, independent of the baseline blood volumes, which are greatest in GL and decreases with depth. We found that the positive BOLD responses were greatest in GL for all three stimulation types (A and C), which is likely reflective of the baseline blood volume conditions rather than the underlying neuronal activity. On the contrary, the largest CBV response was discretely segregated to the layer of evoked synaptic activity: greatest in GL for odor, EPL for LOT, and GCL for AC stimulations, respectively (B and D). Most importantly, despite the least baseline blood volume in GCL, the CBV signal during AC stimulation was greatest in this layer. These layer-specific fMRI responses independent of the baseline blood volume strongly suggest that the vasculature, the blood volume in particular, is regulated within layers. Therefore, high-resolution fMRI can be used to measure local network activity occurring in individual layers, but caution must be taken when interpreting BOLD data. Although the positive BOLD response was poorly localized to the site of neuronal activity, there was some evidence, in the laminar BOLD signal time courses during odor and AC stimulations, that the poststimulus BOLD undershoot was highly localized (Poplawsky et al., 2015). Here it is likely that the CBF contributions returned to baseline sooner than the CBV, and possibly the oxygen metabolism, that unmasked these more neural-specific BOLD components. In addition, if the underlying CBF contributions to the BOLD signal are also greatest in the neural-specific layers, then the competing effects of CBV will decrease the

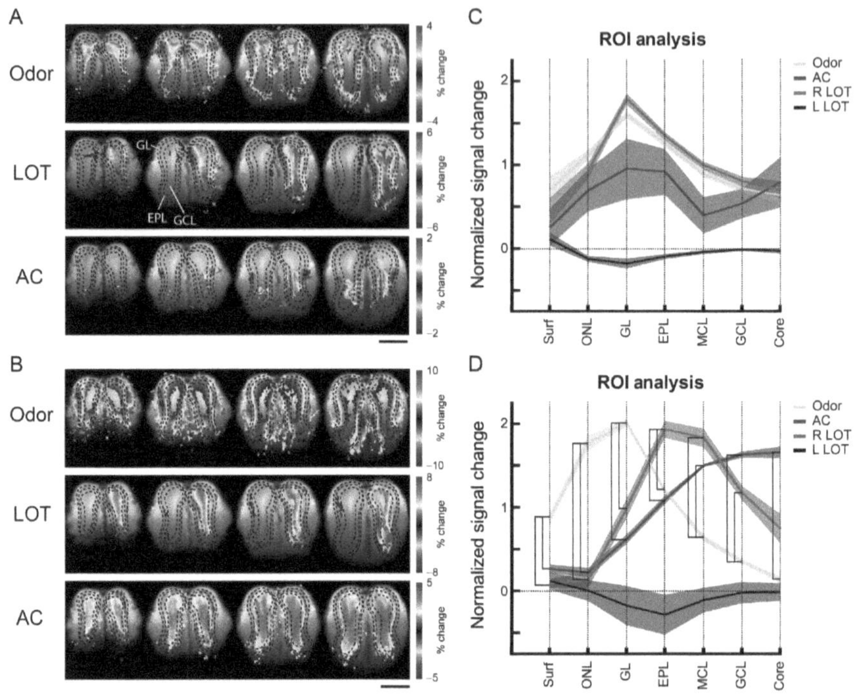

FIG. 6

Maps of layer-specific fMRI responses in the olfactory bulb. (A and B), Group activation maps ($n=6$ rats) for BOLD (A) and CBV fMRI (B) during odor, LOT and AC stimulations, respectively. One rat was removed from BOLD analyses during LOT stimulation only due to inconsistent stimulus delivery. Odor stimulation (5% amyl acetate in mineral oil) evoked the greatest bilateral responses in superficial layers (GL) for both BOLD and CBV fMRI. LOT stimulation produced a unilateral response in superficial and middle layers for BOLD (GL and EPL) and middle layers (EPL) only for CBV fMRI, the latter showing a characteristic thin, ring-like pattern especially in posterior slices. AC stimulation induced a weak BOLD response, while CBV fMRI produced a strong bilateral response in deep layers (eg, GCL). (C and D), ROI analyses show the relative laminar activations for each stimulus (mean $\pm$SEM) in BOLD (C) and CBV fMRI (D). The mean activations between stimuli were compared at each layer with one-way ANOVA and two-sample t-tests (*black brackets* represent significance, $p<0.001$, uncorrected). No thresholds were applied in the ROI analyses. Because almost no activation is expected in the left bulb during right LOT stimulation, the left and right bulbs were reported separately and the activity in the left bulb was normalized by that of the right. The mean fMRI signal change at each bulb layer was normalized by the average absolute response across layers to control for variation in the fMRI response across stimuli and bulbs. For BOLD, the odor (*green, light gray* in the print version; $n=12$ bulbs), LOT (*red, gray* in the print version; $n=5$ right bulbs), and AC (*blue, dark gray* in the print version; $n=12$ bulbs) laminar patterns peak in the superficial glomerular layer (GL). On the contrary, the laminar patterns of CBV changes peak in the same layers as the

BOLD changes in the same layers and obscure the accurate localization of BOLD. Because BOLD fMRI is endogenous and is favored over contrast-enhanced techniques like CBV fMRI, it is necessary to further examine the physiological basis of BOLD and to continue to improve the spatial specificity of the BOLD fMRI.

6 NEUROVASCULAR COUPLING IN OLFACTORY BULB

The layer-specific fMRI responses to LOT and AC stimulations also provide further clues toward the neural interpretability of fMRI. First, LOT and AC stimulations primarily excite the distal and proximal dendrites, respectively, of the same GC population. Thus, the hemodynamic responses measured by CBV fMRI are regulated at the evoked synapse rather than the cell body. Time courses of the fMRI signal changes evoked by these two stimulations were also very different. While AC stimulation evoked a fast hemodynamic response, LOT stimulation evoked a very slow hemodynamic response (Poplawsky et al., 2015). This difference may reflect different vascular compositions in EPL and GCL layers. However, different time courses observed even within the same layer preclude this possibility. Thus, these differences likely reflect different synaptic properties. Second, these GCs are inhibitory neurons, indicating that inhibitory neuronal activity may increase the hemodynamic responses. However, it is not clear yet whether these fMRI responses are coupled to either the presynaptic glutamate release or the postsynaptic inhibitory neuronal activation including the excitation of neurons and GABA release. Further pharmacological studies must be performed.

In addition, the glomerulus in GL has been used to examine relationships of pre- and postsynaptic activity to hemodynamic responses. Studying the pre- and postsynaptic issue in the thalamocortical pathway in layer 4 of the cortex by pharmacologically suppressing postsynaptic activity (Harris et al., 2010; Radhakrishnan et al., 2011) was not conclusive. Neurons in layer 4 receive only weak thalamic input, but dominantly receive inputs from other layer 4 neurons (Bruno and Sakmann, 2006). Therefore, it is not clear whether the decreased hemodynamic response observed following pharmacological blockade is due to the elimination of a substantial

directly evoked neuronal activity for all three stimulus conditions. Specifically, the CBV response to odor (green, *light gray* in the print version; $n=12$ bulbs) peaks in superficial (olfactory nerve layer (ONL) and GL), right LOT (*red, gray* in the print version; $n=6$ right bulbs) in right middle (EPL and mitral cell (MCL) layers), and AC (*blue, dark gray* in the print version; $n=12$ bulbs) in deep layers (GCL and core). Unexpectedly, small contralateral (left) bulb decreases in BOLD (*black*, $n=5$ left bulbs) and CBV (*black*, $n=6$ left bulbs) responses were observed during LOT stimulation. Surf: bulb surface.

Panels C and D are adapted from fig. 7 in Poplawsky, A.J., Fukuda, M., Murphy, M., Kim, S.G., 2015. Layer-specific fMRI responses to excitatory and inhibitory neuronal activities in the olfactory bulb. J. Neurosci. 35, 15263–15275.

number of inputs from other neurons (ie, postsynaptic activity of these neurons is blocked) or due to an insignificant contribution of presynaptic activity to the hemodynamic response. Thus, the glomerulus in GL is an ideal model to examine neural input and output contributions to the hemodynamic response because postsynaptic MCs receive excitatory inputs only from the convergent projections of a single ORN type. The MCs also receive inhibitory input from PGs. However, it is rather tonic and does not correlate with the strength of input to the olfactory bulb (Pirez and Wachowiak, 2008), suggesting that excitatory neuronal activity mainly controls the odor-evoked hemodynamic responses.

Spors and Grinvald (2002) showed that odor stimulation evoked glomerular activities mapped with a voltage-sensitive dye had a postsynaptic origin and were identical to glomerular maps obtained by the early negative deoxy-Hb-weighted OIS. Surprisingly, Gurden et al. (2006) showed that odor stimulation evoked OIS specific to glomeruli was not suppressed by blocking postsynaptic ionotropic glutamate receptors (AMPA and NMDA) and group I metabotropic glutamate receptors, such as mGluR1. Instead, they found that the glomerular-specific OIS was suppressed by blocking glutamate uptake through glutamate transporters (GLT1 and GLAST) of astrocytes. This result suggests that the vascular response does not require postsynaptic activity. Following this study, Petzold et al. (2008) demonstrated that odor stimulation evoked blood flow responses were also intact during the local injection of the AMPA and NMDA glutamate receptor antagonists into a single active glomerulus, but were suppressed by blocking the astrocytic glutamate transporter. Further, they showed that an additional pathway, metabotropic glutamate receptor 5 (mGluR5, a group I metabotropic glutamate receptor) on astrocytes was crucial for the odor stimulation evoked CBF responses. More recently, however, Otsu et al. (2015) demonstrated that astrocyte activity was involved with glomerular neurovascular coupling even in adult mice that lacked mGluR5. These studies suggest that activation of glutamate uptake by astrocytes can be a crucial step for neurovascular coupling in olfactory bulb glomeruli. Thus, presynaptic activity alone seems to account for the hemodynamic responses in the olfactory bulb glomeruli (Nawroth et al., 2007). However, Chaigneau et al. (2007) demonstrated that the odor-evoked blood flow responses were not affected by local injections of antagonists for AMPA and NMDA glutamate receptors, and mGluR1 into a single glomerulus, but were suppressed by injecting the antagonists beyond the single glomerulus. Thus, they concluded that blood flow responses are regulated by a cooperation of several glomeruli, particularly with a strong stimulation, such as a high concentration of odor, and that activation of postsynaptic glutamate receptors are essential for neurovascular coupling in the olfactory bulb glomeruli. In addition, they discussed that glomerular astrocyte transporters are activated by glutamate release from both the presynaptic olfactory nerve terminals and the secondary dendrites of the postsynaptic MCs following spike discharges (De Saint Jan and Westbrook, 2005) (see also Fig. 5A inset top). However, this local vs global drug application does not seem to explain the discrepancy between their and the OIS results (Gurden et al., 2006) because both topically applied the antagonists over the entire bulb. Although the metabolic activity may have a presynaptic origin, the vascular response may require postsynaptic activity. Alternatively, the early negative deoxy-Hb-weighted OIS may not be caused by an increased deoxy-Hb content due to increased oxygen consumption, although odor stimulation evoked an initial decrease in the tissue partial oxygen pressure that resulted from postsynaptic activity (Lecoq et al., 2009), but rather it may be associated with activity-dependent light scattering changes (Meister and Bonhoeffer, 2001; Vincis et al., 2015). In any event, these studies demonstrate the importance of glutamatergic synaptic transmission to odor-evoked hemodynamic responses. Together, these observations make the olfactory bulb an intriguing model to study neurovascular coupling and the neuronal origins of fMRI.

REFERENCES

Adams, D.L., Piserchia, V., Economides, J.R., Horton, J.C., 2015. Vascular supply of the cerebral cortex is specialized for cell layers but not columns. Cereb. Cortex 25, 3673–3681.

Attwell, D., Buchan, A.M., Charpak, S., Lauritzen, M., Macvicar, B.A., Newman, E.A., 2010. Glial and neuronal control of brain blood flow. Nature 468, 232–243.

Berwick, J., Johnston, D., Jones, M., Martindale, J., Martin, C., Kennerley, A.J., Redgrave, P., Mayhew, J.E., 2008. Fine detail of neurovascular coupling revealed by spatiotemporal analysis of the hemodynamic response to single whisker stimulation in rat barrel cortex. J. Neurophysiol. 99, 787–798.

Blasdel, G.G., Salama, G., 1986. Voltage-sensitive dyes reveal a modular organization in monkey striate cortex. Nature 321, 579–585.

Blinder, P., Tsai, P.S., Kaufhold, J.P., Knutsen, P.M., Suhl, H., Kleinfeld, D., 2013. The cortical angiome: an interconnected vascular network with noncolumnar patterns of blood flow. Nat. Neurosci. 16, 889–897.

Bonhoeffer, T., Grinvald, A., 1993. The layout of iso-orientation domains in area 18 of cat visual cortex: optical imaging reveals a pinwheel-like organization. J. Neurosci. 13, 4157–4180.

Borowsky, I.W., Collins, R.C., 1989. Metabolic anatomy of brain: a comparison of regional capillary density, glucose metabolism, and enzyme activities. J. Comp. Neurol. 288, 401–413.

Boyd, A.M., Sturgill, J.F., Poo, C., Isaacson, J.S., 2012. Cortical feedback control of olfactory bulb circuits. Neuron 76, 1161–1174.

Bozza, T., Mcgann, J.P., Mombaerts, P., Wachowiak, M., 2004. In vivo imaging of neuronal activity by targeted expression of a genetically encoded probe in the mouse. Neuron 42, 9–21.

Bruno, R.M., Sakmann, B., 2006. Cortex is driven by weak but synchronously active thalamocortical synapses. Science 312, 1622–1627.

Buxton, R.B., 2001. The elusive initial dip. Neuroimage 13, 953–958.

Chaigneau, E., Oheim, M., Audinat, E., Charpak, S., 2003. Two-photon imaging of capillary blood flow in olfactory bulb glomeruli. Proc. Natl. Acad. Sci. U. S. A. 100, 13081–13086.

Chaigneau, E., Tiret, P., Lecoq, J., Ducros, M., Knopfel, T., Charpak, S., 2007. The relationship between blood flow and neuronal activity in the rodent olfactory bulb. J. Neurosci. 27, 6452–6460.

Chen, G., Wang, F., Gore, J.C., Roe, A.W., 2013. Layer-specific BOLD activation in awake monkey V1 revealed by ultra-high spatial resolution functional magnetic resonance imaging. Neuroimage 64, 147–155.

Cheng, K., Waggoner, R.A., Tanaka, K., 2001. Human ocular dominance columns as revealed by high-field functional magnetic resonance imaging. Neuron 32, 359–374.

Das, A., Gilbert, C.D., 1995. Long-range horizontal connections and their role in cortical reorganization revealed by optical recording of cat primary visual cortex. Nature 375, 780–784.

Dechent, P., Frahm, J., 2000. Direct mapping of ocular dominance columns in human primary visual cortex. Neuroreport 11, 3247–3249.

De Saint Jan, D., Westbrook, G.L., 2005. Detecting activity in olfactory bulb glomeruli with astrocyte recording. J. Neurosci. 25, 2917–2924.

Devor, A., Dunn, A.K., Andermann, M.L., Ulbert, I., Boas, D.A., Dale, A.M., 2003. Coupling of total hemoglobin concentration, oxygenation, and neural activity in rat somatosensory cortex. Neuron 39, 353–359.

Devor, A., Tian, P., Nishimura, N., Teng, I.C., Hillman, E.M., Narayanan, S.N., Ulbert, I., Boas, D.A., Kleinfeld, D., DALE, A.M., 2007. Suppressed neuronal activity and concurrent arteriolar vasoconstriction may explain negative blood oxygenation level-dependent signal. J. Neurosci. 27, 4452–4459.

Drew, P.J., Shih, A.Y., Kleinfeld, D., 2011. Fluctuating and sensory-induced vasodynamics in rodent cortex extend arteriole capacity. Proc. Natl. Acad. Sci. U. S. A. 108, 8473–8478.

Duong, T.Q., Kim, D.S., Ugurbil, K., Kim, S.G., 2000. Spatiotemporal dynamics of the BOLD fMRI signals: toward mapping submillimeter cortical columns using the early negative response. Magn. Reson. Med. 44, 231–242.

Duong, T.Q., Kim, D.S., Ugurbil, K., Kim, S.G., 2001. Localized cerebral blood flow response at submillimeter columnar resolution. Proc. Natl. Acad. Sci. U. S. A. 98, 10904–10909.

Duvernoy, H.M., Delon, S., Vannson, J.L., 1981. Cortical blood vessels of the human brain. Brain Res. Bull. 7, 519–579.

Frostig, R.D., Lieke, E.E., Ts'o, D.Y., Grinvald, A., 1990. Cortical functional architecture and local coupling between neuronal activity and the microcirculation revealed by in vivo high-resolution optical imaging of intrinsic signals. Proc. Natl. Acad. Sci. U. S. A. 87, 6082–6086.

Fukuda, M., Rajagopalan, U.M., Homma, R., Matsumoto, M., Nishizaki, M., Tanifuji, M., 2005. Localization of activity-dependent changes in blood volume to submillimeter-scale functional domains in cat visual cortex. Cereb. Cortex 15, 823–833.

Fukuda, M., Moon, C.H., Wang, P., Kim, S.G., 2006a. Mapping iso-orientation columns by contrast agent-enhanced functional magnetic resonance imaging: reproducibility, specificity, and evaluation by optical imaging of intrinsic signal. J. Neurosci. 26, 11821–11832.

Fukuda, M., Wang, P., Moon, C.H., Tanifuji, M., Kim, S.G., 2006b. Spatial specificity of the enhanced dip inherently induced by prolonged oxygen consumption in cat visual cortex: implication for columnar resolution functional MRI. Neuroimage 30, 70–87.

Fukuda, M., Vazquez, A.L., Zong, X., Kim, S.G., 2013. Effects of the α_2-adrenergic receptor agonist dexmedetomidine on neural, vascular and BOLD fMRI responses in the somatosensory cortex. Eur. J. Neurosci. 37 (1), 80–95.

Gardner, J.L., 2010. Is cortical vasculature functionally organized? Neuroimage 49, 1953–1956.

Goense, J., Merkle, H., Logothetis, N.K., 2012. High-resolution fMRI reveals laminar differences in neurovascular coupling between positive and negative BOLD responses. Neuron 76, 629–639.

Goodyear, B.G., Menon, R.S., 2001. Brief visual stimulation allows mapping of ocular dominance in visual cortex using fMRI. Hum. Brain Mapp. 14, 210–217.

Grinvald, A., Lieke, E., Frostig, R.D., Gilbert, C.D., Wiesel, T.N., 1986. Functional architecture of cortex revealed by optical imaging of intrinsic signals. Nature 324, 361–364.

Guibert, R., Fonta, C., Plouraboue, F., 2010. Cerebral blood flow modeling in primate cortex. J. Cereb. Blood Flow Metab. 30, 1860–1873.

Gurden, H., Uchida, N., Mainen, Z.F., 2006. Sensory-evoked intrinsic optical signals in the olfactory bulb are coupled to glutamate release and uptake. Neuron 52, 335–345.

Guthrie, K.M., Anderson, A.J., Leon, M., Gall, C., 1993. Odor-induced increases in c-fos mRNA expression reveal an anatomical "unit" for odor processing in olfactory bulb. Proc. Natl. Acad. Sci. U. S. A. 90, 3329–3333.

Harel, N., Lin, J., Moeller, S., Ugurbil, K., Yacoub, E., 2006. Combined imaging-histological study of cortical laminar specificity of fMRI signals. Neuroimage 29, 879–887.

Harris, S., Jones, M., Zheng, Y., Berwick, J., 2010. Does neural input or processing play a greater role in the magnitude of neuroimaging signals? Front. Neuroenergetics 2, 1–7 (article 15). http://dx.doi.org/10.3389/fnene.2010.00015.

Hess, A., Stiller, D., Kaulisch, T., Heil, P., Scheich, H., 2000. New insights into the hemodynamic blood oxygenation level-dependent response through combination of functional magnetic resonance imaging and optical recording in gerbil barrel cortex. J. Neurosci. 20, 3328–3338.

Hillman, E.M., 2014. Coupling mechanism and significance of the BOLD signal: a status report. Annu. Rev. Neurosci. 37, 161–181.

Hirsch, S., Reichold, J., Schneider, M., Szekely, G., Weber, B., 2012. Topology and hemodynamics of the cortical cerebrovascular system. J. Cereb. Blood Flow Metab. 32, 952–967.

Howarth, C., Gleeson, P., Attwell, D., 2012. Updated energy budgets for neural computation in the neocortex and cerebellum. J. Cereb. Blood Flow Metab. 32, 1222–1232.

Hubel, D.H., Wiesel, T.N., 1962. Receptive fields, binocular interaction and functional architecture in the cat's visual cortex. J. Physiol. 160, 106–154.

Hubel, D.H., Wiesel, T.N., 1963. Shape and arrangement of columns in cat's striate cortex. J. Physiol. 165, 559–568.

Jezzard, P., Rauschecker, J.P., Malonek, D., 1997. An in vivo model for functional MRI in cat visual cortex. Magn. Reson. Med. 38, 699–705.

Jin, T., Kim, S.G., 2008. Cortical layer-dependent dynamic blood oxygenation, cerebral blood flow and cerebral blood volume responses during visual stimulation. Neuroimage 43, 1–9.

Johnson, B.A., Leon, M., 2000. Modular representations of odorants in the glomerular layer of the rat olfactory bulb and the effects of stimulus concentration. J. Comp. Neurol. 422, 496–509.

Keller, A.L., Schuz, A., Logothetis, N.K., Weber, B., 2011. Vascularization of cytochrome oxidase-rich blobs in the primary visual cortex of squirrel and macaque monkeys. J. Neurosci. 31, 1246–1253.

Kennan, R.P., Scanley, B.E., Innis, R.B., Gore, J.C., 1998. Physiological basis for BOLD MR signal changes due to neuronal stimulation: separation of blood volume and magnetic susceptibility effects. Magn. Reson. Med. 40, 840–846.

Kim, S.G., Fukuda, M., 2008. Lessons from fMRI about mapping cortical columns. Neuroscientist 14, 287–299.

Kim, T., Kim, S.G., 2011. Temporal dynamics and spatial specificity of arterial and venous blood volume changes during visual stimulation: implication for BOLD quantification. J. Cereb. Blood Flow Metab. 31, 1211–1222.

Kim, S.G., Ogawa, S., 2012. Biophysical and physiological origins of blood oxygenation level-dependent fMRI signals. J. Cereb. Blood Flow Metab. 32, 1188–1206.
Kim, D.S., Duong, T.Q., Kim, S.G., 2000. High-resolution mapping of iso-orientation columns by fMRI. Nat. Neurosci. 3, 164–169.
Kim, T., Hendrich, K.S., Masamoto, K., Kim, S.G., 2007. Arterial versus total blood volume changes during neural activity-induced cerebral blood flow change: implication for BOLD fMRI. J. Cereb. Blood Flow Metab. 27, 1235–1247.
Kleinfeld, D., Mitra, P.P., Helmchen, F., Denk, W., 1998. Fluctuations and stimulus-induced changes in blood flow observed in individual capillaries in layers 2 through 4 of rat neocortex. Proc. Natl. Acad. Sci. U. S. A. 95, 15741–15746.
Kohl, M., Lindauer, U., Royl, G., Kuhl, M., Gold, L., Villringer, A., Dirnagl, U., 2000. Physical model for the spectroscopic analysis of cortical intrinsic optical signals. Phys. Med. Biol. 45, 3749–3764.
Lancet, D., Greer, C.A., Kauer, J.S., Shepherd, G.M., 1982. Mapping of odor-related neuronal activity in the olfactory bulb by high-resolution 2-deoxyglucose autoradiography. Proc. Natl. Acad. Sci. U. S. A. 79, 670–674.
Lecoq, J., Tiret, P., Najac, M., Shepherd, G.M., Greer, C.A., Charpak, S., 2009. Odor-evoked oxygen consumption by action potential and synaptic transmission in the olfactory bulb. J. Neurosci. 29, 1424–1433.
Lindauer, U., Royl, G., Leithner, C., Kuhl, M., Gold, L., Gethmann, J., Kohl-Bareis, M., Villringer, A., Dirnagl, U., 2001. No evidence for early decrease in blood oxygenation in rat whisker cortex in response to functional activation. Neuroimage 13, 988–1001.
Logothetis, N.K., 2008. What we can do and what we cannot do with fMRI. Nature 453, 869–878.
Lu, H., Patel, S., Luo, F., Li, S.J., Hillard, C.J., Ward, B.D., Hyde, J.S., 2004. Spatial correlations of laminar BOLD and CBV responses to rat whisker stimulation with neuronal activity localized by Fos expression. Magn. Reson. Med. 52, 1060–1068.
Maier, A., Adams, G.K., Aura, C., Leopold, D.A., 2010. Distinct superficial and deep laminar domains of activity in the visual cortex during rest and stimulation. Front. Syst. Neurosci. 4.
Malonek, D., Grinvald, A., 1996. Interactions between electrical activity and cortical microcirculation revealed by imaging spectroscopy: implications for functional brain mapping. Science 272, 551–554.
Malonek, D., Dirnagl, U., Lindauer, U., Yamada, K., Kanno, I., Grinvald, A., 1997. Vascular imprints of neuronal activity: relationships between the dynamics of cortical blood flow, oxygenation, and volume changes following sensory stimulation. Proc. Natl. Acad. Sci. U. S. A. 94, 14826–14831.
Mandeville, J.B., Marota, J.J., Kosofsky, B.E., Keltner, J.R., Weissleder, R., Rosen, B.R., Weisskoff, R.M., 1998. Dynamic functional imaging of relative cerebral blood volume during rat forepaw stimulation. Magn. Reson. Med. 39, 615–624.
Markopoulos, F., Rokni, D., Gire, D.H., Murthy, V.N., 2012. Functional properties of cortical feedback projections to the olfactory bulb. Neuron 76, 1175–1188.
Masamoto, K., Vazquez, A., Wang, P., Kim, S.G., 2008. Trial-by-trial relationship between neural activity, oxygen consumption, and blood flow responses. Neuroimage 40, 442–450.
Mayhew, J., Hu, D., Zheng, Y., Askew, S., Hou, Y., Berwick, J., Coffey, P.J., Brown, N., 1998. An evaluation of linear model analysis techniques for processing images of microcirculation activity. Neuroimage 7, 49–71.
Mayhew, J., Zheng, Y., Hou, Y., Vuksanovic, B., Berwick, J., Askew, S., Coffey, P., 1999. Spectroscopic analysis of changes in remitted illumination: the response to increased neural activity in brain. Neuroimage 10, 304–326.

Mayhew, J., Johnston, D., Berwick, J., Jones, M., Coffey, P., Zheng, Y., 2000. Spectroscopic analysis of neural activity in brain: increased oxygen consumption following activation of barrel cortex. Neuroimage 12, 664–675.

Mchedlishvili, G., Kuridze, N., 1984. The modular organization of the pial arterial system in phylogeny. J. Cereb. Blood Flow Metab. 4, 391–396.

Meister, M., Bonhoeffer, T., 2001. Tuning and topography in an odor map on the rat olfactory bulb. J. Neurosci. 21, 1351–1360.

Menon, R.S., Goodyear, B.G., 1999. Submillimeter functional localization in human striate cortex using BOLD contrast at 4 Tesla: implications for the vascular point-spread function. Magn. Reson. Med. 41, 230–235.

Menon, R.S., Ogawa, S., Strupp, J.P., Ugurbil, K., 1997. Ocular dominance in human V1 demonstrated by functional magnetic resonance imaging. J. Neurophysiol. 77, 2780–2787.

Moon, C.H., Fukuda, M., Park, S.H., Kim, S.G., 2007. Neural interpretation of blood oxygenation level-dependent fMRI maps at submillimeter columnar resolution. J. Neurosci. 27, 6892–6902.

Moon, C.H., Fukuda, M., Kim, S.G., 2013. Spatiotemporal characteristics and vascular sources of neural-specific and -nonspecific fMRI signals at submillimeter columnar resolution. Neuroimage 64, 91–103.

Nagaoka, T., Zhao, F., Wang, P., Harel, N., Kennan, R.P., Ogawa, S., Kim, S.G., 2006. Increases in oxygen consumption without cerebral blood volume change during visual stimulation under hypotension condition. J. Cereb. Blood Flow Metab. 26, 1043–1051.

Nawroth, J.C., Greer, C.A., Chen, W.R., Laughlin, S.B., Shepherd, G.M., 2007. An energy budget for the olfactory glomerulus. J. Neurosci. 27, 9790–9800.

Nemoto, M., Nomura, Y., Sato, C., Tamura, M., Houkin, K., Koyanagi, I., Abe, H., 1999. Analysis of optical signals evoked by peripheral nerve stimulation in rat somatosensory cortex: dynamic changes in hemoglobin concentration and oxygenation. J. Cereb. Blood Flow Metab. 19, 246–259.

Nguyen, J., Nishimura, N., Fetcho, R.N., Iadecola, C., Schaffer, C.B., 2011. Occlusion of cortical ascending venules causes blood flow decreases, reversals in flow direction, and vessel dilation in upstream capillaries. J. Cereb. Blood Flow Metab. 31, 2243–2254.

Nishimura, N., Schaffer, C.B., Friedman, B., Lyden, P.D., Kleinfeld, D., 2007. Penetrating arterioles are a bottleneck in the perfusion of neocortex. Proc. Natl. Acad. Sci. U. S. A. 104, 365–370.

O'kusky, J., Colonnier, M., 1982. A laminar analysis of the number of neurons, glia, and synapses in the adult cortex (area 17) of adult macaque monkeys. J. Comp. Neurol. 210, 278–290.

Onoda, N., 1992. Odor-induced fos-like immunoreactivity in the rat olfactory bulb. Neurosci. Lett. 137, 157–160.

Orbach, H.S., Cohen, L.B., 1983. Optical monitoring of activity from many areas of the in vitro and in vivo salamander olfactory bulb: a new method for studying functional organization in the vertebrate central nervous system. J. Neurosci. 3, 2251–2262.

Otsu, Y., Couchman, K., Lyons, D.G., Collot, M., Agarwal, A., Mallet, J.M., Pfrieger, F.W., Bergles, D.E., Charpak, S., 2015. Calcium dynamics in astrocyte processes during neurovascular coupling. Nat. Neurosci. 18, 210–218.

Petzold, G.C., Murthy, V.N., 2011. Role of astrocytes in neurovascular coupling. Neuron 71, 782–797.

Petzold, G.C., Albeanu, D.F., Sato, T.F., Murthy, V.N., 2008. Coupling of neural activity to blood flow in olfactory glomeruli is mediated by astrocytic pathways. Neuron 58, 897–910.

Pirez, N., Wachowiak, M., 2008. In vivo modulation of sensory input to the olfactory bulb by tonic and activity-dependent presynaptic inhibition of receptor neurons. J. Neurosci. 28, 6360–6371.

Poplawsky, A.J., Kim, S.G., 2014. Layer-dependent BOLD and CBV-weighted fMRI responses in the rat olfactory bulb. Neuroimage 91, 237–251.

Poplawsky, A.J., Fukuda, M., Murphy, M., Kim, S.G., 2015. Layer-specific fMRI responses to excitatory and inhibitory neuronal activities in the olfactory bulb. J. Neurosci. 35, 15263–15275.

Radhakrishnan, H., Wu, W., Boas, D., Franceschini, M.A., 2011. Study of neurovascular coupling by modulating neuronal activity with GABA. Brain Res. 1372, 1–12.

Ress, D., Glover, G.H., Liu, J., Wandell, B., 2007. Laminar profiles of functional activity in the human brain. Neuroimage 34, 74–84.

Rubin, B.D., Katz, L.C., 1999. Optical imaging of odorant representations in the mammalian olfactory bulb. Neuron 23, 499–511.

Sallaz, M., Jourdan, F., 1993. C-fos expression and 2-deoxyglucose uptake in the olfactory bulb of odour-stimulated awake rats. Neuroreport 4, 55–58.

Sato, C., Nemoto, M., Tamura, M., 2002. Reassessment of activity-related optical signals in somatosensory cortex by an algorithm with wavelength-dependent path length. Jpn. J. Physiol. 52, 301–312.

Schoppa, N.E., Urban, N.N., 2003. Dendritic processing within olfactory bulb circuits. Trends Neurosci. 26, 501–506.

Sharp, F.R., Kauer, J.S., Shepherd, G.M., 1977. Laminar analysis of 2-deoxyglucose uptake in olfactory bulb and olfactory cortex of rabbit and rat. J. Neurophysiol. 40, 800–813.

Shepherd, G.M., 1972. Synaptic organization of the mammalian olfactory bulb. Physiol. Rev. 52, 864–917.

Sheth, S.A., Nemoto, M., Guiou, M., Walker, M., Pouratian, N., Hageman, N., Toga, A.W., 2004. Columnar specificity of microvascular oxygenation and volume responses: implications for functional brain mapping. J. Neurosci. 24, 634–641.

Shih, A.Y., Blinder, P., Tsai, P.S., Friedman, B., Stanley, G., Lyden, P.D., Kleinfeld, D., 2013. The smallest stroke: occlusion of one penetrating vessel leads to infarction and a cognitive deficit. Nat. Neurosci. 16, 55–63.

Shipley, M.T., Ennis, M., 1996. Functional organization of olfactory system. J. Neurobiol. 30, 123–176.

Shmuel, A., Yacoub, E., Pfeuffer, J., Van de Moortele, P.F., Adriany, G., Hu, X., Ugurbil, K., 2002. Sustained negative BOLD, blood flow and oxygen consumption response and its coupling to the positive response in the human brain. Neuron 36, 1195–1210.

Shmuel, A., Augath, M., Oeltermann, A., Logothetis, N.K., 2006. Negative functional MRI response correlates with decreases in neuronal activity in monkey visual area V1. Nat. Neurosci. 9, 569–577.

Shtoyerman, E., Arieli, A., Slovin, H., Vanzetta, I., Grinvald, A., 2000. Long-term optical imaging and spectroscopy reveal mechanisms underlying the intrinsic signal and stability of cortical maps in V1 of behaving monkeys. J. Neurosci. 20, 8111–8121.

Sirotin, Y.B., Hillman, E.M., Bordier, C., Das, A., 2009. Spatiotemporal precision and hemodynamic mechanism of optical point spreads in alert primates. Proc. Natl. Acad. Sci. U. S. A. 106, 18390–18395.

Smirnakis, S.M., Schmid, M.C., Weber, B., Tolias, A.S., Augath, M., Logothetis, N.K., 2007. Spatial specificity of BOLD versus cerebral blood volume fMRI for mapping cortical organization. J. Cereb. Blood Flow Metab. 27, 1248–1261.

Snodderly, D.M., Gur, M., 1995. Organization of striate cortex of alert, trained monkeys (Macaca fascicularis): ongoing activity, stimulus selectivity, and widths of receptive field activating regions. J. Neurophysiol. 74, 2100–2125.

Spors, H., Grinvald, A., 2002. Spatio-temporal dynamics of odor representations in the mammalian olfactory bulb. Neuron 34, 301–315.

Stewart, W.B., Kauer, J.S., Shepherd, G.M., 1979. Functional organization of rat olfactory bulb analysed by the 2-deoxyglucose method. J. Comp. Neurol. 185, 715–734.

Takano, T., Tian, G.F., Peng, W., Lou, N., Libionka, W., Han, X., Nedergaard, M., 2006. Astrocyte-mediated control of cerebral blood flow. Nat. Neurosci. 9, 260–267.

Thompson, J.K., Peterson, M.R., Freeman, R.D., 2003. Single-neuron activity and tissue oxygenation in the cerebral cortex. Science 299, 1070–1072.

Tootell, R.B., Hamilton, S.L., Silverman, M.S., Switkes, E., 1988. Functional anatomy of macaque striate cortex. I. Ocular dominance, binocular interactions, and baseline conditions. J. Neurosci. 8, 1500–1530.

Tsai, P.S., Kaufhold, J.P., Blinder, P., Friedman, B., Drew, P.J., Karten, H.J., Lyden, P.D., Kleinfeld, D., 2009. Correlations of neuronal and microvascular densities in murine cortex revealed by direct counting and colocalization of nuclei and vessels. J. Neurosci. 29, 14553–14570.

Ts'o, D.Y., Frostig, R.D., Lieke, E.E., Grinvald, A., 1990. Functional organization of primate visual cortex revealed by high resolution optical imaging. Science 249, 417–420.

Turner, R., 2002. How much cortex can a vein drain? Downstream dilution of activation-related cerebral blood oxygenation changes. Neuroimage 16, 1062–1067.

Uchida, N., Takahashi, Y.K., Tanifuji, M., Mori, K., 2000. Odor maps in the mammalian olfactory bulb: domain organization and odorant structural features. Nat. Neurosci. 3, 1035–1043.

Van Bruggen, N., Busch, E., Palmer, J.T., Williams, S.P., De Crespigny, A.J., 1998. High-resolution functional magnetic resonance imaging of the rat brain: mapping changes in cerebral blood volume using iron oxide contrast media. J. Cereb. Blood Flow Metab. 18, 1178–1183.

Vanzetta, I., Grinvald, A., 1999. Increased cortical oxidative metabolism due to sensory stimulation: implications for functional brain imaging. Science 286, 1555–1558.

Vanzetta, I., Grinvald, A., 2001. Evidence and lack of evidence for the initial dip in the anesthetized rat: implications for human functional brain imaging. Neuroimage 13, 959–967.

Vanzetta, I., Slovin, H., Omer, D.B., Grinvald, A., 2004. Columnar resolution of blood volume and oximetry functional maps in the behaving monkey; implications for FMRI. Neuron 42, 843–854.

Vanzetta, I., Hildesheim, R., Grinvald, A., 2005. Compartment-resolved imaging of activity-dependent dynamics of cortical blood volume and oximetry. J. Neurosci. 25, 2233–2244.

Vazquez, A.L., Masamoto, K., Fukuda, M., Kim, S.G., 2010. Cerebral oxygen delivery and consumption during evoked neural activity. Front. Neuroenergetics 2, 11.

Vazquez, A.L., Fukuda, M., Crowley, J.C., Kim, S.G., 2014. Neural and hemodynamic responses elicited by forelimb- and photo-stimulation in channelrhodopsin-2 mice: insights into the hemodynamic point spread function. Cereb. Cortex 24, 2908–2919.

Vincis, R., Lagier, S., Van de Ville, D., RODRIGUEZ, I., Carleton, A., 2015. Sensory-evoked intrinsic imaging signals in the olfactory bulb are independent of neurovascular coupling. Cell Reports 12, 313–325.

Wachowiak, M., Cohen, L.B., 2003. Correspondence between odorant-evoked patterns of receptor neuron input and intrinsic optical signals in the mouse olfactory bulb. J. Neurophysiol. 89, 1623–1639.

Wagner, H.J., Hoffmann, K.P., Zwerger, H., 1981. Layer-specific labelling of cat visual cortex after stimulation with visual noise: a [3H]2-deoxy-d-glucose study. Brain Res. 224, 31–43.

Wang, Z., Roe, A.W., 2012. Columnar specificity of microvascular oxygenation and blood flow response in primary visual cortex: evaluation by local field potential and spiking activity. J. Cereb. Blood Flow Metab. 32, 6–16.

Weber, B., Keller, A.L., Reichold, J., Logothetis, N.K., 2008. The microvascular system of the striate and extrastriate visual cortex of the macaque. Cereb. Cortex 18, 2318–2330.

Woolsey, T.A., Van Der Loos, H., 1970. The structural organization of layer IV in the somatosensory region (SI) of mouse cerebral cortex. The description of a cortical field composed of discrete cytoarchitectonic units. Brain Res. 17, 205–242.

Woolsey, T.A., Rovainen, C.M., Cox, S.B., Henegar, M.H., Liang, G.E., Liu, D., Moskalenko, Y.E., Sui, J., Wei, L., 1996. Neuronal units linked to microvascular modules in cerebral cortex: response elements for imaging the brain. Cereb. Cortex 6, 647–660.

Xu, F., Kida, I., Hyder, F., Shulman, R.G., 2000. Assessment and discrimination of odor stimuli in rat olfactory bulb by dynamic functional MRI. Proc. Natl. Acad. Sci. U. S. A. 97, 10601–10606.

Xu, F., Liu, N., Kida, I., Rothman, D.L., Hyder, F., Shepherd, G.M., 2003. Odor maps of aldehydes and esters revealed by functional MRI in the glomerular layer of the mouse olfactory bulb. Proc. Natl. Acad. Sci. U. S. A. 100, 11029–11034.

Yacoub, E., Shmuel, A., Logothetis, N., Ugurbil, K., 2007. Robust detection of ocular dominance columns in humans using Hahn Spin Echo BOLD functional MRI at 7 Tesla. Neuroimage 37, 1161–1177.

Yacoub, E., Harel, N., Ugurbil, K., 2008. High-field fMRI unveils orientation columns in humans. Proc. Natl. Acad. Sci. U. S. A. 105, 10607–10612.

Yang, X., Renken, R., Hyder, F., Siddeek, M., Greer, C.A., Shepherd, G.M., Shulman, R.G., 1998. Dynamic mapping at the laminar level of odor-elicited responses in rat olfactory bulb by functional MRI. Proc. Natl. Acad. Sci. U. S. A. 95, 7715–7720.

Zhao, F., Wang, P., Hendrich, K., Kim, S.G., 2005. Spatial specificity of cerebral blood volume-weighted fMRI responses at columnar resolution. Neuroimage 27, 416–424.

Zhao, F., Wang, P., Hendrich, K., Ugurbil, K., Kim, S.-G., 2006. Cortical layer-dependent BOLD and CBV responses measured by spin-echo and gradient-echo fMRI: insights into hemodynamic regulation. Neuroimage 30, 1149–1160.

Zheng, D., Lamantia, A.S., Purves, D., 1991. Specialized vascularization of the primate visual cortex. J. Neurosci. 11, 2622–2629.

CHAPTER

Hemodynamic signals in fNIRS

7

Y. Hoshi[1]

Institute for Medical Photonics Research, Preeminent Medical Photonics Education & Research Center, Hamamatsu University School of Medicine, Hamamatsu, Japan

[1]*Corresponding author: Tel./Fax: +81-53-435-2329, e-mail address: yhoshi@hama-med.ac.jp*

Abstract

Near-infrared spectroscopy (NIRS) was originally designed for clinical monitoring of tissue oxygenation, and it has also been developed into a useful tool in neuroimaging studies, with the so-called functional NIRS (fNIRS). With NIRS, cerebral activation is detected by measuring the cerebral hemoglobin (Hb), where however, the precise correlation between NIRS signal and neural activity remains to be fully understood. This can in part be attributed to the situation that NIRS signals are inherently subject to contamination by signals arising from extracerebral tissue. In recent years, several approaches have been investigated to distinguish between NIRS signals originating in cerebral tissue and signals originating in extracerebral tissue. Selective measurements of cerebral Hb will enable a further evolution of fNIRS. This chapter is divided into six sections: first a summary of the basic theory of NIRS, NIRS signals arising in the activated areas, correlations between NIRS signals and fMRI signals, correlations between NIRS signals and neural activities, and the influence of a variety of extracerebral tissue on NIRS signals and approaches to this issue are reviewed. Finally, future prospects of fNIRS are described.

Keywords

Oxy-Hb, Deoxy-Hb, t-Hb, fMRI, EEG, NIRS, DOT, Spontaneous fluctuations, Extracerebral tissues

1 BASIC THEORY OF NIRS

1.1 WHAT IS NIRS/fNIRS?

Near-infrared spectroscopy (NIRS) utilizes light in the near-infrared (NIR) region, from 700 to 2500 nm and is a noninvasive technique for analysis of various substances including agricultural products and food. Biological tissue is relatively transparent from 700 to 1300 nm, where absorption spectra vary with the

Progress in Brain Research, Volume 225, ISSN 0079-6123, http://dx.doi.org/10.1016/bs.pbr.2016.03.004

oxygenation–deoxygenation states of hemoglobin (Hb) and myoglobin and the redox state of cytochrome *c* oxidase in mitochondria. Based on these spectroscopic characteristics, NIRS can measure concentration changes in oxygenated (oxy-Hb) and deoxygenated Hb (deoxy-Hb) in the human brain, and the technique was originally designed for clinical monitoring of tissue oxygenation (Jöbsis, 1977). Since the early 1990s, it has also been developed into a useful tool in neuroimaging studies, with the so-called functional NIRS (fNIRS) (Chance et al., 1993; Hoshi and Tamura, 1993a, Kato et al., 1993; Villringer et al., 1993). In this chapter, except where description is restricted to fNIRS, NIRS is used as an umbrella term for this technique.

1.2 NIRS INSTRUMENTATION

A wide range of NIRS instruments have been developed, including some that are commonly commercially available for continuous wave (CW) measurements based on the modified Beer–Lambert law (MBL) (Delpy et al., 1988) (CW-type instruments).

The MBL is expressed as:

$$A = -\log I/I_0 = \epsilon CL + S \tag{1}$$

where A is the attenuation measured in OD (optical density), I and I_0 are the intensities of the detected and illuminated light, ϵ is the molar absorption coefficient ($M^{-1}\,cm^{-1}$), C is the concentration (M) of chromophore (eg, Hb), and L is the length of the optical path (cm), also known as the "total pathlength" (t-PL), and S denotes the optical attenuation mainly due to scattering (Fig. 1). However, CW-type instruments cannot measure the optical pathlength, instruments of this type do not provide absolute values of concentration changes, and here the NIRS signal is expressed as the product of concentration changes and optical pathlength ($M \times cm$). To overcome this problem with quantification, time-resolved spectroscopy (TRS) (Chance et al., 1988; Delpy et al., 1988) and frequency-domain spectroscopy (FDS) (Duncan et al., 1993; Lakowicz and Berndt, 1990), which can determine the t-PL, have been developed. However, functional brain activation is accompanied by localized Hb concentration changes within the cerebral tissue. This requires the determination of the partial optical pathlengths (p-PL) in the cerebral tissue for the quantification, something which is, however, not feasible at present. Since both the t-PL and p-PL vary with the region of measurement, the amplitudes of NIRS signals for different regions do not reflect only the concentration changes in the cerebral Hb of the regions compared.

1.3 NIR LIGHT PROPAGATION IN THE HEAD

Light propagation in biological tissue is generally approximated by the diffusion equation (DE) and can be predicted by Monte Carlo (MC) simulations. Theoretical analyses with MC simulations of the head models consisting of three- or four-layered

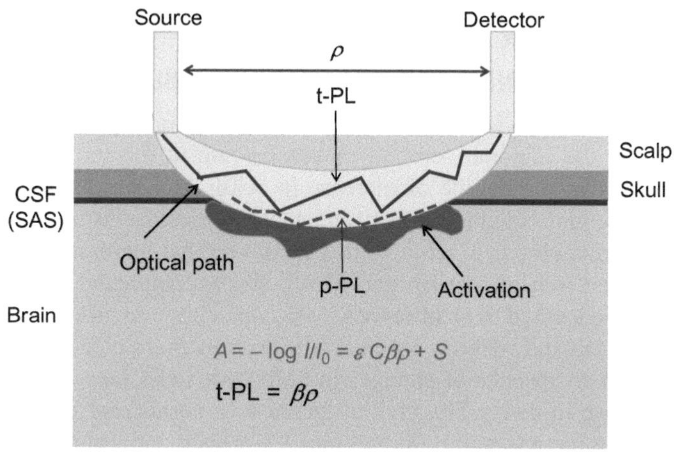

FIG. 1

Schematic representation of near-infrared light traveling through the head and the modified Beer–Lambert law (MBL). The *red* (*dark gray* in the print version) area denotes a region of brain activation. Total pathlength (t-PL) is β times longer than the source–detector distance (ρ) due to scattering. *Solid* and *dashed lines* in the optical path denote schematic mean t-PL and p-PL, respectively. *CSF*, cerebrospinal fluid; *SAS*, subarachnoid space.

slabs, the latter incorporating a clear cerebrospinal fluid (CSF) layer, have demonstrated that light propagation in the adult head is strongly affected by the presence of a low-scattering CSF layer (Okada et al., 1997). This indicates that the light penetration in the adult brain may be limited to the outer cortical gray matter, which has been confirmed by later studies where the light propagation in head models generated by magnetic resonance imaging (MRI) was predicted by MC simulations (Firbank et al., 1998). A subsequent study has reported that a large source–detector (SD) distance only broadens the sampling region on the brain surface and affects the penetration depth in the adult head to a lesser degree, whereas the intensely sensitive region in neonatal heads is confined to the gray matter, and the deeper regions of the white matter can be sampled with a large SD distance (Fukui et al., 2003).

Our group's MC simulations have been performed on four-layered human head models based on MRI scans and have indicated that at the most common SD distance of 30 mm very few photons penetrating deeper than about 25 mm are detected, while the number of photons that travel through the brain surface layer is the largest among the detected photons penetrating into the brain tissue (Hoshi et al., 2005). A more recent study, where the MC simulations were performed on multiple brain regions in a large number of head models, has reported that increasing the SD distance from 20 to 45 mm monotonically increased sensitivity to gray matter, while increasing the separation further resulted in diminishing returns and the sensitivity decreased exponentially with depth (Strangman et al., 2013).

2 NIRS SIGNALS IN ACTIVATED AREAS

2.1 NIRS SIGNALS AND REGIONAL CEREBRAL BLOOD FLOW

Regional brain activation is accompanied by increases in regional cerebral blood flow (rCBF) and the regional cerebral metabolic rate of oxygen ($rCMRO_2$). It is widely accepted that the degree of increase in rCBF exceeds that of the increase in $rCMRO_2$ (Fox and Raichle, 1986), and that this results in a decrease in the deoxy-Hb in venous blood. A result of this would be that increases in oxy-Hb and total Hb (t-Hb, the sum of oxy-Hb and deoxy-Hb) accompanied by a decrease in deoxy-Hb can be expected to be observed in activated areas in NIRS measurements. However, deoxy-Hb and t-Hb do not necessarily show these changes: the observations have shown an absence of changes in t-Hb with an increase in oxy-Hb and a reciprocal decrease in deoxy-Hb, and an increase or absence of change in deoxy-Hb accompanying increases in t-Hb and oxy-Hb (Hoshi and Tamura, 1993b; Kato et al., 1993; Kleinschmidt et al., 1996). Using a novel perfused rat brain model, we examined the direct effects of the changes of each of the components in CBF and $CMRO_2$ on the cerebral Hb oxygenation to become better able to interpret NIRS signals (Fig. 2A and B) (Hoshi et al., 2001). We have confirmed that the directions of the changes in oxy-Hb are always the same as the changes in CBF, while the direction of the changes in deoxy-Hb is determined by changes in the venous blood oxygenation and volume. It has also been confirmed that small changes in CBF are not accompanied by changes in t-Hb. Overall this suggests that oxy-Hb is the most sensitive indicator of changes in rCBF in NIRS measurements.

2.2 VASCULAR SPECIFICITY OF NIRS

The NIRS is sensitive to small vessels including arterioles, capillaries, and venules, because light propagated into large vessels is almost completely absorbed (Chance et al., 1992; Mancini et al., 1994). However, no precise vascular specificity of NIRS is established, one argument is that NIRS exclusively corresponds to capillary oxygenation (Nielsen et al., 2001; Rasmussen et al., 2007), and another is that NIRS reflects Hb oxygenation in venous, capillary, and arterial blood within the sampling volume (Murkin and Arango, 2009). Similarly, the origin of the fNIRS signals has been questioned, and it is not established which fNIRS reflects capillaries or venules (Fig. 3). However, it is well known that cortical activation induces pial artery dilatation (Ngai et al., 1988), which is explained by the putative mechanism that activation-related vasodilation propagates from arterioles and capillaries in the activated site to the pial arteries in a retrograde fashion (Iadecola, 2004). This makes it reasonable to hypothesize that fNIRS reflects concentration changes in Hb within the sampling volume which result from vasodilation from pial arteries to capillaries (Hall et al., 2014) and venules (Buxton et al., 1998), and oxygenation changes in venous blood (see Chapter "New horizons in neurometabolic and neurovascular coupling from calibrated fMRI" by C.Y. Shu, B.G. Sanganahalli, D. Coman, P. Herman, and F. Hyder).

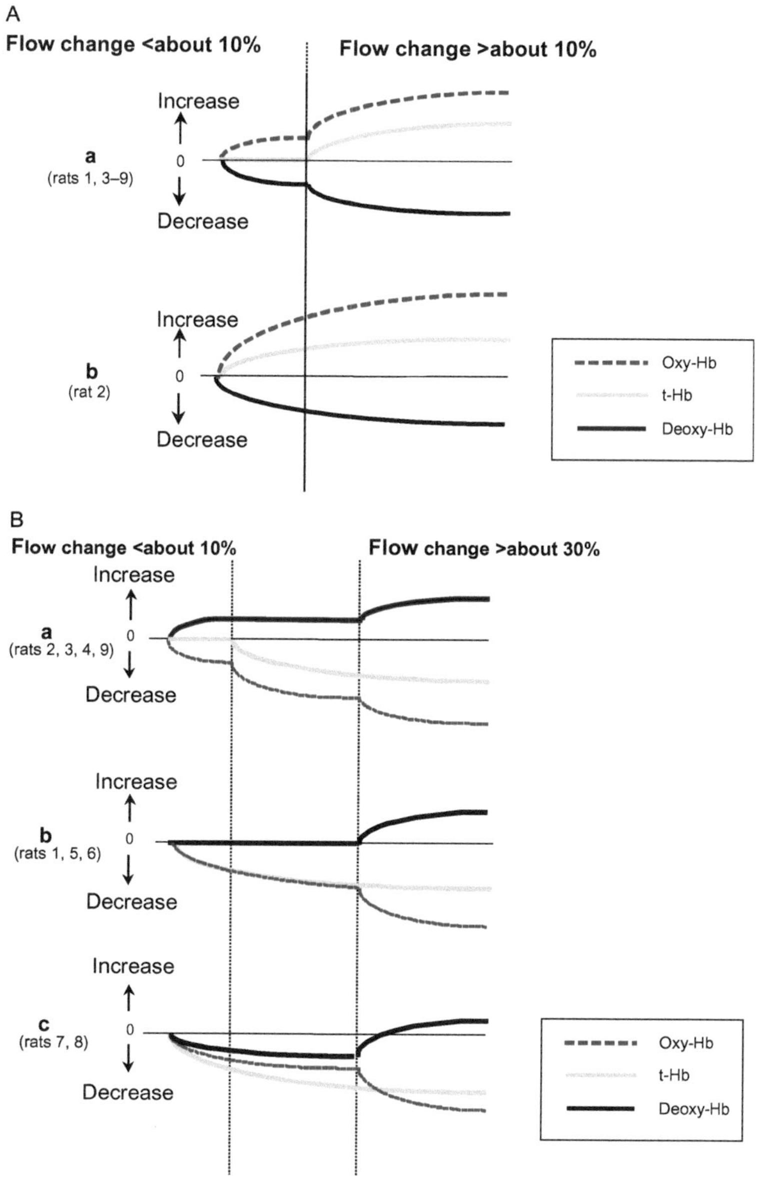

FIG. 2

(A) Pattern of NIRS signals while the flow rate was increasing. The pattern of changes in NIRS signals was divided into A and B according to the direction of changes in t-Hb when the flow change was less than ∼10%. (a) Absence of change in t-Hb and (b) an increase in t-Hb were observed. (B) Pattern of changes in NIRS signals while the flow rate was decreasing. The pattern was divided into a, b, and c, according to the direction of changes in t-Hb and deoxy-Hb when the flow change was less than ∼10%. (a) An increase in deoxy-Hb and a decrease in oxy-Hb with absence of change in t-Hb, (b) absence of changes in deoxy-Hb and t-Hb with a decrease in oxy-Hb, and (c) decreases in all the three Hbs were observed. *From Hoshi, Y., Kobayashi, N., Tamura., M., 2001. Interpretation of near-infrared spectroscopy signals: a study with a newly developed perfused rat brain model. J. Appl. Physiol. 90, 1657–1662, with permission.*

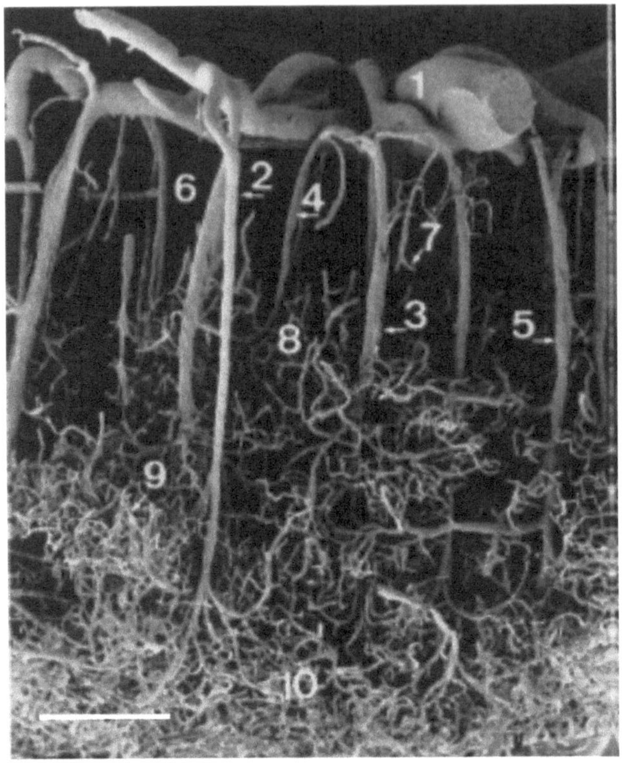

FIG. 3

Vascular structure in the human temporal cortex obtained by vascular casting and electron scanning microscopy. 1, pial artery; 2, long cortical artery; 3, middle cortical artery; 4, short cortical artery; 5, cortical vein; 6, subpial zone; 7, precapillary vessels with blind endings; 8, superficial capillary zone; 9, middle capillary zone; 10, deep capillary zone. Scale bar = 375 μm.

From Reina-De La Torre, F., Rodriguez-Baeza, A., Sahuquillo-Barris, J., 1998. Morphological characteristics and distribution pattern of the arterial vessels in human cerebral cortex: a scanning electron microscopy study. Anat. Rec. 251, 87–96, with permission.

3 NIRS SIGNALS AND fMRI SIGNALS

3.1 COMPARISON OF NIRS SIGNALS AND fMRI SIGNALS

A number of simultaneous NIRS and functional MRI (fMRI) recordings have been performed, by which it is expected that NIRS obtains anatomical information from the latter fMRI and provides complementary information in addition to further insights into the physiological basis of the fMRI signals. Originally, spatial correspondences between fNIRS and blood oxygenation level-dependent (BOLD) fMRI was examined qualitatively by a single-channel NIRS instrument (Kleinschmidt et al., 1996), and then various more elaborate investigations followed, where multichannel NIRS instruments and arterial spin labeling (ASL) MRI were introduced (Hoge et al., 2005; MacIntosh et al., 2003; Strangman et al., 2002; Toronov et al., 2001; Yamamoto and Kato, 2002). Most of these studies demonstrated high correlations between fNIRS and fMRI signals, while only a few studies showed a higher correlation between BOLD and deoxy-Hb than that between BOLD and oxy-Hb (MacIntosh et al., 2003; Toronov et al., 2001). Huppert et al. (2006) argued that the long-duration tasks used in previous studies made it difficult to distinguish subtle temporal differences between oxy-Hb and deoxy-Hb. Employing a short-duration, event-related motor task for temporal comparisons of hemodynamic responses measured by fMRI and NIRS, they reported that deoxy-Hb was better correlated to BOLD signals, while t-Hb and oxy-Hb showed a high correlation with ASL-measured CBF. These studies have contributed to a better understanding of the origins of the BOLD signals. However, the spatial resolution and signal-to-noise ratio of NIRS are lower than those of fMRI, and statistical methods used in NIRS data analysis are generally different from those used with MRI, which makes a precise spatial and temporal comparison difficult. More recent studies have taken these differences into account when comparing NIRS and fMRI signals (Cui et al., 2011; Gagnon et al, 2012a; Toronov et al., 2007), while the exact correlations between NIRS and fMRI signals remain inconclusive.

3.2 NEGATIVE BOLD RESPONSES AND NIRS SIGNALS

3.2.1 Prolonged negative signal

Negative BOLD responses (NBRs), including the prolonged negative signals, the initial dip, and the poststimulus undershoot, have increasingly been the focus of research attention (see Chapter “New horizons in neurometabolic and neurovascular coupling from calibrated fMRI” by C.Y. Shu, B.G. Sanganahalli, D. Coman, P. Herman, and F. Hyder). The prolonged NBR has been observed in various areas, such as visual, somatosensory, and motor areas (Devor et al., 2007; Hamzei et al., 2002; Shmuel et al., 2002). Several mechanisms behind these responses have been proposed (Kim and Ogawa, 2012), and the current general consensus is that the prolonged NBR is accompanied by decreases in the CBF and the $CMRO_2$, which indicates that neural activities are decreased (Huber et al., 2014). A recent study with a multiecho, BOLD-corrected vascular space occupancy (VASO) fMRI, where BOLD signals, cerebral blood volume (CBV), and CBF were measured during stimuli eliciting positive and NBRs in humans at 7T, has shown that the inhibitory stimuli are associated with maximum NBRs in deeper cortical layers with depth-dependent decreases in CBV due to constriction of surface arteries, while the excitatory stimuli are accompanied by a maximum in the positive BOLD responses at the cortical surface with increases in CBV in cortex and pial vasculature (Huber et al., 2014). These results support the current interpretation for NBRs and further suggest that underlying hemodynamic mechanisms for NBRs are different from those of the positive BOLD responses.

An early simultaneous measurement with NIRS and fMRI demonstrated that electrical stimulation on the right median nerve induced prolonged NBRs in the left supramarginal region, which was accompanied by an increase in deoxy-Hb and a decrease in oxy-Hb with no change in t-Hb (Seiyama et al., 2004). A recent study combining NIRS and fMRI has reported that NBRs to photic stimulation in the extrastriate visual cortex are related to a decrease in oxy-Hb and a lower increase in deoxy-Hb, resulting in a decrease in t-Hb (Maggioni et al., 2015). The difference in t-Hb changes between the two studies may have reflected the aforementioned depth-dependent CBV changes (Huber et al., 2014). Alternatively, in the former study (Seiyama et al., 2004), decreases in CBF could have been small, resulting in the absence of change in t-Hb (Hoshi et al., 2001).

It has been reported that visual stimulation induces NBRs in the visual cortex of the infant brain (Born et al, 1998; Yamada et al., 1997). These fMRI findings were supported by some NIRS studies, which demonstrated that oxy-, deoxy-, and t-Hb increased in the visual cortex of the infant and neonate brains during visual stimulation (Hoshi et al., 2000; Meek et al., 1995). The neonatal and early infantile periods correspond to the period of rapid formation of synapses (Bourgeois et al., 1989; Huttenlocher and Dabholkar, 1997), and this stimulus-related increase in deoxy-Hb, which is not specific to the visual cortex (Anderson et al., 2001), has been explained by a lower increase in rCBF compared to an increase in $CMRO_2$. However, this explanation has not yet been verified, and different hemodynamic response patterns have also been observed (Hoshi et al., 2000; Kozberg et al., 2013; Liao et al., 2010). A recent multispectral optical imaging study, where the evolution of hemodynamic responses to hind-paw stimulation was examined in the somatosensory cortex of neonatal rats, has suggested the possibility that the reported neonatal responses could have been influenced by stimulus-induced increases in the blood pressure (Kozberg et al., 2013). Measurements by high-density NIRS arrays with multiple SD distances (see Section 6.2) have shown that the "adult pattern," increases in oxy-Hb and t-Hb with a decrease in deoxy-Hb, was observed in six out eight neonates during a visual stimulation (Liao et al., 2010). It has also been observed that the response pattern to a visual stimulation varies with measurements even in the same neonate (Hoshi et al., 2001). Thus, NBRs in neonates and infants cannot be explained by only the $CBF/CMRO_2$ response mismatch.

Increases in deoxy-Hb accompanying increases in oxy-Hb and t-Hb were also observed in the activation areas of patients with cerebrovascular disease or brain tumors (Fujiwara et al., 2004; Murata et al., 2002). The underlying physiological mechanisms for this are still unknown. However, it must be noted that BOLD-fMRI can fail to detect brain activation if only positive BOLD signals are examined.

3.2.2 Initial dip and poststimulus undershoot

Since the initial dip, a small and transient decrease in BOLD signal compared to the baseline, was first observed in the human brain (Ernst and Hennig, 1994), its existence and origins have been investigated by many researchers but no generally accepted explanation for it has been established (Hu and Yacoub, 2012). The initial dip may be attributable to an increase in deoxy-Hb due to a mismatch between $CMRO_2$ and CBF (Kim and Kim, 2011), increases in CBV (Buxton et al., 1998), or combined changes in CBV and the oxygen extraction fraction (Buxton, 2001). Since NIRS can measure CBV as well as Hb oxygenation, it appears suitable for investigation of the initial dip. However, NIRS has poor spatial resolution, and the behavior of Hb changes is influenced by several assumptions on which NIRS algorithms are based. That is, it is assumed that the MBL is valid and the optical pathlength is independent of wavelength (Lindauer et al., 2001). Thus, it is very difficult to detect subtle Hb changes with NIRS (Jasdzewski et al., 2003).

The poststimulus undershoot of oxy-Hb and overshoot of deoxy-Hb after the cessation of the stimulation have often been observed in NIRS measurements (Jasdzewski et al., 2003; Roche-Labarbe et al., 2010; Schroeter et al., 2006) and is thought to correspond to the poststimulus undershoot of BOLD signals. Three possible physiological factors behind the poststimulus undershoot of BOLD signal have been suggested: (1) a delayed recovery of CBV (Mandeville et al., 1998), (2) a slow return of $CMRO_2$ to the baseline (Frahm et al., 1996), and (3) decreases in CBF due to cortical inhibition (Mullinger et al., 2013). Simultaneous measurements with NIRS and fMRI showed that the poststimulus period was accompanied by an undershoot of the BOLD signal, a decrease in oxy-Hb, and an increase in deoxy-Hb with no change in t-Hb, which indicated that there is a prolonged high-level $CMRO_2$ (Schroeter et al., 2006). However, more recent BOLD, CBF, and CBV fMRI studies have suggested that more than one factor contribute to the poststimulus undershoot of the BOLD signal (Chen and Pike, 2009; Hua et al., 2011), whereas it is supposed that the relative contribution of each factor may vary with stimulus conditions (Kim and Ogawa, 2012).

4 NIRS SIGNALS AND NEURAL ACTIVITIES

4.1 ELECTROENCEPHALOGRAM OSCILLATIONS AND HEMODYNAMIC RESPONSES

4.1.1 Alpha activity

The electroencephalogram (EEG) is presumed to reflect synchronized synaptic activity of large numbers of neurons (Buzsáki et al., 2012). The EEG oscillations have been subdivided into five frequency bands: delta (0.5–4 Hz), theta (4–8 Hz), alpha (8–12 Hz), beta (12–30 Hz), and gamma ($\sim$30 Hz). It has been suggested that ongoing activity in the EEG, the so-called background activity, plays a crucial role in functional communication in the brain (Başar et al., 2001); however, the physiological roles of each EEG rhythm remain to be fully understood. Since NIRS measures regional hemodynamic responses to changes in neural activities, simultaneous measurements of EEG and NIRS are expected to help interpret neural processes reflected by the EEG oscillations.

Alpha activity is the dominant rhythm in the EEG in the awake state and is predominantly observed in the occipital regions. It has long been thought that the alpha rhythm reflects "cortical idling" (Adrian and Matthews, 1934). However, one early EEG-NIRS study, where the occipital cortex was measured during visual stimulation with a checkerboard, eyes-open and eyes-close, and resting state, demonstrated a positive cross-correlation between the alpha activity and changes in deoxy-Hb (Moosman et al., 2003). Since increases in deoxy-Hb and decreases in oxy-Hb had been observed in the occipital cortex during saccadic suppression (Wenzel et al., 2000), it was concluded that alpha activity in the occipital cortex was associated with metabolic deactivation (Moosman et al., 2003). Several recent EEG studies have also indicated that this rhythm is related to functional inhibition (Jensen and Mazaheri, 2010; Klimesch et al., 2007). As mentioned in an earlier section, however, an increase in deoxy-Hb does not necessarily mean deactivation (Hoshi et al., 2001). More recent EEG studies have indicated that the increases in alpha activity reflect inhibition of task-irrelevant cortical areas (Klimesch et al., 2007), active processing required for unified cognitive operations (Palva and Palva, 2007), or gaiting of information by inhibiting task-irrelevant regions (Jensen and Mazaheri, 2010).

4.1.2 Gamma activity

High-frequency brain activity in the gamma band has recently been attracting attention in EEG/MEG (magnetoencephalography) studies. It has been suggested that the gamma rhythm contributes to cortical information processing and reflects brain excitation (Cardin et al., 2009; Gray and Singer, 1989; Ray and Maunsell, 2011). However, the question has also been raised whether the EEG/MEG gamma rhythms are attributed to artifacts of muscle activity (Yuval-Greenberg et al., 2008), and various approaches to reducing muscle artifacts have been recommended (Muthukumaraswamy, 2013). Several studies have suggested that cerebral hemodynamic responses can be attributed to the local field potential (LFP) in the gamma band, although it remains to be fully understood which of LFP and MUA (multiunit activity) contributes to hemodynamic responses (see Chapter "Neurogenic control of parenchymal arterioles in the cerebral cortex" by H. Hotta). It has been reported that the degree of hemodynamic responses to sensory and visual stimulation are positively correlated to the intensity of LFPs in the gamma band in the somatosensory cortex of rats (Nemoto et al., 2012) and in the visual cortex of cats (Niessing et al., 2005), respectively. An fMRI study has also reported that BOLD signals induced by auditory stimulation are closely correlated to the LFP in the gamma band in the auditory cortex of neurosurgical patients (Mukamel et al., 2005). A study combining NIRS and EEG, focusing on the enhancement in top-down processing of the gamma activity (Engel et al., 2001), examined the effects of stimulus strength and internal state on sustained gamma activity and hemodynamic responses (Koch et al., 2009). This study has demonstrated that the strength of a visual stimulation and attentive state modulate the gamma activity, something that is closely correlated to hemodynamic responses in the human visual cortex (Koch et al., 2009).

4.2 FLUCTUATIONS IN NIRS SIGNALS AND EEG OSCILLATIONS IN THE RESTING STATE

Spontaneous low-frequency (<0.1 Hz) BOLD signal fluctuations in the resting state (Biswal et al., 1995) and correlation network analysis of such data (Raichle and Synder, 2007) are attracting growing attention in neuroimaging studies. It has been demonstrated that the correlation structure of spontaneous BOLD fluctuations corresponds to that of slow (<4 Hz) and infra-slow (<0.1 Hz) cortical potentials

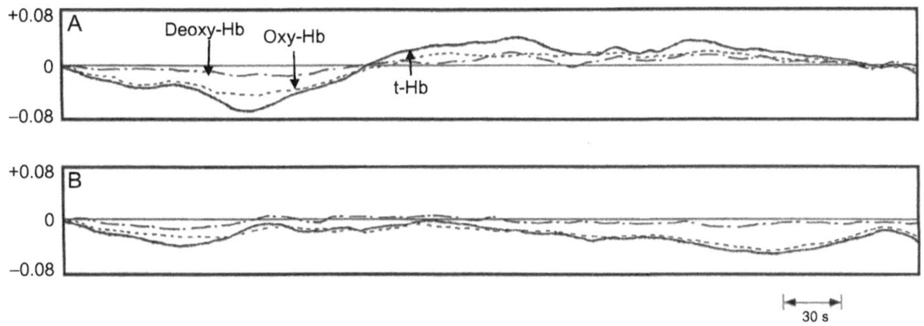

FIG. 4

Fluctuations in oxy-Hb (*dotted line*), deoxy-Hb (*broken line*), and t-Hb (*solid line*) in the frontal (A) and occipital (B) regions during the resting state. The two regions were measured simultaneously.

From Hoshi, Y., 2005. Functional near-infrared spectroscopy: potential and limitations in neuroimaging studies. In: Glabus, M.F. (Ed.), Neuroimaging Part A: International Review of Neurobiology. Elsevier Academic Press, New York, pp. 238–266, with permission.

recorded by electrocorticography in the human brain (He et al., 2008). It has also been found that the LFP in infra-slow band is correlated to resting BOLD signals in the rat brain (Pan et al., 2013).

Similar fluctuations have been observed in NIRS measurements (Fig. 4) (Hoshi and Tamura, 1997; Obrig et al. 2000). One early NIRS-EEG study suggested that the fluctuations in oxy-Hb in the frontal region during a 15-min resting period could be attributed to brain activities reflected by the alpha rhythm (Hoshi et al., 1998). A recent NIRS-EEG study, in which blood pressure, respiratory rate, and heart rate were also simultaneously recorded during a 5-min resting period and cross-spectral and sliding cross-correlation analyses were performed, has demonstrated a short-lasting coupling between prefrontal oxy-Hb in the frequency band between 0.07 and 0.13 Hz and central EEG in the alpha and/or beta bands (Pfurtscheller et al., 2012). Although a good correlation between BOLD signals and NIRS signals at rest has been reported (Sasai et al., 2012), here unlike with BOLD signals there was no definite correlation observed between slow/infra-cortical slow potentials and NIRS signal fluctuation (Nikulin et al., 2014).

5 NIRS SIGNALS ORIGINATING FROM CEREBRAL AND EXTRACEREBRAL TISSUE

5.1 INFLUENCES OF THE SKIN BLOOD FLOW ON NIRS SIGNALS DURING BRAIN ACTIVATION

During the past 38 years since Jöbsis first reported biomedical applications of NIRS (Jöbsis, 1977), difficulties in selective and quantitative measurements of the cerebral Hb have been central issues in the NIRS field. The light detected on the skin contains information about both cerebral and extracerebral tissue, and changes in the skin blood flow can influence NIRS signals. Based on this, two assumptions have been made in applying NIRS to neuroimaging studies. The first assumption is that the contribution of the skin blood flow changes on NIRS signals is far smaller than that of the CBF changes. The second assumption is that even though the skin blood flow changes are too large to be ignored, the distribution of these changes is homogeneous. Thus, if Hb changes are localized, these changes are considered to reflect changes in the brain activity. However, the first assumption has been questioned by a recent study (Takahashi et al., 2011), in which the prefrontal region was measured at the SD distance of 30 mm (FAR channels) by a multichannel CW instrument during a verbal fluency task. Hemodynamic changes in the skin were also simultaneously measured at the short SD distance of 5 mm (NEAR channels) and by using a laser Doppler flow meter (LDF). The results suggested that a major part of the task-related changes in oxy-Hb can be attributed to task-related changes in the skin blood flow (Takahashi et al., 2011). A study combining NIRS, fMRI, and MR-angiography, where changes in Hb, intracranial and extracranial fMRI signals, and the superficial veins in the prefrontal regions were measured during a continuous performance task, has demonstrated that skin blood volume is strongly correlated to the cognitive state, and suggested that task-evoked hemodynamic changes of veins draining the skin can induce artifacts in NIRS signals (Kirillina et al., 2012). In response to the arguments that the prefrontal NIRS signals are influenced by superficial hemodynamic changes, a more recent study, comparing simultaneously measured NIRS, fMRI, and LDF signals, has reported that the working memory task-induced Hb changes mainly reflect hemodynamic changes in the gray matter (Sato et al., 2013a). There is no doubt that NIRS can detect brain activity-related hemodynamic changes, while influences of superficial hemodynamic changes, especially the skin blood flow changes, have to be considered.

However, the mechanisms behind the regulation of the scalp and facial skin blood flows are complex. When the sympathetic nervous system is activated by performing a task, systemic blood pressure and heart rate elevate increasing the skin blood flow. Whereas sympathetic activation mediated by α-adrenoreceptors induces vasoconstriction of the skin blood vessels, resulting in reductions in skin blood flow (Drummond, 1997). Further, β-adrenoreceptors also control the skin blood flow and its activation induces vasodilation, increasing the blood flow (Kenney et al., 1991). Since the skin blood flow is determined by all these factors, an increase, decrease, or absence of change in the blood flow may occur during the performance of a task. However, it has not been possible to determine which of the three changes will occur.

5.2 INFLUENCES OF THE SKIN BLOOD FLOW ON NIRS SIGNALS IN THE RESTING STATE

As mentioned in the previous section, currently, the resting-state fMRI is a focus of research interests, and NIRS has also been employed in investigating functional connectivity (Homae et al., 2010; Medvedev, 2014). However, it is well known that skin blood flow spontaneously fluctuates and five characteristic frequencies (1.0, 0.3, 0.1, 0.04, and 0.01 Hz) originate from the heart beat, respiration, the myogenic activity of vascular smooth muscles, neurogenic activity, and vascular endothelial metabolic

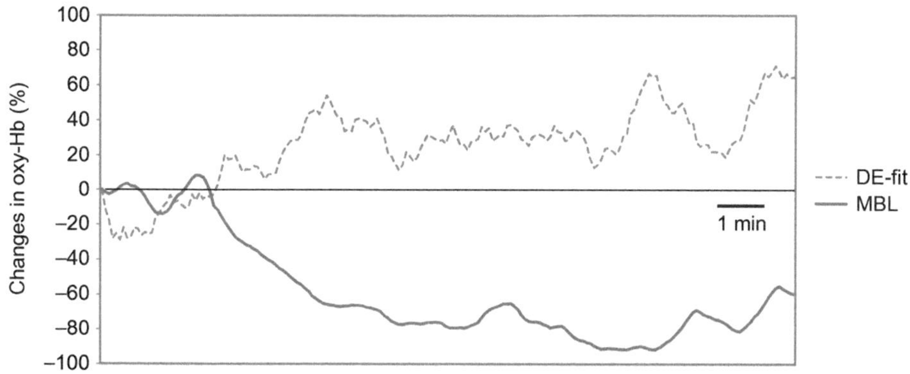

FIG. 5

Changes in oxy-Hb in the left frontal region of a healthy adult in the resting state. *Dotted line*, oxy-Hb changes calculated by the MBL method; *solid line*, oxy-Hb changes calculated by the DE-fit method. Changes in oxy-Hb were expressed as percentage of the maximum change during the observation period.

From Hoshi, Y., 2011. Toward the next generation of NIRS. Phil. Trans. R. Soc. A 369, 4425–4439, with permission.

activity (Jan et al., 2009; Liao et al., 2013; Stefanovska et al., 1999). This makes it likely that resting NIRS signals also include signals arising from the skin, although in infants the contribution of extracerebral hemodynamics has been estimated as very small (~22%) at the SD of 30 mm (Funane et al., 2014a). Fig. 3 shows oxy-Hb changes in the left frontal region of an adult subject in the resting state measured by TRS. The TRS data can be analyzed by two methods: one is based on the MBL, which is analogous to CW measurements (MBL method), and the other is based on diffusion theory (DE-fit method) (see Section 6.2). In the case in Fig. 5, the two methods produced different results: the upper and lower traces are the oxy-Hb changes calculated by the DE-fit and the MBL method, respectively. Such discrepancies are occasionally observed in resting-state TRS measurements. It has been reported that the DE-fit method shows higher sensitivity to the deeper regions of the head, while the MBL method is more sensitive to absorbance changes in the superficial layers (Hielscher et al., 1996; Sato et al., 2007). From this it may be assumed that the lower trace (the MBL method) mainly reflects the skin blood flow in the case of Fig. 5 (Hoshi, 2011). This indicates that NIRS signals in the resting state measured by CW instruments in adults could be influenced by skin blood flow as well as in the activation state.

5.3 INFLUENCES OF THE SUBARACHNOID SPACE AND THE SKULL ON NIRS SIGNALS

Hemodynamic changes in the subarachnoid space (SAS) and the skull during functional activation and resting are assumed to be negligible. However, the CSF in the SAS and the skull has a strong influence on the light propagation in the head (Sato et al., 2013b; Wang et al., 2010). A simulation study using

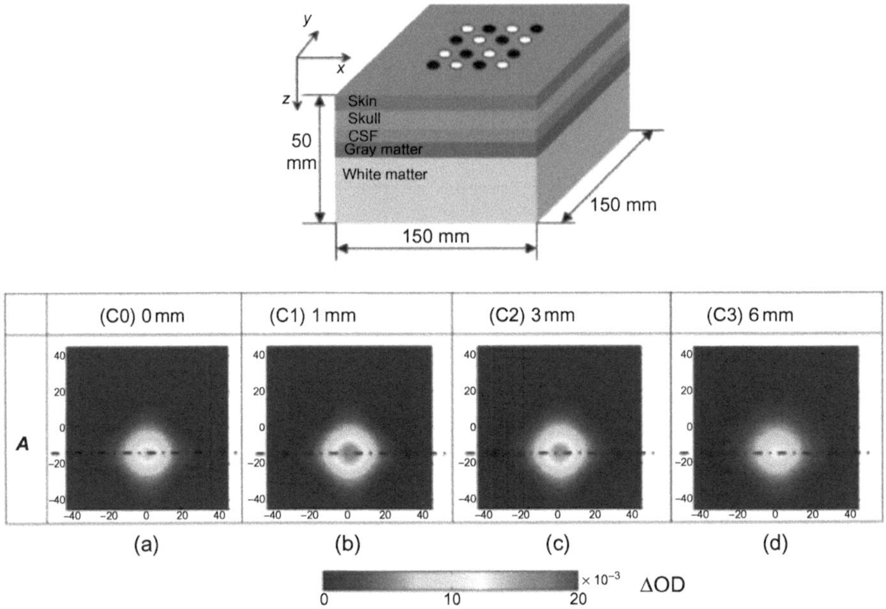

FIG. 6

Influences of the thickness of subarachnoid space (SAS) on optical mapping images of a single cortical activation region. *Upper*: A five-layered human head model consisting of the skin, skull, SAS (CSF), gray matter, and white matter. The thickness of CSF layer was changed from 0 to 6 mm. A single activated region ($10 \times 10 \times 3.5$ mm^3) was placed in the gray matter and its center was at coordinates x=75 mm and y=60 mm. *Open circle*, source position; *filled circle*, detector position. *Lower*: simulated topographic images of optical density (OD) changes caused by brain activation under the conditions of the CSF layer thickness of 0, 1, 3, and 6 mm.

From Wang, S., Shibahara, N., Kuramashi, D., Okawa, S., Okada, E., Maki, A., Yamada, Y. 2010. Effects of spatial variation of skull and cerebrospinal fluid layers on optical mapping of brain activities. Opt. Rev. 17, 410–420, with permission.

the finite element method on a five-layered human head model has shown that the sensitivity of the topographic images to an activation in the gray matter is negatively correlated to the thickness of the skull, while the thickness of the SAS is positively correlated to the sensitivity in the range of 0–3 mm (Fig. 6) (Wang et al., 2010). It has been reported that most of the developmental changes in brain-scalp distance in children can be attributed to increases in the SAS and inner table of the cranium (Beauchamp et al., 2011). These findings mean that interregional comparisons of NIRS signal amplitudes are not valid to those of hemodynamic changes. Also, it should be borne in mind that age-dependent changes in NIRS signals can be attributed to developmental changes in the brain-scalp distance.

6 SELECTIVE AND QUANTITATIVE NIRS MEASUREMENTS OF CEREBRAL Hb

6.1 SELECTIVE MEASUREMENTS OF CEREBRAL Hb

Elaborate experimental designs have been used to prevent inducing hemodynamic changes in the extracerebral tissue; however, the results are still insufficient and further work is required. It is necessary to distinguish the NIRS signals originating in the cerebral tissue from those coming from the superficial layers, mainly the skin. Several approaches to achieve this have been tried, and in recent years various methods with a multidistance probe arrangement have been proposed. Here short SD separation measurements are used as regressors to eliminate the skin blood flow signals from the longer SD measurements (Gagnon et al., 2011; Saager and Berger, 2005; Zhang et al., 2007). Data-driven approaches, including principal component analysis (PCA) and independent component analysis (ICA), have been used to remove extracerebral interference from NIRS signals (Kohno et al, 2007; Virtannen et al., 2009; Zhang et al., 2005). Further, methods combining multidistance probe arrangements and data-driven approaches have also been proposed (Funane et al., 2014b). These methods have been reported to improve the quality of NIRS signals, but the key assumption that hemodynamic changes in the extracerebral tissue are homogeneous has recently been questioned (Gagnon et al., 2012b; Kohno and Hoshi, 2016). Further refining of the algorithms for routine applications to NIRS measurements is required.

6.2 QUANTITATIVE AND SELECTIVE MEASUREMENTS OF CEREBRAL Hb

In TRS, ultrashort (picosecond order) laser pulses are irradiated on the tissue, and the intensity of the emerging light is recorded over time (the temporal point spread function, TPSF) with picosecond resolution (Fig. 7). The mean t-PL is determined by multiplying the light speed in the media by the mean transit time of the scattered photons, which is calculated with the TPSF. The TRS data carry information about depth-dependent attenuation based on the correlation of the detection time to the penetration depth of photons. Changes in cerebral Hb concentrations can be more selectively and quantitatively determined by analyzing the TPSF, because this function contains information about the optical properties of the media, that is, absorption (μ_a) and reduced scattering (μ_s') coefficients.

Several different methods for analyzing the TPSF have been proposed (Liebert et al., 2004; Sato et al. 2005; Steinbrink et al., 2001; Torricelli et al., 2014). One method is a curve-fitting method based on the DE (DE-fit method) described earlier (see Section 5.2). In this method, optical properties are determined by fitting the TPSF, the temporal probability function for all detected photons, obtained from the analytical solution to the DE for the observed TPSF on the assumption that the human head is a semiinfinite homogeneous medium. It is still not conclusively

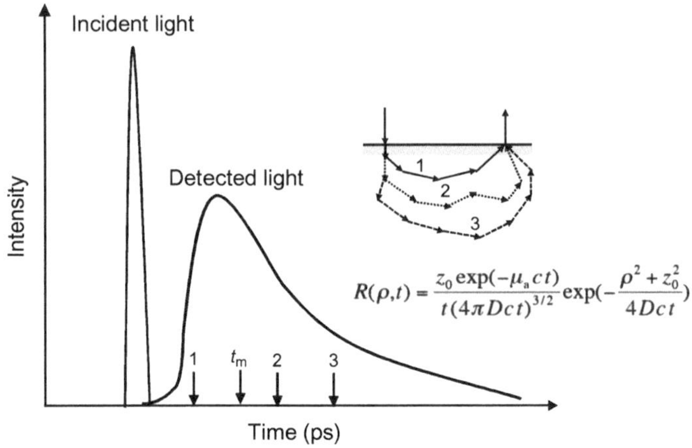

FIG. 7

Temporal profiles of incident light and intensity of detected light. t_m, mean transit time; 1, 2, and 3 denote detection times of photons traveling through the optical paths 1, 2, and 3 (inserted figure), respectively. $R(r,t)$ is a flux (TPSF) derived from analytical solution of the DE. D is the diffusion coefficient and expressed as $1/3\mu_s'$. The optical properties are determined by fitting $R(r,t)$ for the measured temporal profile of detected light.

established whether the human head can be considered as a homogeneous and/or a simple slab model, however. Nevertheless, it has been reported that the estimated μ_a by the DE-fit method is nearly equal to the deeper-layer μ_a under the condition where the μ_a of the upper layer is larger than that of the lower layer in two-layered slab models (Sato et al., 2005). We have also confirmed that the DE-fit method is more sensitive to μ_a changes in the cerebral tissue, by measuring adult heads with TRS during carotid endarterectomy (Fig. 8) (Sato et al., 2007). This is explained by the fact that the later (falling) part of the TPSF, which includes more photons traveling through the cerebral tissue, is critical for the determination of the μ_a in the fitting process. Thus, changes in cerebral Hb concentrations can be more selectively measured by TRS, although the selective and quantitative accuracy are not perfect.

6.3 DIFFUSE OPTICAL TOMOGRAPHY

Diffuse optical tomography (DOT) is a very promising approach to further a resolution of the aforementioned issues. With DOT, 3D images of the optical properties are reconstructed from boundary measurements, and this could enable a quantitative determination of regional cerebral Hb details (Fig. 9) (Arridge et al., 1992; Singer et al., 1990). The DOT algorithm essentially consists of two parts: one is a forward model to calculate the light propagation and the resultant outward reemissions at the boundary of the tissue. The other is an inverse model to search for the distribution of optical properties minimizing the differences between the calculated and the measured data.

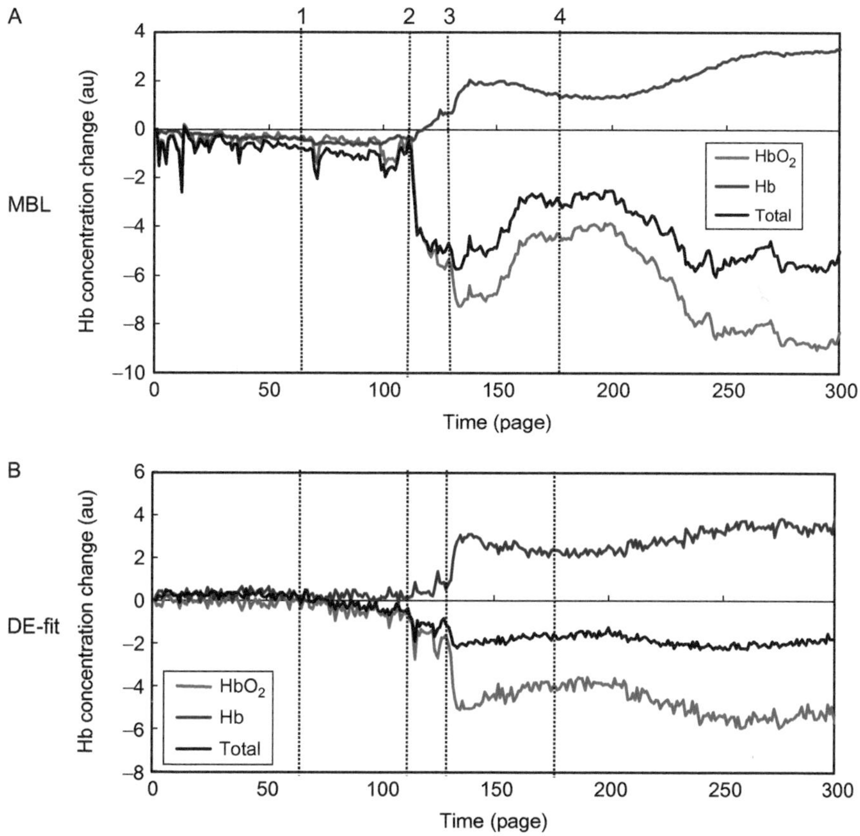

FIG. 8

Changes in oxy-, deoxy-, and t-Hb in the frontal region of a patient with internal carotid artery stenosis during carotid endarterectomy. (A) Time course of Hb changes calculated by the MBL method. (B) Time course of changes in Hb calculated by the DE-fit method. Before the dissection of the carotid artery for atherectomy, (1) the superior thyroid artery, (2) the external carotid artery, (3) the common carotid artery, and (4) the internal carotid artery were clamped in this order. The *vertical lines* in the graphs represent the time point of clamping each artery. Oxy-Hb decreased by clamping of the external carotid artery in the MBL method, while a large decrease in oxy-Hb was observed at clamping the internal carotid artery in the DE-fit method.

From Sato, C., Yamaguchi, T., Seida, M., Ota, Y., Yu, I., Iguchi, Y., Nemoto, M., Hoshi, Y., 2007. Intraoperative monitoring of depth-dependent hemoglobin concentration changes during carotid endarterectomy by time-resolved spectroscopy. Appl. Opt. 46, 2785–2792, with some modifications, with permission.

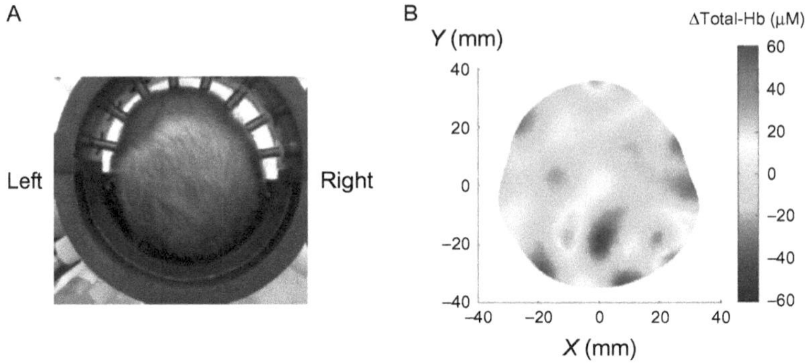

FIG. 9

DOT image of total Hb concentrations of the brain in an extremely low birth weight infant under hypocapnic conditions. (A) Sixteen coaxial optical fibers, consisting of a single fiber for illumination in the center and outer bundle fibers for detection, were placed on the scalp of the infant. (B) Image reconstruction was performed by using the diffusion equation-based algorithm. The axial image shows the total Hb concentration changes cause by hypocapnia in comparison with normocapnic conditions.

From Fukuzawa, R., 2009. Diffuse optical tomography for premature neonate head (Master thesis). The University of Electro-Communications, Tokyo, Japan, with permission.

The DOT image reconstruction can be approximately divided into two kinds: one is a linear single-step image reconstruction; the other is a nonlinear iterative image reconstruction. The former was first proposed in 1995 (O'Leary et al., 1995) and can now provide high quality images with multichannel CW-type instruments (Eggebrecht et al., 2014; Zeff et al., 2007). This approach reconstructs only qualitative images of measured changes, which is, however, sufficient for neuroimaging studies. In clinical use, quantitative images of steady-state Hb are further useful with diagnostic optical imaging, and DOT based on the nonlinear approach has been being developed to provide this. The DOT of this type can be performed with CW measurements, FDS, and TRS, where TRS provides more of the information required for image reconstruction.

The nonlinear iterative image reconstruction has been applied to imaging of organs and tissue, including the brain and breasts (Deng et al., 2015; Hebden et al., 2002; Montejo et al., 2013; Zhao et al., 2005). However, the image quality is still inadequate. The radiative transfer equation (RTE) correctly describes photon propagation in living tissue; however, the computational load in solving the RTE is very heavy. Thus, the DE is widely used as a forward model, but the DE is invalid in low-scattering regions (eg, the SAS), highly absorbing regions (eg, large vessels) and in the vicinity of light sources. Further, the inverse model is an ill-posed and ill-conditioned problem. It is very difficult to develop accurate and efficient algorithms based on the time-domain RTE for DOT image reconstruction. However, DOT of this type will help contribute to establish an optical imaging diagnosis.

7 NEXT-GENERATION NIRS

Although NIRS provides only information of the functioning of the cerebral cortex facing the skull convexity, NIRS has become widely applied in a variety of neuroimaging studies because of its advantages: (1) measurements are noninvasive and can be performed with fewer motion restrictions, (2) the temporal resolution is high, and (3) simultaneous measurements with other neuroimaging modalities are possible. In addition, NIRS enables neuroimaging studies of children, the elderly, and patients with psychoneurological complaints where it is difficult to conduct measurements by other neuroimaging techniques. However, now that the basic assumption that the contribution of the extracerebral tissues to NIRS signals can be ignored does not always hold true, removing signals arising from the extracerebral tissue from NIRS signals is critical to the future development of fNIRS. Selective measurements of cerebral Hb is a challenging object, and it is important for the potentially useful approaches mentioned earlier to preserve the advantages of NIRS, its ability to provide a high temporal resolution (<1 s). Even linear single-step CW-DOT cannot yet reconstruct images in real time. Although TRS provides depth information, it takes a few seconds to obtain a temporal profile with the detected photons (TPSF). However, recent advances in optical and computer technologies will be able to overcome these difficulties. Next-generation NIRS, such as DOT, can be expected to enable further evolution of fNIRS.

REFERENCES

Adrian, E.D., Matthews, B.H., 1934. The Berger rhythm: potential changes from the occipital lobes in man. Brain 57, 355–385.

Anderson, A.W., Marois, R., Colson, E.R., Peterson, B.S., Duncan, C.C., Ehrenkranz, R.A., Schneider, K.C., Gore, J.C., Ment, L.R., 2001. Neonatal auditory activation detected by functional magnetic resonance imaging. Magn. Reson. Imaging 19, 1–5.

Arridge, S.R., Schweiger, M., Delpy, D.T., 1992. Iterative reconstruction of near infra-red absorption images. Proc. SPIE 1767, 372–383.

Başar, E., Başar-Erouglu, C., Karakaş, S., Schürmann, M., 2001. Gamma, alpha, delta, theta oscillations govern cognitive process. Int. J. Psychophysiol. 39, 241–248.

Beauchamp, M.S., Beurlot, M.R., Fava, E., Nath, A.R., Parkin, N.A., Saad, Z.S., Bortfeld, H., Oghalai, J.S., 2011. The developmental trajectory of brain-scalp distance from birth through childhood: implications for functional neuroimaging. PLoS One 6, e24981.

Biswal, B., Yetkin, F.Z., Haughton, V.M., Hyde, J.S., 1995. Functional connectivity in the motor cortex of resting human brain using echo-planar MRI. Magn. Reson. Med. 34, 537–541.

Born, P., Leth, H., Miranda, M.J., Rostrup, E., Stensqaard, A., Peitersen, B., Larsson, H.B., Lou, C., 1998. Visual activation in infants and young children studies by functional magnetic resonance imaging. Pediatr. Res. 44, 578–583.

Bourgeois, J.-P., Jastreboff, P.J., Rakic, Pasko, 1989. Synaptogenesis in visual cortex of normal and preterm monkeys: evidence for intrinsic regulation of synaptic overproduction. Proc. Natl. Acad. Sci. U.S.A. 86, 4297–4301.

Buxton, R.B., 2001. The elusive initial dip. NeuroImage 13, 953–958.
Buxton, R.B., Wong, E.C., Frank, L.R., 1998. Dynamics of blood flow and oxygenation changes during brain activation: the balloon model. Magn. Reson. Med. 39, 855–864.
Buzsáki, G., Anastassiou, C.A., Koch, C., 2012. The origin of extracellular fields and currents—EEG, ECoG, and spikes. Nat. Rev. 13, 407–420.
Cardin, J.A., Carlèn, M., Meletis, K., Knoblich, U., Zhang, F., Deisseroth, K., Tsai, L.H., Moore, C.I., 2009. Driving fast-spiking cells induces gamma rhythm and controls sensory responses. Nature 459, 663–668.
Chance, B., Leigh, J.S., Miyake, H., Smith, D.S., Nioka, S., Greenfeld, R., Finander, M., Kaufmann, K., Levy, W., Young, M., Choen, P., Yoshioka, H., Boretsky, R., 1988. Comparison of time-resolved and -unresolved measurements of deoxyhemoglobin in brain. Proc. Natl. Acad. Sci. U.S.A. 85, 4971–4975.
Chance, B., Dait, M.T., Zhang, C., Hamaoka, T., Hargerman, F., 1992. Recovery from exercise-induced desaturation in the quadriceps muscle of elite competitive rowers. Am. J. Physiol. 262, C766–C775.
Chance, B., Zhuang, Z., Unah, C., Alter, C., Lipton, L., 1993. Cognition-activated low frequency modulation of light absorption in human brain. Proc. Natl. Acad. Sci. U.S.A. 90, 3770–3774.
Chen, J.J., Pike, G.B., 2009. Origins of the BOLD poststimulus undershoot. NeuroImage 45, 559–568.
Cui, X., Bray, S., Bryant, D.M., Glover, G.H., Reiss, A., 2011. A quantitative comparison of NIRS and fMRI across multiple cognitive tasks. NeuroImage 54, 2808–2821.
Delpy, D.T., Cope, M., van der Zee, P., Arridge, S., Wray, S., Wyatt, J., 1988. Estimation of optical pathlength through tissue from direct time of flight measurement. Phys. Med. Biol. 33, 1433–1442.
Deng, B., Brooks, D.H., Boas, D.A., Lundqvist, M., Fang, Q., 2015. Characterization of structural-prior guided optical tomography using realistic breast models derived from dual-energy X-ray mammography. Biomed. Opt. Express 6, 2366–2379.
Devor, A., Tian, P., Nishimura, N., Teng, I.C., Hillman, E.M., Narayanan, S.N., Ulbert, I., Boas, D.A., Kleinfeld, D., Dale, A.M., 2007. Suppressed neuronal activity and concurrent arteriolar vasoconstriction may explain negative blood oxygenation level-dependent signal. J. Neurosci. 27, 4452–4459.
Drummond, P.D., 1997. The effect of adrenergic blockade on blushing and facial flushing. Psychophysiology 34, 163–168.
Duncan, A., Whitlock, T.L., Cope, M., Delpy, D.T., 1993. A multiwavelength, wideband, intensity modulated optical spectrometer for near infrared spectroscopy and imaging. Proc. SPIE 1888, 248–257.
Eggebrecht, A.T., Ferradal, S.L., Robichaux-Viehoever, A., Hassanpour, M.S., Dehghani, H., Snyder, A.Z., Hershe, T., Culver, J.P., 2014. Mapping distributed brain function and networks with diffuse optical tomography. Nat. Photonics 8, 448–454.
Engel, A.K., Fries, P., Singer, W., 2001. Dynamic predictions: oscillations and synchrony in top-down processing. Nat. Rev. Neurosci. 2, 704–716.
Ernst, T., Hennig, J., 1994. Observation of a fast response in functional MR. Magn. Reson. Med. 32, 146–149.
Firbank, M., Okada, E., Delpy, D.T., 1998. A theoretical study of the signal contribution of regions of the adult head to near-infrared spectroscopy studies of visual evoked responses. NeuroImage 8, 69–78.

Fox, P., Raichle, M.E., 1986. Focal physiological uncoupling of cerebral blood flow and oxidative metabolism during somatosensory stimulation in human subjects. Proc. Natl. Acad. Sci. U.S.A. 83, 1140–1144.

Frahm, J., Kruger, G., Merboldt, K.D., Kleinschmidt, A., 1996. Dynamic uncoupling and recoupling of perfusion and oxidative metabolism during focal brain activation in man. Magn. Reson. Med. 35, 143–148.

Fujiwara, N., Sakatani, K., Katayama, Y., Murata, Y., Hoshino, T., Fukaya, C., Yamamoto, T., 2004. Evoked-cerebral blood oxygenation changes in false-negative activations in BOLD contrast functional MRI of patients with brain tumors. NeuroImage 21, 1464–1471.

Fukui, Y., Ajichi, Y., Okada, E., 2003. Monte Carlo prediction of near-infrared light propagation in realistic adult and neonatal head models. Appl. Opt. 42, 2881–2887.

Funane, T., Homae, F., Watanabe, H., Kiguchi, M., Taga, G., 2014a. Greater contribution of cerebral than extracerebral hemodynamics to near-infrared spectroscopy signals for functional activation and resting-state connectivity in infants. Neurophotonics 1, 025003.

Funane, T., Atsumori, H., Katura, T., Obata, A.N., Sato, H., Tanikawa, Y., Okada, E., Kiguchi, M., 2014b. Quantitative evaluation of deep and shallow tissue layers' contribution to NIRS signal using multi-distance optodes and independent component analysis. NeuroImage 85, 150–165.

Gagnon, L., Perdue, K., Greve, D.N., Goldenholz, D., Kaskhedikar, G., Boas, D.A., 2011. Improved recovery of the hemodynamic response in diffuse optical imaging using short optode separations and state-space modeling. NeuroImage 56, 1362–1371.

Gagnon, L., Yücel, M.A., Dehaes, M., Cooper, R.J., Perdue, K.L., Selb, J., Huppert, T.J., Hoge, R.D., Boas, D.A., 2012a. Quantification of the cortical contribution to the NIRS signal over the motor cortex using concurrent NIRS-fMRI measurements. NeuroImage 15, 3933–3940.

Gagnon, L., Cooper, R.J., Yucel, M.A., Perdue, K.L., Greve, D.N., Boas, D.A., 2012b. Short separation channel location impacts the performance of short channel regression in NIRS. NeuroImage 59, 2518–2528.

Gray, C.M., Singer, W., 1989. Stimulus-specific neuronal oscillations in orientation columns of cat visual cortex. Proc. Natl. Acad. Sci. U.S.A. 86, 1698–1702.

Hall, C.N., Reynell, C., Gesslein, B., Hamilton, N.B., Mishra, A., Sutherland, B.A., O'Farrell, F.M., Buchan, A.M., Lauritzen, M., Attwell, D., 2014. Capillary pericytes regulate cerebral blood flow in health and disease. Nature 508, 55–60.

Hamzei, F., Dettmers, C., Rzanny, R., Liepert, J., Büchel, C., Weiller, C., 2002. Reduction of excitability ("inhibition") in the ipsilateral primary motor cortex is mirrored by fMRI signal decreases. NeuroImage 17, 490–496.

He, B.J., Synder, A.Z., Zempel, J.M., Smyth, M.D., Raichle, M.E., 2008. Electrophysiological correlates of the brain's intrinsic large-scale functional architecture. Proc. Natl. Acad. Sci. U.S.A. 41, 16039–16044.

Hebden, J.C., Gibson, A., Yusof, R.M., Everdell, N., Hillman, E.M.C., Delpy, D.T., Arridge, S.R., Austin, T., Meek, J.H., Wyatt, J.S., 2002. Three-dimensional optical tomography of the premature infant brain. Phys. Med. Biol. 47, 4155–4166.

Hielscher, A.H., Liu, H., Chance, B., Tittel, F.K., Jacques, S.L., 1996. Time-resolved photon emission from layered turbid media. Appl. Opt. 35, 719–728.

Hoge, R.D., Franceschini, M.A., Covolan, R.J.M., Huppert, T., Mandeville, J.B., Boas, D.A., 2005. Simultaneous recording of task-induced changes in blood oxygenation, volume, and flow using diffuse optical imaging and arterial spin-labeling MRI. NeuroImage 25, 701–707.

Homae, F., Watanabe, H., Otobe, T., Nakano, T., Go, T., Konishi, Y., Taga, G., 2010. Development of global cortical networks in early infancy. J. Neurosci. 30, 4877–4882.

Hoshi, Y., 2011. Toward the next generation of NIRS. Phil. Trans. R. Soc. A 369, 4425–4439.

Hoshi, Y., Tamura, M., 1993a. Detection of dynamic changes in cerebral oxygenation coupled to neuronal function during mental work in man. Neurosci. Lett. 150, 5–8.

Hoshi, Y., Tamura, Y., 1993b. Dynamic multichannel near-infrared optical imaging of human brain activity. J. Appl. Physiol. 75, 1842–1846.

Hoshi, Y., Tamura, M., 1997. Fluctuation in the cerebral oxygenation state during the resting period in functional mapping studies of the human brain. Med. Biol. Eng. Comput. 35, 328–330.

Hoshi, Y., Kosaka, S., Xie, Y., Kohri, S., Tamura, M., 1998. Relationship between fluctuations in the cerebral hemoglobin oxygenation state and neuronal activity under resting conditions in man. Neurosci. Lett. 245, 147–150.

Hoshi, Y., Kohri, S., Matsumoto, Y., Cho, K., Matsuda, T., Okajima, S., Fujimoto, S., 2000. Hemodynamic responses to photic stimulation in neonates. Pediatr. Neurol. 23, 323–327.

Hoshi, Y., Kobayashi, N., Tamura, M., 2001. Interpretation of near-infrared spectroscopy signals: a study with a newly developed perfused rat brain model. J. Appl. Physiol. 90, 1657–1662.

Hoshi, Y., Shimada, M., Sato, C., Iguchi, Y., 2005. Reevaluation of near-infrared propagation in the adult human head: implications for functional near-infrared spectroscopy. J. Biomed. Opt. 10, 064032.

Hu, X., Yacoub, E., 2012. The story of the initial dip in fMRI. NeuroImage 62, 1103–1108.

Hua, J., Stevens, R.D., Huang, A.J., Pekar, J.J., van Zijl, C.M.P., 2011. Physiological origin for the BOLD poststimulus undershoot in human brain: vascular compliance versus oxygen metabolism. J. Cereb. Blood Flow Metab. 31, 1599–1611.

Huber, L., Goense, J., Kennerley, A.J., Ivanov, D., Krieger, S.N., Lepsien, J., Trampel, R., Turner, R., Möller, H.E., 2014. Investigation of the neurovascular coupling in positive and negative BOLD responses in human brain at 7T. NeuroImage 97, 349–362.

Huppert, T.J., Hoge, R.D., Diamond, S.G., Franceschini, M.A., Boas, D.A., 2006. A temporal comparison of BOLD, ASL, and NIRS hemodynamic responses to motor stimuli in adult humans. NeuroImage 29, 368–382.

Huttenlocher, P.R., Dabholkar, A.S., 1997. Regional differences in synaptogenesis in human cerebral cortex. J. Comp. Neurol. 387, 167–178.

Iadecola, C., 2004. Neurovascular regulation in the normal brain and in Alzheimer's disease. Nat. Rev. Neurosci. 5, 347–360.

Jan, Y.K., Struck, B.D., Foreman, R.D., Robinson, C., 2009. Wavelet analysis of sacral skin blood flow oscillations to assess soft tissue viability in older adults. Microvasc. Res. 78, 162–168.

Jasdzewski, G., Strangman, G., Wager, J., Kwong, K.K., Poldrack, R.A., Boas, D.A., 2003. Differences in hemodynamic response to event-related motor and visual paradigms as measured by near-infrared spectroscopy. NeuroImage 20, 479–488.

Jensen, O., Mazaheri, A., 2010. Shaping functional architecture by oscillatory alpha activity: gating by inhibition. Front. Hum. Neurosci. 4, 186.

Jöbsis, F.F., 1977. Noninvasive infrared monitoring of cerebral and myocardial oxygen sufficiency and circulatory parameters. Science 198, 1264–1267.

Kato, T., Kamei, A., Takashima, S., Ozaki, T., 1993. Human visual cortical function during photic stimulation monitoring by means of near-infrared spectroscopy. J. Cereb. Blood Flow Metab. 13, 516–520.

Kenney, W.L., Tankersley, C.G., Newswanger, D.L., Puhl, S.M., 1991. α_1-Adrenergic blockade does not alter control of skin blood flow during exercise. Am. J. Physiol. 260, H855–H861 (Heart Circ. Physiol. 29).

Kim, T., Kim, S., 2011. Temporal dynamics and spatial specificity of arterial and venous blood volume changes during visual stimulation: implication for BOLD quantification. J. Cereb. Blood Flow Metab. 31, 1211–1222.

Kim, S.G., Ogawa, S., 2012. Biophysical and physiological origins of blood oxygenation level-dependent fMRI signals. J. Cereb. Blood Flow Metab. 32, 1188–1206.

Kirillina, E., Jelzow, A., Heine, A., Niessing, M., Wabnitz, H., Brühl, R., Ittermann, B., Jacobs, A.M., Tachitsidis, I., 2012. The physiological origin of task-evoked systemic artefacts in functional near infrared spectroscopy. NeuroImage 61, 70–81.

Kleinschmidt, A., Obrig, H., Requardt, M., Merboldt, K.-D., Dirnagl, D., Villringer, A., Frahm, J., 1996. Simultaneous recording of cerebral blood oxygenation changes during human brain activation by magnetic resonance imaging and near-infrared spectroscopy. J. Cereb. Blood Flow Metab. 16, 817–826.

Klimesch, W., Sauseng, P., Hanslmayr, S., 2007. EEG alpha oscillations: the inhibition-timing hypothesis. Brain Res. Rev. 53, 63–88.

Koch, S.P., Werner, P., Steinbrink, J., Fries, P., Obrig, H., 2009. Stimulus-induced and state-dependent sustained gamma activity is tightly coupled to the hemodynamic response in humans. J. Neurosci. 29, 13962–13970.

Kohno, S., Hoshi, Y., 2016. Spatial distributions of hemoglobin signals in superficial layers on the forehead during a verbal fluency task. J. Biomed. Opt. (in submission).

Kohno, S., Miyai, I., Seiyama, A., Oda, I., Ishikawa, A., Tsuneishi, S., Amita, T., Shimizu, K., 2007. Removal of the skin blood flow artifact in functional near-infrared spectroscopic imaging data through independent component analysis. J. Biomed. Opt. 12, 062111.

Kozberg, M.G., Chen, B.R., Deleo, S.E., Bouchard, M.B., Hillman, E.M.C., 2013. Resolving the transition from negative to positive blood oxygen level-dependent responses in the developing brain. Proc. Natl. Acad. Sci. U.S.A. 110, 4380–4385.

Lakowicz, J.R., Berndt, K., 1990. Frequency domain measurement of photon migration in tissues. Chem. Phys. Lett. 166, 246–252.

Liao, S.M., Gregg, N.M., White, B.R., Zeff, B.W., Inder, T.E., Culver, J.P., 2010. Neonatal hemodynamic response to visual cortex activity: high-density near-infrared spectroscopy study. J. Biomed. Opt. 15, 026010.

Liao, F., O'Brien Jr., W.D., Jan, Y.K., 2013. Assessing complexity of skin blood flow oscillations in response to locally applied heating and pressure in rats: implications for pressure ulcer risk. Physica A 392, 4905–4915.

Liebert, A., Wabnitz, H., Steinbrink, J., Obrig, H., Moller, M., Macdonald, R., Villringer, A., Rinnerberg, H., 2004. Time-resolved multidistance near-infrared spectroscopy of the adult head: intracerebral and extracerebral absorption changes from moments of distribution of times of flight of photons. Appl. Opt. 43, 3037–3047.

Lindauer, U., Royl, G., Leithner, C., Kühl, M., Gold, L., Gethmann, J., Kohl-Bareis, M., Villringer, A., Dirnagl, U., 2001. No evidence for early decrease in blood oxygenation in rat whisker cortex in response to functional activation. NeuroImage 13, 988–1001.

MacIntosh, B.J., Klassen, L.M., Menon, R.S., 2003. Transient hemodynamics during a breath hold challenge in a two part functional imaging study with simultaneous near-infrared spectroscopy in adult humans. NeuroImage 20, 1246–1252.

Maggioni, E., Molteni, E., Zucca, C., Reni, G., Cerutti, S., Triulzi, F.M., Arrigoni, F., Bianchi, A.M., 2015. Investigation of negative BOLD responses in human brain through NIRS technique. A visual stimulation study. NeuroImage 108, 410–422.

Mancini, D.M., Bolinger, L., Li, H., Kendrick, K., Chance, B., Wilson, J.R., 1994. Validation of near-infrared spectroscopy in humans. J. Appl. Physiol. 77, 2740–2747.

Mandeville, J.B., Marota, J.J., Kosofsky, B.E., Keltner, J.R., Weissleder, R., Rosen, B.R., Weisskoff, R.M., 1998. Dynamic functional imaging of relative cerebral blood volume during rat forepaw stimulation. Magn. Reson. Med. 39, 615–624.

Medvedev, A., 2014. Does the resting state connectivity have hemispheric asymmetry? A near-infrared spectroscopy study. NeuroImage 85, 400–407.

Meek, J.H., Elwell, C.E., Khan, M.J., Romaya, J., Wyatt, J.S., Delpy, D.T., Zeki, S., 1995. Regional changes in cerebral haemodynamics as a result of a visual stimulus measured by near infrared spectroscopy. Proc. Biol. Sci. 261, 351–356.

Montejo, L.D., Jia, J., Kim, H.K., Netz, U.J., Blaschke, S., Müller, G.A., Hielscher, A.H., 2013. Computer-aided diagnosis of rheumatoid arthritis with optical tomography. Part 1: feature extraction. J. Biomed. Opt. 18, 076001.

Moosman, M., Ritter, P., Krastel, I., Brink, A., Thees, S., Blankenburg, F., Taskin, B., Obrig, H., Villringer, A., 2003. Correlates of alpha rhythm in functional magnetic resonance imaging and near infrared spectroscopy. NeuroImage 20, 145–158.

Mukamel, R., Gelbard, H., Arieli, A., Hasson, U., Fried, I., Malach, R., 2005. Coupling between neuronal firing, field potentials, and fMRI in human auditory cortex. Science 309, 951–954.

Mullinger, K.J., Mayhew, S.D., Bagshaw, A.P., Bowtell, R., Francis, S., 2013. poststimulus undershoots in cerebral blood flow and BOLD fMRI responses are modulated by poststimulus neuronal activity. Proc. Natl. Acad. Sci. U.S.A. 110, 13636–13641.

Murata, Y., Sakatani, K., Katayama, Y., Fukaya, C., 2002. Increase in focal concentration of deoxyhaemoglobin during neuronal activity in cerebral ischaemic patients. J. Neurol. Neurosurg. Psychiatry 73, 182–184.

Murkin, J.M., Arango, M., 2009. Near-infrared spectroscopy as an index of brain and tissue oxygenation. Br. J. Anaesth. 103, i3–i13.

Muthukumaraswamy, S.D., 2013. High-frequency brain activity and muscle artifacts in MEG/EEG: a review and recommendations. Front. Hum. Neurosci. 7, 138.

Nemoto, M., Hoshi, Y., Sato, C., Iguchi, Y., Hashimoto, I., Kohno, E., Hirano, T., Terakawa, S., 2012. Diversity of neural-hemodynamic relationships associated with differences in cortical processing during bilateral somatosensory activation in rats. NeuroImage 59, 3325–3338.

Ngai, A.L., Ko, K.R., Morii, S., Winn, R., 1988. Effect of sciatic nerve stimulation on pial arterioles in rats. Am. J. Physiol. 254, H133–H139.

Nielsen, H.B., Boesen, M., Secher, N.H., 2001. Near-infrared spectroscopy determined brain and muscle oxygenation during exercise with normal and resistive breathing. Acta Physiol. Scand. 171, 63–70.

Niessing, J., Ebisch, B., Schmidt, K.E., Niessing, M., Singer, W., Galuske, R.A.W., 2005. Hemodynamic signals correlate tightly with synchronized gamma oscillations. Science 309, 948–951.

Nikulin, V.V., Fedele, T., Mehnert, J., Lipp, A., Noack, C., Steinbrink, J., Curio, G., 2014. Monochromatic ultra-slow (~0.1 Hz) oscillations in the human electroencephalogram and their relation to hemodynamics. NeuroImage 97, 71–80.

O'Leary, M.A., Boas, D.A., Chance, B., Yodh, A.G., 1995. Experimental images of heterogeneous turbid media by frequency-domain diffusing-photon tomography. Opt. Lett. 20, 426–428.

Obrig, H., Neufang, M., Wenzel, R., Kohl, M., Steinbrink, J., Einhäupl, K., Villringer, A., 2000. Spontaneous low frequency oscillations of cerebral hemodynamics and metabolism in human adults. NeuroImage 12, 623–639.

Okada, E., Firbank, M., Schweiger, M., Arridge, S.R., Cope, M., Delpy, D.T., 1997. Theoretical and experimental investigation of near-infrared light propagation in a model of the adult head. Appl. Opt. 36, 21–31.

Palva, S., Palva, J.M., 2007. New vistas for alpha-frequency band oscillations. Trends Neurosci. 30, 150–158.

Pan, W.-J., Thompson, G.J., Magnuson, M.E., Jaeger, D., Keilholz, S., 2013. Infraslow LFP correlates to resting-state fMRI BOLD signals. NeuroImage 74, 288–297.

Pfurtscheller, G., Daly, I., Bauernfeind, G., Müller-Puz, G.R., 2012. Coupling between intrinsic prefrontal HbO_2 and central EEG beta power oscillations in the resting brain. PLoS One 7, e43640.

Raichle, M.E., Synder, A.Z., 2007. A default mode of brain function: a brief history of an evolving idea. NeuroImage 37, 1083–1090.

Rasmussen, P., Dawson, E.A., Nybo, L., van Lieshout, J.J., Secher, N.H., Gjedde, A., 2007. Capillary-oxygenation-level-dependent near-infrared spectrometer in frontal lobe of humans. J. Cereb. Blood Flow Metab. 27, 1082–1093.

Ray, S., Maunsell, J.H.R., 2011. Different origins of gamma rhythm and high-gamma activity in macaque visual cortex. PLoS Biol. 9, e1000610.

Roche-Labarbe, N., Zaaimi, B., Mahmoudzadeh, M., Osharina, V., Wallois, A., Nehlig, A., Grebe, R., Wallois, F., 2010. NITS-measured oxy- and deoxyhemoglobin changes associated with EEG spike-and-wave discharges in a genetic model of absence epilepsy: the GAERS. Epilepsia 51, 1374–1384.

Saager, R.B., Berger, A.J., 2005. Direct characterization and removal of interfering absorption trends in two-layer turbid media. J. Opt. Soc. Am. A Opt. Image Sci. Vis. 22, 1874–1882.

Sasai, S., Homae, F., Watanabe, H., Sasaki, A.T., Tanabe, H.C., Sadato, N., Taga, G., 2012. A NIRS-fMRI study of resting state network. NeuroImage 63, 179–193.

Sato, C., Shimada, M., Hoshi, Y., Yamada, Y., 2005. Extraction of depth-dependent signals from time-resolved reflectance in layered turbid media. J. Biomed. Opt. 10, 064008.

Sato, C., Yamaguchi, T., Seida, M., Ota, Y., Yu, I., Iguchi, Y., Nemoto, M., Hoshi, Y., 2007. Intraoperative monitoring of depth-dependent hemoglobin concentration changes during carotid endarterectomy by time-resolved spectroscopy. Appl. Opt. 46, 2785–2792.

Sato, H., Yahata, N., Funane, T., Takizawa, R., Katura, T., Atsumori, H., Nishimura, Y., Kinoshita, A., Kiguchi, M., Koizumi, H., Fukuda, M., kasai, K., 2013a. A NIRS-fMRI investigation of prefrontal cortex activity during a working memory task. NeuroImage 83, 158–173.

Sato, C., Shimada, M., Tanikawa, Y., Hoshi, Y., 2013b. Estimating the absorption coefficient of the bottom layer in four-layered turbid mediums based on the time-domain depth sensitivity of near-infrared light reflectance. J. Biomed. Opt. 18, 09705.

Schroeter, M.L., Kupka, T., Mildner, T., Uludağ, K., Yves von Cramon, D., 2006. Investigating the poststimulus undershoot of the BOLD signal—a simultaneous fMRI and fNIRS study. NeuroImage 30, 349–358.

Seiyama, A., Seki, J., Tanabe, H.C., Sase, I., Takatsuki, A., Miyauchi, S., Eda, H., Hayashi, S., Imaruoka, T., Iwakura, T., Yanagida, T., 2004. Circulatory basis of fMRI signals: relationship between changes in the hemodynamic parameters and BOLD signal intensity. NeuroImage 21, 1204–1214.

Shmuel, A., Yacoub, E., Pfeuffer, J., de Moortele, P.-F.V., Adriany, G., Hu, X., Ugurbil, K., 2002. Sustained negative BOLD, blood flow and oxygen consumption response and its coupling to the positive response in the human brain. Neuron 36, 1195–1210.

Singer, J.R., Grünbaum, F.A., Kohn, P., Zubelli, J.P., 1990. Image reconstruction of the interior of bodies that diffusion radiation. Science 248, 990–993.

Stefanovska, A., Bracic, M., Kvernmo, H.D., 1999. Wavelet analysis of oscillations in the peripheral blood circulation measured by laser Doppler technique. IEEE Trans. Biomed. Eng. 46, 1230–1239.

Steinbrink, J., Wabnitz, H., Obrig, H., Villringer, A., Rinneberg, H., 2001. Determining changes in NIR absorption using a layered model of the human head. Phys. Med. Biol. 46, 879–896.

Strangman, G., Culver, J.P., Thomson, J.H., Boas, D.A., 2002. A quantitative comparison of simultaneous BOLD fMRI and NIRS recordings during functional brain activation. NeuroImage 17, 719–731.

Strangman, G.E., Li, Z., Zhang, Q., 2013. Depth sensitivity and source-detector separations for near infrared spectroscopy based on the Colin 27 brain template. PLoS One 8, e66319.

Takahashi, T., Takikawa, Y., Kawagoe, R., Shibuya, S., Iwano, T., Kitazawa, S., 2011. Influence of skin blood flow on near-infrared spectroscopy signals measured on the forehead during a verbal fluency task. NeuroImage 57, 991–1002.

Toronov, V., Webb, A., Choi, J.H., 2001. Investigation of human brain hemodynamics by simultaneous near-infrared spectroscopy and functional magnetic resonance imaging. Med. Phys. 28, 521–527.

Toronov, V.Y., Zhang, X., Webb, A.G., 2007. A spatial and temporal comparison of hemodynamic signals measured using optical and functional magnetic resonance imaging during activation in the human primary visual cortex. NeuroImage 34, 1136–1148.

Torricelli, A., Contini, D., Pifferi, A., Caffini, M., Re, R., Zucchelli, L., Spinelli, L., 2014. Time domain functional NIRS imaging for human brain mapping. NeuroImage 85, 28–50.

Villringer, A., Plank, J., Hock, C., Schleikofer, L., Dirnagl, U., 1993. Near-infrared spectroscopy (NIRS): a new tool to study hemodynamic changes during activation of brain function in human adults. Neurosci. Lett. 154, 101–104.

Virtannen, J., Noponen, T., Meriläinen, P., 2009. Comparison of principal and independent component analysis in removing extracerebral interference from near-infrared spectroscopy signals. J. Biomed. Opt. 14, 054032.

Wang, S., Shibahara, N., Kuramashi, D., Okawa, S., Kakuts, N., Okada, E., Maki, A., Yamada, Y., 2010. Effects of spatial variation of skull and cerebrospinal fluid layers on optical mapping of brain activities. Opt. Rev. 17, 410–420.

Wenzel, R., Wobst, P., Heekeren, H.H., Kwong, K.K., Brandt, S.A., Kohl, M., Obrig, H., Dirnagl, U., Villringer, A., 2000. Saccadic suppression induced focal hypooxygenation in the occipital cortex. J. Cereb. Blood Flow Metab. 20, 1103–1110.

Yamada, H., Sadato, N., Konishi, Y., Kimura, K., Tanaka, M., Yonekura, Y., Ishii, Y., 1997. A rapid metabolic change in infants detected by fMRI. Neuroreport 8, 3775–3778.

Yamamoto, T., Kato, T., 2002. Paradoxical correlation between signal in functional magnetic resonance imaging and deoxygenated haemoglobin content in capillaries: a new theoretical explanation. Phys. Med. Biol. 47, 1121–1141.

Yuval-Greenberg, S., Tomer, O., Keren, A.S., Nelken, I., Deouell, L.Y., 2008. Transient induced gamma-band response in EEG as a manifestation of miniature saccades. Neuron 58, 429–441.

Zeff, B.W., White, B.R., Dehghani, H., Schlaggar, B.L., Culver, J.P., 2007. Retinotopic mapping of adult human visual cortex with high-density diffuse optical tomography. Proc. Natl. Acad. Sci. U.S.A. 104, 12169–12174.

Zhang, Y., Brooks, D.H., Franceschini, M.A., Boas, D.A., 2005. Eigenvector-based spatial filtering for reduction of physiological interference in diffuse optical imaging. J. Biomed. Opt. 10, 011014.

Zhang, Q., Brown, E.N., Strangman, G.E., 2007. Adaptive filtering to reduce global interference in evoked brain activity detection: a human subject case study. J. Biomed. Opt. 12, 064009.

Zhao, H., Gao, F., Tanikawa, Y., Homma, K., Yamada, Y., 2005. Time-resolved diffuse optical tomographic imaging for the provision of both anatomical and functional information about biological tissue. Appl. Opt. 44, 1905–1916.

PART

III

Plastic Changes in Neurovascular Coupling

CHAPTER

Mechanisms of cellular plasticity in cerebral perivascular region

8

N. Egawa*, J. Lok*,†, K. Arai*,1

*Neuroprotection Research Laboratory, Massachusetts General Hospital and Harvard Medical School, Charlestown, MA, United States

†Massachusetts General Hospital and Harvard Medical School, Charlestown, MA, United States

[1]Corresponding author: Tel.: +1-617-724-9503; Fax: +1-617-726-7830, e-mail address: karai@partners.org

Abstract

Brain vasculature acts in synergism with neurons to maintain brain function. This neurovascular coupling, or trophic coupling between cerebral endothelium and neuron, is now well accepted as a marker for mapping brain activity. Neurovascular coupling is most active in the perivascular region, in which there are ample opportunities for cell–cell interactions within the neurovascular unit. This trophic coupling between cells maintains neurovascular function and cellular plasticity. Recent studies have revealed that even adult brains contain multiple stem cells of various lineages, which may provide cellular plasticity through the process of differentiation among these stem cell populations. In this chapter, we provide an overview of the process by which neurovascular components contribute to cellular plasticity in the cerebral perivascular regions, focusing on mechanisms of cell–cell interaction in adult brain.

Keywords

Perivascular region, Neurovascular coupling, Neurogenesis, Oligodendrogenesis, Neurovascular niche, Oligovascular niche, Neurovascular unit

1 INTRODUCTION

Trophic coupling between different brain cells plays an essential role in maintaining normal brain function. In this regard, the concept of the neurovascular unit is a useful depiction of brain physiology and pathophysiology. The neurovascular unit provides a conceptual framework in which cell–cell interactions between neuronal, glial, and vascular elements constitute a functional entity (del Zoppo, 2009; Hawkins and Davis, 2005; Iadecola, 2004; Lo et al., 2003, 2004; Zacchigna et al., 2008; Zlokovic, 2008). One important phenomenon within the neurovascular unit is that

Progress in Brain Research, Volume 225, ISSN 0079-6123, http://dx.doi.org/10.1016/bs.pbr.2016.03.005

of neurovascular coupling, wherein neural activity evokes local blood flow changes. Imaging techniques including PET, fMRI, and fNIRS utilize the cerebrovascular–neuron interaction as a principle mechanism to map brain activity (see Chapters "Bridging macroscopic and microscopic methods for the measurements of cerebral blood flow: Toward finding the determinants in maintaining the CBF homeostasis" by I. Kanno and K. Masamoto; "New horizons in neurometabolic and neurovascular coupling from calibrated fMRI" by C.Y. Shu, B.G. Sanganahalli, D. Coman, P. Herman and F. Hyder; "Submillimeter-resolution fMRI: Toward understanding local neural processing" by M. Fukuda, A.J. Poplawsky and S.-G. Kim).

Neurovascular coupling takes place most prominently in the perivascular region (the microenvironment surrounding the microvasculature). This perivascular pocket may also provide an opportunity for trophic coupling between other types of cells. For example, the perivascular region gives rise to the neurovascular niche and gliovascular niche, wherein signaling pathways between cerebral endothelial cells and neuronal/glial precursor cells help mediate pockets of ongoing neurogenesis, gliogenesis, and angiogenesis in adult brain (Kokovay et al., 2010; Ohab et al., 2006). These intricate interactions are maintained during the remodeling phase after brain injury. Angiogenic signals are known to enhance neurogenesis after stroke (Ohab et al., 2006; Taguchi et al., 2004). In turn, neuroblasts migrate along perivascular routes, and the process of neurogenesis enhances vascular regrowth (Thored et al., 2007). This cellular plasticity is an important mechanism in brain remodeling after injury.

Recent studies have revealed that even in the adult brain, mechanisms of cellular plasticity are retained to some extent. Adult brains possess stem cells from multiple lineages in the perivascular region, and coordination of self-renewal and differentiation in those stem cell populations maintains brain homeostasis. Notably, cell–cell trophic coupling may contribute to the cellular plasticity in the perivascular region, which plays an important role in repair responses in the adult brain after injury. In this chapter, we review mechanisms of cellular plasticity in the neurovascular unit, focusing on cell–cell interaction in the perivascular region.

2 CELLULAR PLASTICITY AND FUNCTION OF NEUROVASCULAR UNIT COMPONENTS

The neurovascular unit is composed of neurons, glial cells (astrocytes, microglia, and oligodendrocytes), and vascular cells (endothelial cells and pericytes). The dynamic interactions among these components regulate complex brain function, an example of which is that of adjustment of local blood flow by neural activity. Recent research advances have shown that these neurovascular unit components have diverse heterogeneity in their morphology, developmental origin, gene expression profile, physiological properties, function, and response to injury and disease (Allen and Barres, 2009; Barres, 2008; Olah et al., 2012; Urich et al., 2012; Zamanian et al., 2012;

Zhang and Barres, 2010). This diversity is a major contributor that underlies the cellular plasticity of neurovascular unit components in the adult brain.

Within the neurovascular unit, cell plasticity may come from differentiation of progenitor cells in the perivascular region. As noted, several kinds of progenitor cells reside even in the adult brain, and with trophic support by neighboring cells, they differentiate as needed for maintenance and/or remodeling of the relevant brain region. In this regard, neuronal stem/progenitor cells (NSPCs) play an essential role because they have the potential to differentiate and mature into all the subtypes of neurons and glial cells. They undergo both self-renewal and diversification into other types of cells; the former through symmetric proliferation and the later through asymmetric proliferation. The cellular plasticity exhibited by NSPCs during differentiation (ie, neurogenesis and gliogenesis) is partly a response to intrinsic or extrinsic mechanisms, which trigger the pathways that determine the eventual cell fate. For example, NSPCs respond to inner or outer stimuli such as epigenetic regulation as in chromatin remodeling (Hirabayashi and Gotoh, 2010), cellular signaling by cell–cell communication (Ables et al., 2011) and the action of local trophic factors.

In this section, we briefly review the function of neurovascular components, focusing on cell differentiation mechanisms. Because mechanisms of adult neurogenesis and gliogenesis may mirror the cell differentiation steps observed during embryonic development, we also overview intrinsic and extrinsic signaling mechanisms for cell differentiation during brain development.

2.1 NEURONS

Radial glia (RG), which is the progenitor cell for neural/glial cells, is morphologically characterized by the projection of its processes from the ventricular zone (VZ) to the meningeal zone with apicobasal polarity during the embryonic stage of brain development. During this period, RG switches its role among three main functions: self-proliferation, neurogenesis, and gliogenesis (astrogenesis). NSPCs, which are originally derived from RG, generate both neurons and astrocytes even in cell culture conditions (Qian et al., 2000). Thus far, several intrinsic signaling pathways, such as Notch, Wnt, and JAK/STAT signaling, have been found to have the ability to maintain NSPCs in undifferentiated states during the self-proliferation phase. In addition, chromatin remodeling of DNA by methylation and histone modification has also been shown to contribute to the plasticity of neurons and glial cells during the developmental stage (Fig. 1).

Notch signaling is one of the major mediators of cell fate determination in cells of neural lineage. Notch signaling is mediated by the Notch receptors activated by ligand Jagged on the surface of the neighboring cells (Wang, 2011). The targets are transcriptional factors Hes (Hes1 and Hes5), which inhibit the neural differentiation of NSPC. The Notch-mediated cell–cell interactions spread to surrounding cells to maintain the undifferentiated conditions in their surrounding, which is known as "lateral inhibition." Notch 1 knockout in mature neurons did not change the

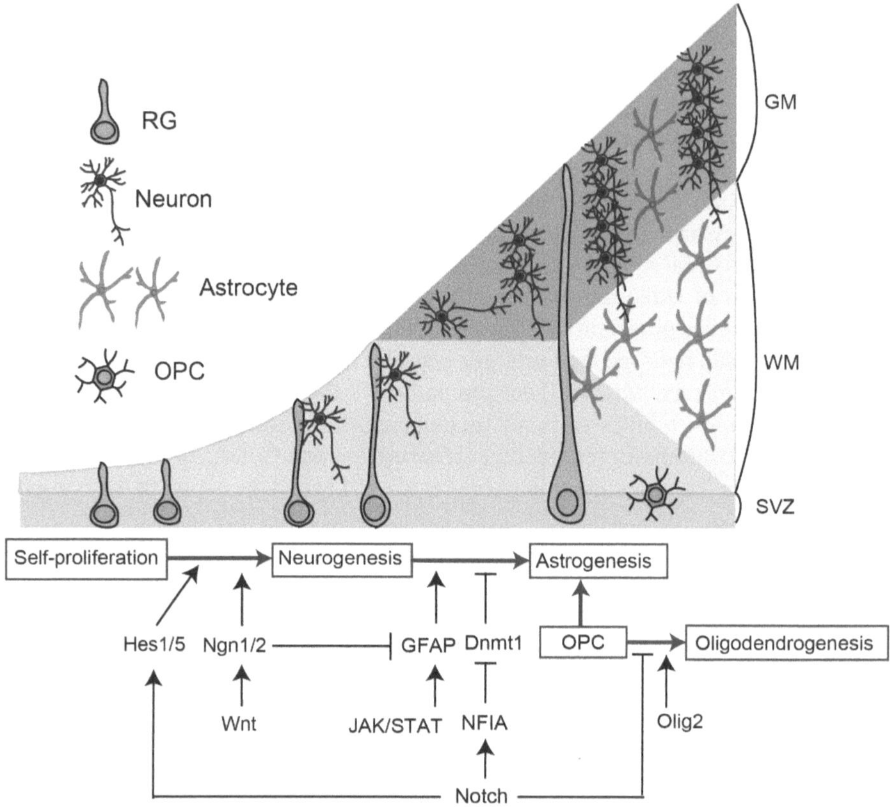

FIG. 1

Cellular signaling in neurogeneis and gliogenesis during brain development. Abbreviation: *RG*, radial glia; *OPC*, oligodendrocyte progenitor cell; *GM*, gray matter; *WM*, white matter; *SVZ*, subventricular zone.

morphology but altered dendritic spines and disrupted long-term potentiation and long-term depression, leading to deficits in performing several memory tasks in mice (Alberi et al., 2011).

Phase-dependent activation of Wnt signaling is another critical regulator of neuronal cell fate. Binding of Wnt to its receptor Frizzled inhibits the degradation of β-catenin, which binds the T-cell factor to upregulate downstream genes. The importance of β-catenin in neurogenesis is illustrated by the fact that β-catenin deletion is associated with impaired neurogenesis from immature neuron progenitor cells (Zechner et al., 2003). Among genes downstream of Wnt signaling, Ngn1 and Ngn2 are key regulators that are highly expressed transcriptional factors during the neurogenesis phase (Hirabayashi et al., 2004; Israsena et al., 2004). Ngn regulates neurogenesis and gliogenesis in a complex way. Ngn strongly promotes neurogenesis to constitute the six layers of brain cortex in an order from deep to superficial layers, and Ngn negatively regulates JAK-STAT-mediated demethylation of the GFAP promoter region and inhibits NSPC differentiation into astrocytes (Sun et al., 2001). Ngn2-knockout mice exhibit prominent decrements in neurogenesis and gliogenesis (Nieto et al., 2001). Although Wnt signaling upregulates the expression of Ngn during the neurogenesis phase, its effects on chromatin remodeling by histone modification of the promoter region of Ngn in the later phase results in decreased neurogenesis and Wnt-induced chromatin remodeling blunts the sensitivity to Wnt signaling, leading to a shift from neurogenesis to gliogenesis (Hirabayashi et al., 2009). Since precise mechanisms as to how Ngn regulates neurogenesis/gliogenesis are still mostly unknown, elucidating phase- and region-dependent mechanisms of Ngn signaling in NSPC function is warranted for future studies.

Taken together, phase-dependent cellular signaling and epigenetic modification in the promoter region of genes related to neurogenesis and gliogenesis are key regulators involved in the shift from neurogenesis to gliogenesis. Further studies will be needed to elucidate how these factors act synergistically to organize the process of neurogenesis by maintaining asymmetrical differentiation, and contribute to phase-dependent cellular plasticity.

2.2 GLIAL CELLS (ASTROCYTES, OLIGODENDROCYTES, AND MICROGLIA)

Astrocytes are mainly generated from three distinct areas in the brain: VZ, subventricular zone (SVZ), and the distributed localized area (Pinto and Gotz, 2007). RG in the VZ generates astrocytes in white matter during the gliogenesis phase. Glial progenitor cells in SVZ migrate to the cortical or subcortical layer of the brain and generate astrocytes in both gray and white matter. These newly generated astrocytes are morphologically categorized into two types—protoplasmic astrocytes in gray matter and fibroblastic astrocytes in white matter.

Several signaling pathways have been identified to modulate NSPC differentiation into astrocytes. As mentioned above, a central one is JAK/STAT signaling, in which JAK phosphorylates STAT1/3 and promotes the dimerization of STAT1/3 to upregulate GFAP expression. Notch signaling is also an important regulator of GFAP gene expression both genetically and epigenetically. Notch intracellular domain (NICD)/CBF1 complex upregulates GFAP via direct binding to the GFAP promoter region. There are CpG-methylated regions on the GFAP promoter to which Notch signaling-activated NFIA binds and prevents Dnmt1 from accessing the promoter region (Namihira et al., 2009). This leads to demethylation of the promoter region and upregulation of GFAP expression.

Besides neurons and astrocytes, oligodendrocyte lineage cells are also generated from NSPCs. In general, oligodendrocytes are derived from so-called oligodendrocyte precursor cells (OPCs), which originate from NSPCs. However, there is ongoing debate as to whether oligodendrocytes should be generated from OPCs with fate-restricted lineage or not, since several studies have shown that OPCs retain context-dependent lineage plasticity in the differentiation period (Kondo and Raff, 2000; Nishiyama et al., 2009).

Historically, OPCs have been purified from rat optic nerves by immunopanning with antibody against A2B5 and identified as bipotential oligodendrocyte-type2 astrocyte progenitor cells (O-2A cells) (Raff et al., 1983), which indicate that OPCs could differentiate into oligodendrocytes as well as astrocytes. In the postnatal period, OPCs appear in the SVZ and diffusely distributed to cover the entire parenchyma of the brain. Even in the adult CNS, OPCs have been found to comprise 2–8% of all the cells (Dawson et al., 2003; Peters, 2004). While majority of SVZ-derived OPCs can differentiate into oligodendrocytes, some may directly differentiate into protoplasmic astrocytes in gray matter. Actually, OPCs contribute to about 40% of protoplasmic astrocytes in the gray matter of the ventral forebrain (Zhu et al., 2008). In addition, when purified OPCs were transplanted into the glia-depleted mice, they were differentiated into both oligodendrocytes and astrocytes (Franklin et al., 1995). Moreover, when transplanted into myelin basic protein (MBP)-depleted conditions, some OPCs were differentiated into GFAP positive cells, while others expressed MBP, a marker of mature oligodendrocytes (Windrem et al., 2004). Astrocyte generation from OPCs is limited during the initial postnatal period. These data suggest that OPCs are capable of differentiating into different types of cells, depending on the condition of the surrounding environment (Hill et al., 2014).

Thus far, several key mechanisms have been identified in cell fate decisions for oligodendrocyte lineage cells. Olig2 is a critical transcriptional regulator, which directs OPCs to become oligodendrocytes during embryogenesis (Lu et al., 2002; Zhou and Anderson, 2002; Zhu et al., 2012). In fact, in Olig2-deficient conditions, OPCs were directed to become astrocytes, and postnatal hypomyelination was observed in the neocortex and corpus callosum (Cai et al., 2007). Epigenetic regulation such as DNA methylation and histone modification associated with chromatin remodeling was also shown to be involved in cell specification for oligodendrocytes (Moyon et al., 2015). Treatment with HDAC inhibitor or genetic ablation of HDAC1/2 in OPCs resulted in failure of differentiation in vivo (Cunliffe and Casaccia-Bonnefil, 2006; Ye et al., 2009). Several groups have confirmed that HDAC activities are required for efficient myelination (Shen et al., 2008). For example, HDACs act in conjunction with the transcription factor YY1 to start myelin gene expression by suppressing Id4 and Tcf4 (He et al., 2007). Neuronal activity (Ishibashi et al., 2006), microRNAs (Dugas et al., 2010), and Notch signaling (Wang et al., 1998) are also well-defined mechanisms for cell fate decisions in oligodendrocyte lineage cells. Working together, the factors and signaling pathways discussed here coordinately regulate oligodendrocyte generation during development.

Microglia, which originates from mesodermal yolk sac tissue during development (Chan et al., 2007; Hirasawa et al., 2005), is a major cell type for the immune response in brain. After infiltrating into the brain during development microglia reside and proliferate locally. In terms of cellular plasticity in the neurovascular unit, microglia plays a major role in synaptic pruning during development. The fractalkine receptor (CX3CR1), a marker of microglia (Harrison et al., 1998), has been shown to be involved in the cytokine-mediated pathway in which synaptic plasticity would dynamically be formed by the interaction between neurons (Kim et al., 2013). Also, microglia responds to neurotransmitter released by neuronal activity and contributes to the synaptic pruning via fractalkine/fractalkine receptors (CX3CL1/CX3CR1) signaling (Pocock and Kettenmann, 2007; Schafer et al., 2012). Genetic fate mapping studies using CX3R1-knockout mice have revealed deficits in multiple learning abilities as well as a remarkable reduction in glutamatergic synapse formation (Paolicelli et al., 2011; Parkhurst et al., 2013). The function of CX3R1 is partially dependent on microglia-derived BDNF activity. The homeostasis of microglia is dependent on CSF-1/CD115 receptor because its inhibition results in complete depletion of brain microglia (Elmore et al., 2014). These findings support the idea that in addition to immune responses, microglia plays an important role in maintaining neurovascular homeostasis including cellular plasticity in neurons.

2.3 VASCULAR CELLS (CEREBRAL ENDOTHELIAL CELLS AND PERICYTES)

Vascular cells constitute a major cellular element of the brain, and cerebral endothelial cells play a central role in the neurovascular unit. The cerebrovascular system was traditionally seen as a passive conduit for blood; however, recent research has revealed that the microvasculature plays an active role in maintaining brain homeostasis. Cerebral endothelial cells are differentiated from endothelial progenitor cells (EPCs). EPCs originate from mesodermal cells, and are responsible for vasculogenesis such as the formation of capillary network in the developing embryo. In addition, EPCs also contribute to angiogenesis—in the setting of remodeling and repairing of the capillary network (Asahara et al., 1997; Augustin, 2001).

In addition to its role as a conduit for blood flow, cerebral endothelial cells act as a source of multiple mediators to regulate neurovascular plasticity. Through secreting soluble factors, cerebral endothelial cells control neuronal regeneration through the processes of proliferation, migration, and differentiation of NSPCs (Goldman and Chen, 2011). Major mediators from cerebral endothelial cells are nitric oxide (NO) (Chen et al., 2005; Packer et al., 2003), SXCL12 (SDF1), vascular endothelial growth factor (VEGF) (Licht and Keshet, 2013), brain-derived neurotrophic factor (BDNF) (Leventhal et al., 1999), and pigmented epithelium-derived factor. These endothelial-derived factors may act in concert with secreting factors from other neurovascular unit components to coordinate neurogenesis, neural plasticity, and NSPC division during the developmental stage. For example, NO inhibits epidermal growth factor receptor-dependent PI3K/Akt signaling (Torroglosa et al., 2007) and acts as a tonic repressor of neurogenesis in vivo. VEGF plays multiple functions in maintaining the fenestration permeability of the vessel (Maharaj et al., 2008), activity-induced neurogenesis (Cao et al., 2004), and synaptic plasticity such as enhancement of the long-term potentiation response (Licht and Keshet, 2013). BDNF is reported to support the recruitment of newly generated neurons in SVZ (Rasika et al., 1999). PDGF acts to promote NSC renewal through potentiating Notch-dependent transcription (Andreu-Agullo et al., 2009). Since cerebral endothelial cells play a central role in the neurovascular unit, future studies are needed to further investigate the mechanisms by which cerebral endothelial cells contribute to cellular plasticity within the neurovascular unit.

Within the neurovascular unit, pericytes play several unique roles. Pericytes are especially abundant in the brain vasculature. Similar to other neurovascular unit components, pericytes also interact with neighboring cells. For example, cerebral vasculature is covered with a 1:1–3 ratio between endothelial cells and pericytes (Armulik et al., 2011; Mathiisen et al., 2010; Sims, 1986), and pericytes are well known to regulate the integrity of blood–brain barrier (Armulik et al., 2010; Bell et al., 2010). In addition, pericytes and endothelial cells reciprocally support each other in the perivascular area through the secretion of soluble factors. Platelet-derived growth factor-receptor β (PDGFRβ) is a constitutive marker of pericytes, and endothelial cells secrete PDGFβ to activate the recruitment and proliferation of pericytes. PDGF-β/PDGFRβ knockout mice suffer from a deficiency pericytes in the brain as well as other organs such as kidney, heart, and lung (Bjarnegard et al., 2004). On the other hand, pericytes secrete angiopoietin-1, the ligand for Tie-1 that is predominantly expressed in endothelial cells, and promotes endothelial maturation and stability (Falcon et al., 2009; Gaengel et al., 2009). In addition to endothelium–pericyte interactions, recent studies demonstrated that in the perivascular region, pericytes may also exchange soluble factors with OPCs to support their function (Maki et al., 2015).

As discussed above, through secreting trophic factors, pericytes regulate cellular plasticity through their involvement in angiogenesis and oligodendrogenesis in the perivascular area. However, pericytes may contribute to cellular plasticity in a more direct way. Early studies have demonstrated that pericytes in the brain originate from mesenchyme cells that localize on the abluminal side of the endothelial cells (Hungerford and Little, 1999). Recent studies in turn have stressed that the pericytes exhibit stem-cell-like properties, which enable them to differentiate into a variety of cell types, such as osteoblasts, adipocytes, chondrocytes, vascular smooth muscle, and skeletal muscle. In fact, under ischemic conditions, pericytes can be reprogrammed into neurons (Karow et al., 2012). As illustrated by these examples, the pericyte is an important cell in the regulation of cell plasticity in the perivascular region.

3 NEUROVASCULAR/OLIGOVASCULAR NICHE MEDIATES NEUROGENESIS AND OLIGODENDROGENESIS IN THE NEUROVASCULAR UNIT

The components of the neurovascular unit coordinately affect cellular plasticity through self-renewal, cell fate, and synaptic plasticity in processes that respond to stimulation by intrinsic and extrinsic factors (Fig. 2). In this regard, cerebral endothelial cells play a central role in cellular plasticity because they produce the neurovascular and oligovascular niches to support neurogenesis and oligodendrogenesis.

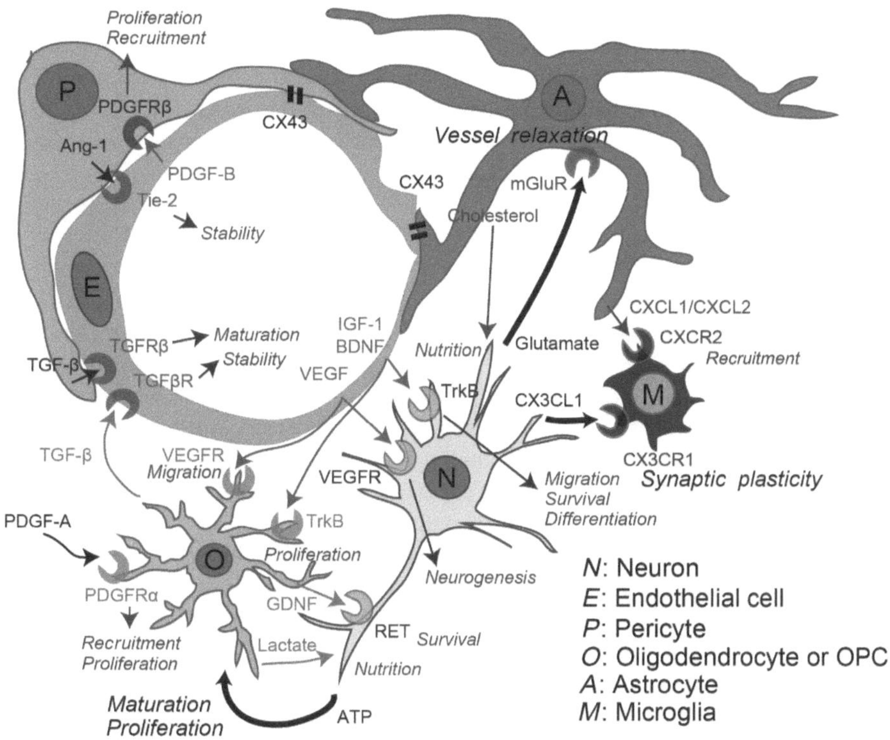

FIG. 2

Interactive behavior in the neurovascular unit for systematic plasticity. Trophic factors are secreted from each neurovascular unit component to affect the function of its target cell. Abbreviation: *A*, astrocyte; *P*, pericyte; *O*, oligodendrocyte or oligodendrocyte progenitor cell; *N*, neuron; *M*, microglia; *E*, endothelial cell.

3.1 NEUROVASCULAR NICHE

As noted in the introduction section, cell–cell interaction in the neurovascular niche is one of the major facets within the neurovascular unit. Cerebral endothelial cells and NSPCs are closely related in the perivascular region in the processes of neurogenesis and angiogenesis (Ohab et al., 2006). In this neurovascular niche, a number of trophic factors such as VEGF, BDNF, EGF, and FGF2 are released from endothelial cells to upregulate NSPC expansion for neurogenesis. This is an important mechanism for adult neurogenesis in the adult brain. Additionally, after brain injury, cerebral endothelial cells respond to ischemic or traumatic stress by releasing necessary soluble factors that are important in the recovery of cellular functions.

In turn, NSPCs (and neurons) can regulate development of functional integrity in the neurovascular unit by secreting the factors such as VEGF that induce endothelial cell migration and guide vessel sprouting (Obermeier et al., 2013). For example, neuroblasts migrate along perivascular routes and the promotion of neurogenesis enhances vascular regrowth (Thored et al., 2007). Besides angiogenesis, neuronal activity also regulates blood flow in the local area of the brain (eg, neurovascular coupling). This is mediated at least partly by the neuron-to-astrocyte signaling in which glutamate-mediated Ca^{2+} elevation in astrocytes promotes dilation of the arterioles (Zonta et al., 2003) (see Chapter "Involvement of astrocytes in neurovascular communication" by M. Nuriya and H. Hirase).

3.2 OLIGOVASCULAR NICHE

Similar to the neurovascular niche, the oligovascular niche mediates interactions between cerebral endothelial cells and OPCs, mostly in cerebral white matter (Arai and Lo, 2009a; Miyamoto et al., 2014). OPCs are widely distributed even in adult brain (Chang et al., 2000; Levine et al., 2001; Nishiyama et al., 1999) and when guided to the injured site, they contribute to myelin repair after brain injury (Tanaka et al., 2003). At the same time, cerebral endothelial cells may promote the survival and proliferation of perivascular OPCs (Maki et al., 2015) via FGF-2 and BDNF (Arai and Lo, 2009b). VEGF-A may also be an important mediator in the oligovascular niche. VEGF-A is secreted from cerebral endothelial cells and its receptor (Flk-1) is expressed in OPCs (Hayakawa et al., 2011). In vitro cell culture studies confirmed that endothelial-derived VEGF-A did promote the migration of OPCs (Hayakawa et al., 2012). Notably, the trophic support from cerebral endothelial cells to OPCs may change under diseased conditions. Even sublethal level of oxidative stress could suppress the secretion of FGF-2/BDNF from cerebral endothelial cells, indicating that cerebral endothelial cells would not provide sufficient support to OPCs under diseased conditions (Arai and Lo, 2009b). Therefore, disruption of OPC–endothelium trophic coupling may hasten the progression of white matter-related diseases.

In turn, cells from the oligodendrocyte lineage may modulate the cerebrovascular system. For example, OPCs produce TGF-β, which maintains the tightness of BBB during development (Seo et al., 2014). However, under diseased conditions such as cerebral hypoperfusion in the adult brain, OPCs could damage the BBB by secreting matrix metalloproteinase-9 (Seo et al., 2013). Importantly, patterns of cell–cell interaction in the oligovascular niche may change again during the chronic (repairing) phase after injury. After white matter damage, oligodendrocyte-derived MMP-9 would work on white matter remodeling by promoting angiogenesis (Pham et al., 2012). Taken together, these findings indicate that the oligovascular niche contributes to cellular plasticity in cerebral white matter.

4 CELL–CELL INTERACTION IN DISEASE

Cell–cell interaction in the perivascular region is an important mechanism for cellular plasticity, which maintains neurovascular homeostasis. Importantly, these cell–cell interactions may change under disease conditions—for example, disruption of cell–cell trophic coupling causes cellular damage, but at the same time, changes in neurovascular interactions after brain injury may promote neurovascular repair. Therefore, an understanding of cell–cell interaction in disease may lead to new therapeutic approaches for CNS diseases. In this section, we will briefly introduce some key findings related to changes in neurovascular components under pathological conditions.

In the acute phase after the cerebral ischemia, injured neurons release damage-associated molecular pattern proteins (DAMPs), leading to the secretion of deleterious mediators such as inflammatory cytokines (eg, IL-1β and TNF-α) and reactive oxygen species. These factors induce BBB breakdown and facilitate the infiltration of circulating monocytes, neutrophils, and lymphocytes (del Zoppo et al., 2000; Jackman and Iadecola, 2015), which then accelerate postischemic inflammation, which may further damage the surviving neurons (Fitch and Silver, 2008). Under ischemic conditions, CX3CL1/CX3CR1 signaling promotes microglial phagocytosis of apoptotic neurons, an important aspect of remodeling (Cardona et al., 2006). This response also promotes neurogenesis in the ipsilateral SVZ—residual NSPCs are activated to migrate to the ischemic lesion and to proliferate at that site (Zhang et al., 2004). This dual effect is also observed in other signaling proteins. For example, HMGB-1, which belongs to DAMP family, induces inflammatory/deleterious effects in the early phase of stroke, while it contributes to neurovascular remodeling by promoting neurogenesis and angiogenesis in the chronic recovery phase (Hayakawa et al., 2010).

Other mechanisms of cell–cell interactions are also affected under conditions of stress. Deficits in axonal energy caused by metabolic disturbance in glial cells are closely related to neurodegenerative disease. Oligodendrocyte dysfunction may cause neuronal/axonal damage. Under normal conditions, oligodendrocytes supply lactate to neurons through monocarboxylate transporter 1 (MCT1) (Morrison et al., 2013). However, in the setting of amyotrophic lateral sclerosis, oligodendrocytes exhibit MCT1 deficiency—a phenomenon observed both in a mouse model and in human subjects (Kang et al., 2010). This deficit would likely lead to a disruption of the trophic support from oligodendrocytes to neurons (axons).

Protein aggregation and degeneration in glial cells are also related to the progression of neuronal disease. Astrocytes take up and degrade Aβ protein, which is deposited in brains of patients with Alzheimer's disease (AD) and thought to cause neuronal damage (Guenette, 2003). Enhancing the autophagy-mediated proteolysis in astrocytes attenuate Aβ deposit (Xiao et al., 2014), suggesting that astrocytes are important in Aβ clearance and that astrocyte dysfunction may accelerate the progression of AD pathology. In addition to Aβ protein, α-synuclein is another important protein in neurodegenerative diseases. α-synuclein deposits mostly in neuronal Lewy bodies, and α-synuclein-aggregated Lewy neurites are observed in brains of Parkinson's disease or dementia patients. On the other hand, in multiple system atrophy patients, α-synuclein deposits largely in oligodendroglial cytoplasmic inclusions (Halliday et al., 2011). These findings suggest that mechanisms of α-synuclein clearance could be deranged in specific cell populations depending on the type of disease.

In general, pathological conditions disturb cell–cell trophic coupling, leading to neurovascular dysfunction. But, as briefly mentioned above, cell–cell interactions may recover to some extent during the chronic phase to promote compensative neurogenesis/angiogenesis/oligodendrogenesis. However, it still remains unknown how those cell–cell interactions contribute to barriergenesis (ie, resealing BBB) after brain injury. During development, cells in the perivascular region, such as pericytes and astrocytes, provide support to cerebral endothelial cells in BBB formation (Siegenthaler et al., 2013). Therefore, to reveal the mechanisms of cellular plasticity in the neurovascular unit, future studies are warranted to investigate the roles of pericyte–endothelium or astrocyte–endothelium trophic coupling in compensative barriergenesis of BBB in adult brains.

5 CONCLUSION

The concept of the neurovascular unit has provided a basic framework for understanding brain physiology and pathology. As discussed in this chapter, cell–cell trophic coupling is important for the regulation of neurovascular function and preservation of brain homeostasis. In this regard, perivascular region may play an essential role because the neurovascular/oligovascular niche mediates the interaction of cerebral endothelial cells with other cells within the neurovascular unit to regulate cellular plasticity. Ultimately, cellular plasticity can be easily affected by neighboring microenvironments in both beneficial and detrimental ways, and therefore, a precise understanding of cell–cell interaction pathways is needed to develop efficient therapeutic treatments for CNS diseases.

REFERENCES

Ables, J.L., Breunig, J.J., Eisch, A.J., Rakic, P., 2011. Not(ch) just development: notch signalling in the adult brain. Nat. Rev. Neurosci. 12, 269–283. http://dx.doi.org/10.1038/nrn3024.

Alberi, L., et al., 2011. Activity-induced Notch signaling in neurons requires Arc/Arg3.1 and is essential for synaptic plasticity in hippocampal networks. Neuron 69, 437–444. http://dx.doi.org/10.1016/j.neuron.2011.01.004.

Allen, N.J., Barres, B.A., 2009. Neuroscience: glia—more than just brain glue. Nature 457, 675–677. http://dx.doi.org/10.1038/457675a. 457675a [pii].

Andreu-Agullo, C., Morante-Redolat, J.M., Delgado, A.C., Farinas, I., 2009. Vascular niche factor PEDF modulates Notch-dependent stemness in the adult subependymal zone. Nat. Neurosci. 12, 1514–1523. http://dx.doi.org/10.1038/nn.2437.

Arai, K., Lo, E.H., 2009a. Oligovascular signaling in white matter stroke. Biol. Pharm. Bull. 32, 1639–1644.

Arai, K., Lo, E.H., 2009b. An oligovascular niche: cerebral endothelial cells promote the survival and proliferation of oligodendrocyte precursor cells. J. Neurosci. 29, 4351–4355. http://dx.doi.org/10.1523/JNEUROSCI.0035-09.2009.

Armulik, A., et al., 2010. Pericytes regulate the blood–brain barrier. Nature 468, 557–561. http://dx.doi.org/10.1038/nature09522. nature09522 [pii].

Armulik, A., Genove, G., Betsholtz, C., 2011. Pericytes: developmental, physiological, and pathological perspectives, problems, and promises. Dev. Cell 21, 193–215. http://dx.doi.org/10.1016/j.devcel.2011.07.001.

Asahara, T., et al., 1997. Isolation of putative progenitor endothelial cells for angiogenesis. Science 275, 964–967.

Augustin, H.G., 2001. Tubes, branches, and pillars: the many ways of forming a new vasculature. Circ. Res. 89, 645–647.

Barres, B.A., 2008. The mystery and magic of glia: a perspective on their roles in health and disease. Neuron 60, 430–440. http://dx.doi.org/10.1016/j.neuron.2008.10.013. S0896-6273(08)00886-6 [pii].

Bell, R.D., et al., 2010. Pericytes control key neurovascular functions and neuronal phenotype in the adult brain and during brain aging. Neuron 68, 409–427. http://dx.doi.org/10.1016/j.neuron.2010.09.043.

Bjarnegard, M., et al., 2004. Endothelium-specific ablation of PDGFB leads to pericyte loss and glomerular, cardiac and placental abnormalities. Development 131, 1847–1857. http://dx.doi.org/10.1242/dev.01080.

Cai, J., et al., 2007. A crucial role for Olig2 in white matter astrocyte development. Development 134, 1887–1899. http://dx.doi.org/10.1242/dev.02847.

Cao, L., et al., 2004. VEGF links hippocampal activity with neurogenesis, learning and memory. Nat. Genet. 36, 827–835. http://dx.doi.org/10.1038/ng1395.

Cardona, A.E., et al., 2006. Control of microglial neurotoxicity by the fractalkine receptor. Nat. Neurosci. 9, 917–924. http://dx.doi.org/10.1038/nn1715.

Chan, W.Y., Kohsaka, S., Rezaie, P., 2007. The origin and cell lineage of microglia: new concepts. Brain Res. Rev. 53, 344–354. http://dx.doi.org/10.1016/j.brainresrev.2006.11.002.

Chang, A., Nishiyama, A., Peterson, J., Prineas, J., Trapp, B.D., 2000. NG2-positive oligodendrocyte progenitor cells in adult human brain and multiple sclerosis lesions. J. Neurosci. 20, 6404–6412.

Chen, J., et al., 2005. Endothelial nitric oxide synthase regulates brain-derived neurotrophic factor expression and neurogenesis after stroke in mice. J. Neurosci. 25, 2366–2375. http://dx.doi.org/10.1523/JNEUROSCI.5071-04.2005.

Cunliffe, V.T., Casaccia-Bonnefil, P., 2006. Histone deacetylase 1 is essential for oligodendrocyte specification in the zebrafish CNS. Mech. Dev. 123, 24–30. http://dx.doi.org/10.1016/j.mod.2005.10.005.

Dawson, M.R., Polito, A., Levine, J.M., Reynolds, R., 2003. NG2-expressing glial progenitor cells: an abundant and widespread population of cycling cells in the adult rat CNS. Mol. Cell. Neurosci. 24, 476–488.

del Zoppo, G.J., 2009. Inflammation and the neurovascular unit in the setting of focal cerebral ischemia. Neuroscience 158, 972–982. http://dx.doi.org/10.1016/j.neuroscience.2008.08.028.

del Zoppo, G., et al., 2000. Inflammation and stroke: putative role for cytokines, adhesion molecules and iNOS in brain response to ischemia. Brain Pathol. 10, 95–112.

Dugas, J.C., et al., 2010. Dicer1 and miR-219 Are required for normal oligodendrocyte differentiation and myelination. Neuron 65, 597–611. http://dx.doi.org/10.1016/j.neuron.2010.01.027.

Elmore, M.R., et al., 2014. Colony-stimulating factor 1 receptor signaling is necessary for microglia viability, unmasking a microglia progenitor cell in the adult brain. Neuron 82, 380–397. http://dx.doi.org/10.1016/j.neuron.2014.02.040.

Falcon, B.L., et al., 2009. Contrasting actions of selective inhibitors of angiopoietin-1 and angiopoietin-2 on the normalization of tumor blood vessels. Am. J. Pathol. 175, 2159–2170. http://dx.doi.org/10.2353/ajpath.2009.090391.

Fitch, M.T., Silver, J., 2008. CNS injury, glial scars, and inflammation: inhibitory extracellular matrices and regeneration failure. Exp. Neurol. 209, 294–301. http://dx.doi.org/10.1016/j.expneurol.2007.05.014. S0014-4886(07)00213-0 [pii].

Franklin, R.J., Bayley, S.A., Milner, R., Ffrench-Constant, C., Blakemore, W.F., 1995. Differentiation of the O-2A progenitor cell line CG-4 into oligodendrocytes and astrocytes following transplantation into glia-deficient areas of CNS white matter. Glia 13, 39–44. http://dx.doi.org/10.1002/glia.440130105.

Gaengel, K., Genove, G., Armulik, A., Betsholtz, C., 2009. Endothelial-mural cell signaling in vascular development and angiogenesis. Arterioscler. Thromb. Vasc. Biol. 29, 630–638. http://dx.doi.org/10.1161/ATVBAHA.107.161521.

Goldman, S.A., Chen, Z., 2011. Perivascular instruction of cell genesis and fate in the adult brain. Nat. Neurosci. 14, 1382–1389. http://dx.doi.org/10.1038/nn.2963.

Guenette, S.Y., 2003. Astrocytes: a cellular player in Abeta clearance and degradation. Trends Mol. Med. 9, 279–280.

Halliday, G.M., Holton, J.L., Revesz, T., Dickson, D.W., 2011. Neuropathology underlying clinical variability in patients with synucleinopathies. Acta Neuropathol. 122, 187–204. http://dx.doi.org/10.1007/s00401-011-0852-9.

Harrison, J.K., et al., 1998. Role for neuronally derived fractalkine in mediating interactions between neurons and CX3CR1-expressing microglia. Proc. Natl. Acad. Sci. U. S. A. 95, 10896–10901.

Hawkins, B.T., Davis, T.P., 2005. The blood–brain barrier/neurovascular unit in health and disease. Pharmacol. Rev. 57, 173–185. http://dx.doi.org/10.1124/pr.57.2.4.

Hayakawa, K., Qiu, J., Lo, E.H., 2010. Biphasic actions of HMGB1 signaling in inflammation and recovery after stroke. Ann. N. Y. Acad. Sci. 1207, 50–57. http://dx.doi.org/10.1111/j.1749-6632.2010.05728.x.

Hayakawa, K., et al., 2011. Vascular endothelial growth factor regulates the migration of oligodendrocyte precursor cells. J. Neurosci. 31, 10666–10670. http://dx.doi.org/10.1523/JNEUROSCI.1944-11.2011. 31/29/10666 [pii].

Hayakawa, K., et al., 2012. Cerebral endothelial derived vascular endothelial growth factor promotes the migration but not the proliferation of oligodendrocyte precursor cells in vitro. Neurosci. Lett. 513, 42–48. http://dx.doi.org/10.1016/j.neulet.2012.02.004.

He, Y., et al., 2007. The transcription factor Yin Yang 1 is essential for oligodendrocyte progenitor differentiation. Neuron 55, 217–230. http://dx.doi.org/10.1016/j.neuron.2007.06.029.

Hill, R.A., Patel, K.D., Goncalves, C.M., Grutzendler, J., Nishiyama, A., 2014. Modulation of oligodendrocyte generation during a critical temporal window after NG2 cell division. Nat. Neurosci. 17, 1518–1527. http://dx.doi.org/10.1038/nn.3815.

Hirabayashi, Y., Gotoh, Y., 2010. Epigenetic control of neural precursor cell fate during development. Nat. Rev. Neurosci. 11, 377–388. http://dx.doi.org/10.1038/nrn2810.

Hirabayashi, Y., et al., 2004. The Wnt/beta-catenin pathway directs neuronal differentiation of cortical neural precursor cells. Development 131, 2791–2801. http://dx.doi.org/10.1242/dev.01165.

Hirabayashi, Y., et al., 2009. Polycomb limits the neurogenic competence of neural precursor cells to promote astrogenic fate transition. Neuron 63, 600–613. http://dx.doi.org/10.1016/j.neuron.2009.08.021.

Hirasawa, T., et al., 2005. Visualization of microglia in living tissues using Iba1-EGFP transgenic mice. J. Neurosci. Res. 81, 357–362. http://dx.doi.org/10.1002/jnr.20480.
Hungerford, J.E., Little, C.D., 1999. Developmental biology of the vascular smooth muscle cell: building a multilayered vessel wall. J. Vasc. Res. 36, 2–27. http://dx.doi.org/10.1159/000025622.
Iadecola, C., 2004. Neurovascular regulation in the normal brain and in Alzheimer's disease. Nat. Rev. Neurosci. 5, 347–360. http://dx.doi.org/10.1038/nrn1387.
Ishibashi, T., et al., 2006. Astrocytes promote myelination in response to electrical impulses. Neuron 49, 823–832. http://dx.doi.org/10.1016/j.neuron.2006.02.006.
Israsena, N., Hu, M., Fu, W., Kan, L., Kessler, J.A., 2004. The presence of FGF2 signaling determines whether beta-catenin exerts effects on proliferation or neuronal differentiation of neural stem cells. Dev. Biol. 268, 220–231. http://dx.doi.org/10.1016/j.ydbio.2003.12.024.
Jackman, K., Iadecola, C., 2015. Neurovascular regulation in the ischemic brain. Antioxid. Redox Signal. 22, 149–160. http://dx.doi.org/10.1089/ars.2013.5669.
Kang, S.H., Fukaya, M., Yang, J.K., Rothstein, J.D., Bergles, D.E., 2010. NG2+ CNS glial progenitors remain committed to the oligodendrocyte lineage in postnatal life and following neurodegeneration. Neuron 68, 668–681. http://dx.doi.org/10.1016/j.neuron.2010.09.009.
Karow, M., et al., 2012. Reprogramming of pericyte-derived cells of the adult human brain into induced neuronal cells. Cell Stem Cell 11, 471–476. http://dx.doi.org/10.1016/j.stem.2012.07.007.
Kim, K.H., Son, S.M., Mook-Jung, I., 2013. Contributions of microglia to structural synaptic plasticity. J. Exp. Neurosci. 7, 85–91. http://dx.doi.org/10.4137/JEN.S11269.
Kokovay, E., et al., 2010. Adult SVZ lineage cells home to and leave the vascular niche via differential responses to SDF1/CXCR4 signaling. Cell Stem Cell 7, 163–173. http://dx.doi.org/10.1016/j.stem.2010.05.019. S1934-5909(10)00279-1 [pii].
Kondo, T., Raff, M., 2000. Oligodendrocyte precursor cells reprogrammed to become multipotential CNS stem cells. Science 289, 1754–1757.
Leventhal, C., Rafii, S., Rafii, D., Shahar, A., Goldman, S.A., 1999. Endothelial trophic support of neuronal production and recruitment from the adult mammalian subependyma. Mol. Cell. Neurosci. 13, 450–464. http://dx.doi.org/10.1006/mcne.1999.0762.
Levine, J.M., Reynolds, R., Fawcett, J.W., 2001. The oligodendrocyte precursor cell in health and disease. Trends Neurosci. 24, 39–47. http://dx.doi.org/10.1016/S0166-2236(00)01691-X. S0166-2236(00)01691-X [pii].
Licht, T., Keshet, E., 2013. Delineating multiple functions of VEGF-A in the adult brain. Cell. Mol. Life Sci. 70, 1727–1737. http://dx.doi.org/10.1007/s00018-013-1280-x.
Lo, E.H., Dalkara, T., Moskowitz, M.A., 2003. Mechanisms, challenges and opportunities in stroke. Nat. Rev. Neurosci. 4, 399–415. http://dx.doi.org/10.1038/nrn1106. nrn1106 [pii].
Lo, E.H., Broderick, J.P., Moskowitz, M.A., 2004. tPA and proteolysis in the neurovascular unit. Stroke 35, 354–356. http://dx.doi.org/10.1161/01.str.0000115164.80010.8a.
Lu, Q.R., et al., 2002. Common developmental requirement for Olig function indicates a motor neuron/oligodendrocyte connection. Cell 109, 75–86.
Maharaj, A.S., et al., 2008. VEGF and TGF-beta are required for the maintenance of the choroid plexus and ependyma. J. Exp. Med. 205, 491–501. http://dx.doi.org/10.1084/jem.20072041.
Maki, T., et al., 2015. Potential interactions between pericytes and oligodendrocyte precursor cells in perivascular regions of cerebral white matter. Neurosci. Lett. 597, 164–169. http://dx.doi.org/10.1016/j.neulet.2015.04.047.

Mathiisen, T.M., Lehre, K.P., Danbolt, N.C., Ottersen, O.P., 2010. The perivascular astroglial sheath provides a complete covering of the brain microvessels: an electron microscopic 3D reconstruction. Glia 58, 1094–1103. http://dx.doi.org/10.1002/glia.20990.

Miyamoto, N., et al., 2014. Crosstalk between cerebral endothelium and oligodendrocyte. Cell. Mol. Life Sci. 71, 1055–1066. http://dx.doi.org/10.1007/s00018-013-1488-9.

Morrison, B.M., Lee, Y., Rothstein, J.D., 2013. Oligodendroglia: metabolic supporters of axons. Trends Cell Biol. 23, 644–651. http://dx.doi.org/10.1016/j.tcb.2013.07.007.

Moyon, S., Liang, J., Casaccia, P., 2015 Jun 17. Epigenetics in NG2 glia cells. Brain Res. pii: S0006-8993(15)00480-1, http://dx.doi.org/10.1016/j.brainres.2015.06.009. [Epub ahead of print] Review. PMID: 26092401.

Namihira, M., et al., 2009. Committed neuronal precursors confer astrocytic potential on residual neural precursor cells. Dev. Cell 16, 245–255. http://dx.doi.org/10.1016/j.devcel.2008.12.014.

Nieto, M., Schuurmans, C., Britz, O., Guillemot, F., 2001. Neural bHLH genes control the neuronal versus glial fate decision in cortical progenitors. Neuron 29, 401–413.

Nishiyama, A., Chang, A., Trapp, B.D., 1999. NG2+ glial cells: a novel glial cell population in the adult brain. J. Neuropathol. Exp. Neurol. 58, 1113–1124.

Nishiyama, A., Komitova, M., Suzuki, R., Zhu, X., 2009. Polydendrocytes (NG2 cells): multifunctional cells with lineage plasticity. Nat. Rev. Neurosci. 10, 9–22. http://dx.doi.org/10.1038/nrn2495. nrn2495 [pii].

Obermeier, B., Daneman, R., Ransohoff, R.M., 2013. Development, maintenance and disruption of the blood–brain barrier. Nat. Med. 19, 1584–1596. http://dx.doi.org/10.1038/nm.3407.

Ohab, J.J., Fleming, S., Blesch, A., Carmichael, S.T., 2006. A neurovascular niche for neurogenesis after stroke. J. Neurosci. 26, 13007–13016. http://dx.doi.org/10.1523/JNEUROSCI.4323-06.2006. 26/50/13007 [pii].

Olah, M., et al., 2012. Identification of a microglia phenotype supportive of remyelination. Glia 60, 306–321. http://dx.doi.org/10.1002/glia.21266.

Packer, M.A., et al., 2003. Nitric oxide negatively regulates mammalian adult neurogenesis. Proc. Natl. Acad. Sci. U. S. A. 100, 9566–9571. http://dx.doi.org/10.1073/pnas.1633579100.

Paolicelli, R.C., et al., 2011. Synaptic pruning by microglia is necessary for normal brain development. Science 333, 1456–1458. http://dx.doi.org/10.1126/science.1202529.

Parkhurst, C.N., et al., 2013. Microglia promote learning-dependent synapse formation through brain-derived neurotrophic factor. Cell 155, 1596–1609. http://dx.doi.org/10.1016/j.cell.2013.11.030.

Peters, A., 2004. A fourth type of neuroglial cell in the adult central nervous system. J. Neurocytol. 33, 345–357. http://dx.doi.org/10.1023/B:NEUR.0000044195.64009.27.

Pham, L.D., et al., 2012. Crosstalk between oligodendrocytes and cerebral endothelium contributes to vascular remodeling after white matter injury. Glia 60, 875–881. http://dx.doi.org/10.1002/glia.22320.

Pinto, L., Gotz, M., 2007. Radial glial cell heterogeneity—the source of diverse progeny in the CNS. Prog. Neurobiol. 83, 2–23. http://dx.doi.org/10.1016/j.pneurobio.2007.02.010.

Pocock, J.M., Kettenmann, H., 2007. Neurotransmitter receptors on microglia. Trends Neurosci. 30, 527–535. http://dx.doi.org/10.1016/j.tins.2007.07.007.

Qian, X., et al., 2000. Timing of CNS cell generation: a programmed sequence of neuron and glial cell production from isolated murine cortical stem cells. Neuron 28, 69–80.

Raff, M.C., Miller, R.H., Noble, M., 1983. A glial progenitor cell that develops in vitro into an astrocyte or an oligodendrocyte depending on culture medium. Nature 303, 390–396.

Rasika, S., Alvarez-Buylla, A., Nottebohm, F., 1999. BDNF mediates the effects of testosterone on the survival of new neurons in an adult brain. Neuron 22, 53–62.

Schafer, D.P., et al., 2012. Microglia sculpt postnatal neural circuits in an activity and complement-dependent manner. Neuron 74, 691–705. http://dx.doi.org/10.1016/j.neuron.2012.03.026.

Seo, J.H., et al., 2013. Oligodendrocyte precursors induce early blood–brain barrier opening after white matter injury. J. Clin. Invest. 123, 782–786. http://dx.doi.org/10.1172/JCI65863. 65863 [pii].

Seo, J.H., et al., 2014. Oligodendrocyte precursor cells support blood–brain barrier integrity via TGF-beta signaling. PLoS One 9, e103174. http://dx.doi.org/10.1371/journal.pone.0103174.

Shen, S., et al., 2008. Age-dependent epigenetic control of differentiation inhibitors is critical for remyelination efficiency. Nat. Neurosci. 11, 1024–1034. http://dx.doi.org/10.1038/nn.2172.

Siegenthaler, J.A., Sohet, F., Daneman, R., 2013. "Sealing off the CNS": cellular and molecular regulation of blood–brain barriergenesis. Curr. Opin. Neurobiol. 23, 1057–1064. http://dx.doi.org/10.1016/j.conb.2013.06.006.

Sims, D.E., 1986. The pericyte—a review. Tissue Cell 18, 153–174.

Sun, Y., et al., 2001. Neurogenin promotes neurogenesis and inhibits glial differentiation by independent mechanisms. Cell 104, 365–376.

Taguchi, A., et al., 2004. Administration of CD34+ cells after stroke enhances neurogenesis via angiogenesis in a mouse model. J. Clin. Invest. 114, 330–338. http://dx.doi.org/10.1172/JCI20622.

Tanaka, K., Nogawa, S., Suzuki, S., Dembo, T., Kosakai, A., 2003. Upregulation of oligodendrocyte progenitor cells associated with restoration of mature oligodendrocytes and myelination in peri-infarct area in the rat brain. Brain Res. 989, 172–179.

Thored, P., et al., 2007. Long-term neuroblast migration along blood vessels in an area with transient angiogenesis and increased vascularization after stroke. Stroke 38, 3032–3039. http://dx.doi.org/10.1161/STROKEAHA.107.488445. STROKEAHA.107.488445 [pii].

Torroglosa, A., et al., 2007. Nitric oxide decreases subventricular zone stem cell proliferation by inhibition of epidermal growth factor receptor and phosphoinositide-3-kinase/Akt pathway. Stem Cells 25, 88–97. http://dx.doi.org/10.1634/stemcells.2006-0131.

Urich, E., Lazic, S.E., Molnos, J., Wells, I., Freskgard, P.O., 2012. Transcriptional profiling of human brain endothelial cells reveals key properties crucial for predictive in vitro blood–brain barrier models. PLoS One 7, e38149. http://dx.doi.org/10.1371/journal.pone.0038149. PONE-D-12-03987 [pii].

Wang, M.M., 2011. Notch signaling and Notch signaling modifiers. Int. J. Biochem. Cell Biol. 43, 1550–1562. http://dx.doi.org/10.1016/j.biocel.2011.08.005.

Wang, S., et al., 1998. Notch receptor activation inhibits oligodendrocyte differentiation. Neuron 21, 63–75.

Windrem, M.S., et al., 2004. Fetal and adult human oligodendrocyte progenitor cell isolates myelinate the congenitally dysmyelinated brain. Nat. Med. 10, 93–97. http://dx.doi.org/10.1038/nm974.

Xiao, Q., et al., 2014. Enhancing astrocytic lysosome biogenesis facilitates Abeta clearance and attenuates amyloid plaque pathogenesis. J. Neurosci. 34, 9607–9620. http://dx.doi.org/10.1523/JNEUROSCI.3788-13.2014.

Ye, F., et al., 2009. HDAC1 and HDAC2 regulate oligodendrocyte differentiation by disrupting the beta-catenin-TCF interaction. Nat. Neurosci. 12, 829–838. http://dx.doi.org/10.1038/nn.2333.

Zacchigna, S., Lambrechts, D., Carmeliet, P., 2008. Neurovascular signalling defects in neurodegeneration. Nat. Rev. Neurosci. 9, 169–181. http://dx.doi.org/10.1038/nrn2336.

Zamanian, J.L., et al., 2012. Genomic analysis of reactive astrogliosis. J. Neurosci. 32, 6391–6410. http://dx.doi.org/10.1523/JNEUROSCI.6221-11.2012. 32/18/6391 [pii].

Zechner, D., et al., 2003. beta-Catenin signals regulate cell growth and the balance between progenitor cell expansion and differentiation in the nervous system. Dev. Biol. 258, 406–418.

Zhang, Y., Barres, B.A., 2010. Astrocyte heterogeneity: an underappreciated topic in neurobiology. Curr. Opin. Neurobiol. 20, 588–594. http://dx.doi.org/10.1016/j.conb.2010.06.005. S0959-4388(10)00102-9 [pii].

Zhang, R., et al., 2004. Activated neural stem cells contribute to stroke-induced neurogenesis and neuroblast migration toward the infarct boundary in adult rats. J. Cereb. Blood Flow Metab. 24, 441–448. http://dx.doi.org/10.1097/00004647-200404000-00009.

Zhou, Q., Anderson, D.J., 2002. The bHLH transcription factors OLIG2 and OLIG1 couple neuronal and glial subtype specification. Cell 109, 61–73.

Zhu, X., Bergles, D.E., Nishiyama, A., 2008. NG2 cells generate both oligodendrocytes and gray matter astrocytes. Development 135, 145–157. http://dx.doi.org/10.1242/dev.004895.

Zhu, X., et al., 2012. Olig2-dependent developmental fate switch of NG2 cells. Development 139, 2299–2307. http://dx.doi.org/10.1242/dev.078873.

Zlokovic, B.V., 2008. The blood–brain barrier in health and chronic neurodegenerative disorders. Neuron 57, 178–201. http://dx.doi.org/10.1016/j.neuron.2008.01.003. S0896-6273(08)00034-2 [pii].

Zonta, M., et al., 2003. Neuron-to-astrocyte signaling is central to the dynamic control of brain microcirculation. Nat. Neurosci. 6, 43–50. http://dx.doi.org/10.1038/nn980.

CHAPTER

Development and pathological changes of neurovascular unit regulated by hypoxia response in the retina

9

T. Kurihara[1]

Keio University School of Medicine, Tokyo, Japan

[1]Corresponding author: Tel.: +81-3-5363-3204; Fax: +81-3-5363-3274, e-mail address: kurihara@z8.keio.jp

Abstract

Retina is a highly vascularized tissue with a high oxygen and metabolic demand receiving light located in the back of the eye. The development and the maintenance of the retinal vasculature are important to regulate the homeostasis in the tissue. α Subunits of hypoxia-inducible factor (HIF) are key molecules in hypoxia response inducing genes required for cell survival such as *vascular endothelial growth factor* under hypoxia. Neurons, glia, and vascular endothelium cells interdependently form neurovascular unit in the retina tightly regulated by hypoxia response via HIF expression. A corruption of the precise hypoxia response in the developmental or matured retinal tissue may lead congenital vascular anomalies or adult neovascular ocular diseases. To regulate hypoxia response through HIF activity would be an ideal therapeutic target for these vision-threatening eye diseases.

Keywords

Retina, Hypoxia-inducible factor, Vascular endothelial growth factor, Von Hippel-Lindau protein, Developmental anomaly, Neovascular ocular disease, Persistent hyperplastic primary vitreous

1 INTRODUCTION

The ocular circulatory system has a unique development and pathophysiology, and the corruption of the system induces severe vision-threatening eye diseases. There are mainly three types of ocular circulatory system, namely, the hyaloidal vascular system which only exist basically in utero and two more different adult circulatory

Progress in Brain Research, Volume 225, ISSN 0079-6123, http://dx.doi.org/10.1016/bs.pbr.2016.03.006

systems: the retinal and the choroidal vascular systems. The hyaloidal vascular system regresses dramatically during the embryonic stage and is taken over by the retinal vascular system. Some of congenital ocular diseases such as persistent hyperplastic primary vitreous (PHPV) may occur if the process is once disturbed. Vascular abnormalities in the mature retina are induced by remodeling of vascular components, eg, the retinal vascular system in proliferative diabetic retinopathy and the choroidal vascular system in the wet form of age-related macular degeneration. Recently we have found that ectopic abnormal hypoxia response in the retina largely affects the normal development and the pathophysiology of the retina. Furthermore, neurons, glia, and blood vessels form a neurovascular unit in the retina affecting each other, and these three components play important roles interdependently in normal and disease development. In this chapter, I would like to describe how the retinal neurovascular unit regulated by hypoxia response.

2 HYPOXIA RESPONSE: HYPOXIA-INDUCIBLE FACTOR AND VON HIPPEL-LINDAU PROTEIN

Retinal angioma, which was described by a German ophthalmologist Eugen von Hippel in his literature titled "a very rare disease of the retina" in 1904 (von Hippel, 1904), was lately found as part of autosomal dominant disease that causes hemangioma in the central nervous system and renal cell carcinoma (von Hippel-Lindau (VHL) disease) (Davison et al., 1936; Lindau, 1927). Although the mechanism of the disease development had been unknown for a long time, the discovery of the causative gene (*Vhl tumor suppressor* gene) in 1993 (Latif et al., 1993) made substantial progress of the research. VHL protein (pVHL) encoded by the *Vhl* gene was further revealed as a component of a ubiquitin ligase complex (Maxwell et al., 1999) required for the specific degradation of hypoxia-inducible factor (HIF) discovered at the same period (Semenza and Wang, 1992).

In 1992, Gregg Semenza at Johns Hopkins University discovered a factor which induces the expression of erythropoietin under hypoxia, and named HIF (Semenza and Wang, 1992; Wang and Semenza, 1995). HIF is composed with α and β subunits. In addition to the first discovered α subunit HIF-1α, there exist at least two other α subunits (HIF-2α and HIF-3α) in mammals. In normoxia, particular proline residues in HIF-α are hydroxylated by enzymes called prolyl hydroxylase domain isoforms (Fig. 1). Hydroxylated HIF-α is recognized by the ubiquitin ligase complex-containing pVHL and is degraded in proteasome. On the other hand, HIF-1α is translocated into nucleus escaping from degradation, and forms a heterodimer with β subunit (HIF-1β) expressed constantly in nucleus, and binds to the targeted genes. HIF prompts not only erythropoiesis as an important hypoxia response by the expression of erythropoietin but also regulates angiogenesis as a representative transcriptional factor for vascular endothelial growth factor (VEGF) (Forsythe et al., 1996; Kurihara et al., 2014). Furthermore, more than 800 genes, which are required for cell

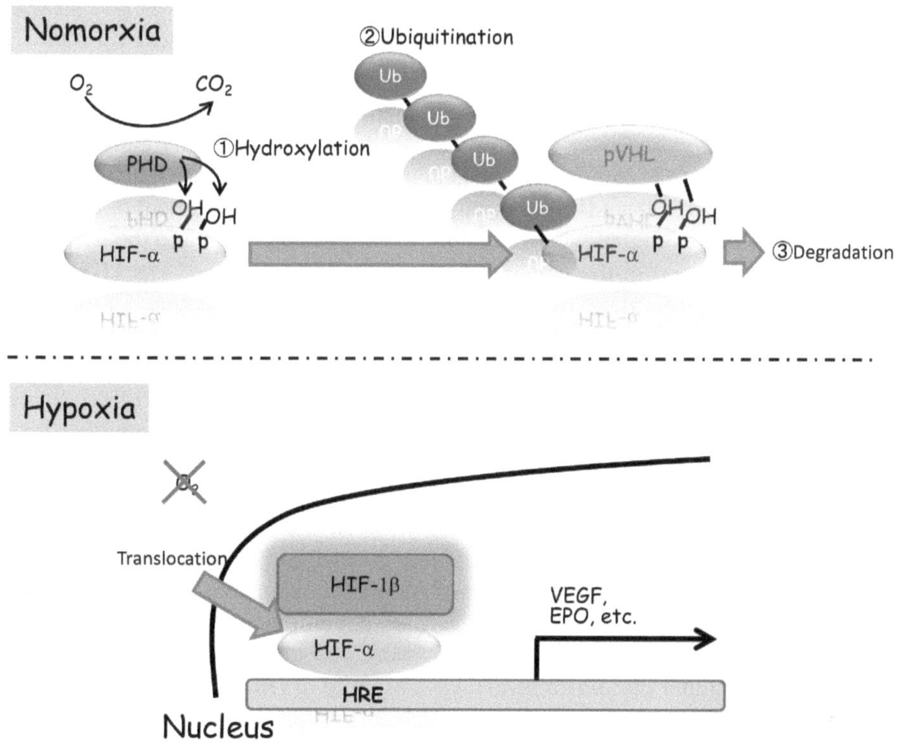

FIG. 1

Regulatory mechanism of hypoxia-inducible factor (HIF). (*Upper panel*) In normoxia, specific proline residues in HIF-α are hydroxylated ①. Hydroxylated HIF-α is multiubiquitinated by ubiquitin ligase complex-containing VHL protein (pVHL) ②. Ubiquitinated HIF-α is degraded in proteasome ③. (*Lower panel*) Under hypoxia, HIF-α is translocated into nucleus, forms a heterodimer with β subunit, and binds to the hypoxia-responsive element (HRE) in the targeted genes.

survival under hypoxia such as cell proliferation, metabolism, and inflammation, are regulated by HIF (Schodel et al., 2011).

HIF transcriptional factor responds to hypoxia using the way of protein degradation to avoid critical situations such as hypoxia immediately. Thus, the HIF-α subunits are always generated and degraded in usual situations. VHL patients carry a heterozygous mutation in the *Vhl* gene and get a mutation in another allele in the cells of some particular tissues inducing constitutive activation of HIF regardless of the oxygen concentration. As a result, angiomas and carcinomas are caused by the induction of targeted genes such as VEGF and the oncogene c-Myc. The etiology of VHL disease was elucidated by the progress of molecular cell biology almost 100 years after its first description.

3 DEVELOPMENT OF THE OCULAR CIRCULATORY SYSTEM

The similar distribution of nervous and vascular system was described by a medieval anatomist Andreas Vesalius in his book De humani corporis fabrica (Vesalius, 1543). In fact, the retinal vasculature distribute along the axons of retinal ganglion cells spread radially from the optic nerve head (Fig. 2). The vasculature extends toward the distal hypoxic area from the proximal side during the establishment of our body. The process of extending retinal vasculature toward avascular area in the peripheral retina can be observed in the ocular fundus of a premature baby. The HIF-VEGF pathway plays an important role in the physiological developmental process of the retinal vasculature. Murine retinae are often used for the experiment of physiological angiogenesis in vivo because the retinal vasculature in rodents unlike in human starts to extend after birth. The *Vegf* gene detected by in situ hybridization is strongly expressed in the avascular area of the murine retina immediate after birth (Fig. 3). In contrast, a lower expression of the gene is observed in the area covered by the vasculature (Kubota et al., 2008; Kurihara et al., 2010). To investigate the regulatory system, we generated retinal neuron-specific *Hif-1α*-deficient mice showing a decreased number of endothelial tip cells located at the edge of the spreading vasculature in the inner retinal layer and of filopodial extension from these tip cells which results a delay of the retinal vascular development (Lang et al., 1994; Nakamura-Ishizu et al., 2012). Moreover, a sparse vascular plexus in the intermediate retinal layer (inner plexiform layer) was observed in *Hif-1α* knockout mice specifically in amacrine and horizontal cells which are neurons located in the layer (Usui et al., 2015). On the other hand, *Vegf*-, *Hif-1α*-, or *Hif-2α*-specific deletion in astrocyte glia, which forms a template for the retinal vasculature in the inner layer, induced no phenotypic change of the retinal vascular development (Weidemann

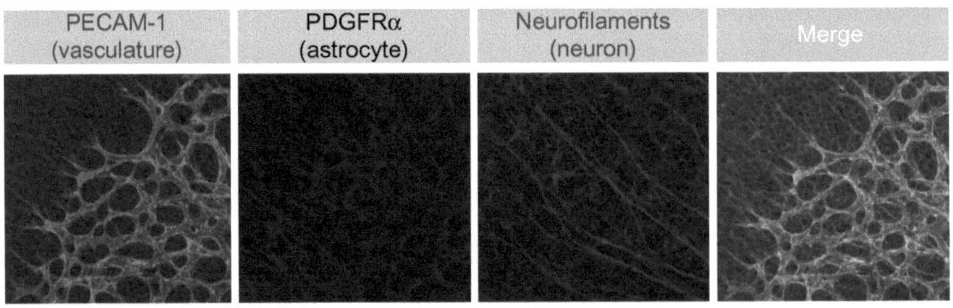

FIG. 2

Neurovascular unit in the retina. The retinal vasculature (*green* (*gray* in the print version)) extends radially toward avascular area from the optic nerve head on the astrocytic template (*blue* (*black* in the print version)) along the axons of retinal ganglion cells (*red* (*dark gray* in the print version)). Wholemount immunostaining for PECAM-1, PDGFRα, and neurofilaments of postnatal day 6 murine retina.

Courtesy of Prof. Yoshiaki Kubota, Keio University School of Medicine.

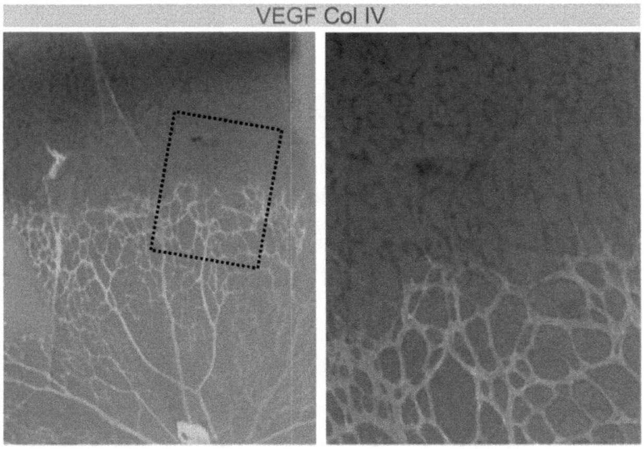

FIG. 3

Vascular extension in the developing murine retina and VEGF expression. *Vegf* mRNA expression (*purple* (dark gray in the print version)) detected by in situ hybridization and the retinal vasculature (*green* (light gray in the print version), Collagen IV) in the inner retinal layer of a postnatal day 6 mouse. Note that *Vegf* expression is high in the avascular area, whereas the expression is low in the area covered by the vasculature. The *dashed rectangle* in the *left panel* is magnified to the *right panel.*

Reprinted with permission from Kurihara, T., Kubota, Y., Ozawa, Y., Takubo, K., Noda, K., Simon, M.C., Johnson, R.S., Suematsu, M., Tsubota, K., Ishida, S., et al., 2010. von Hippel-Lindau protein regulates transition from the fetal to the adult circulatory system in retina. Development 137, 1563–1571.

et al., 2010). Thus, hypoxia response mainly in neurons is important during the physiological retinal vascular development inducing appropriate *Vegf* expression regulated by HIF in the avascular area.

Another important process during the retinal development is the regression of the hyaloidal vasculature. The hyaloidal vascular system, which provides blood and oxygen supply during the embryonic stage, starts to regress by the beginning of the retinal vascular extension from optic nerve head. The regression process completes till the seventh month of pregnancy in human. The failure of the transition from embryonic circulation (hyaloidal vascular system) to adult (retinal vascular system) may result in a persistence of the embryonic hyaloidal vasculature in the eye called PHPV suppressing normal development of the other tissues in the eye and causing severe visual disturbance. We confirmed that decreased HIF-α protein expression in retinal cells during normal developmental process in the murine retina along with increasing oxygen tension in the tissue (Kurihara et al., 2010) (Fig. 4). Therefore, we generated retinal neuron- or astrocyte-specific *Vhl* knockout mice revealing phenotypes similar to human PHPV because HIF-α degradation and proper hypoxia response is disturbed (Kurihara et al., 2010, 2011). This indicates that precise hypoxia response in neurons and glia as components of neurovascular unit during retinal development is important for the ocular circulatory transition from the embryo to the adult.

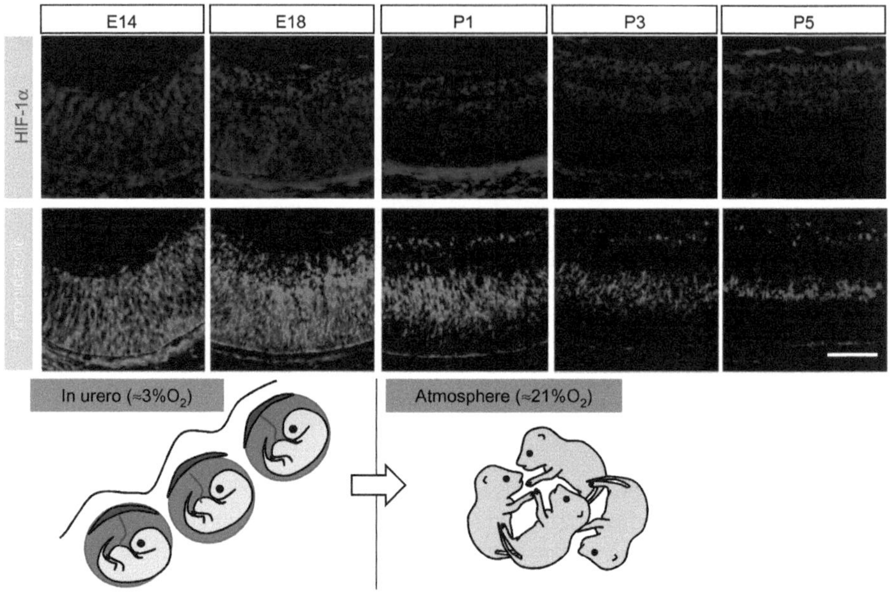

FIG. 4

The change of HIF-1α protein expression before and after birth in the murine retina. The oxygen tension in utero is markedly low compared to that in atomorsphere. Retinal HIF-1α expression is rapidly decreased after birth. The expression pattern of HIF-1α matches the distribution of hypoxia probe (pimonidazole) taken into the hypoxic tissues. Scale bar: 100 μm.

Reprinted with permission from Kurihara, T., Kubota, Y., Ozawa, Y., Takubo, K., Noda, K., Simon, M.C., Johnson, R.S., Suematsu, M., Tsubota, K., Ishida, S., et al., 2010. von Hippel-Lindau protein regulates transition from the fetal to the adult circulatory system in retina. Development 137, 1563–1571.

4 PATHOPHYSIOLOGY OF RETINAL DISEASES AND HYPOXIA RESPONSE

The end stages of ischemic retinal diseases such as diabetic retinopathy, retinopathy of prematurity, and retinal vein occlusion are led by pathological neovascularization. Although a decrease of visual function is observed in early stages of diabetic retinopathy (Kurihara et al., 2008), tractional retinal detachment or neovascular glaucoma associated with proliferative retinopathy leads blindness. The wet form of age-related macular degeneration is also a representative neovascular eye disease as a leading cause of blindness. Recently, various animal models of ocular neovascularization revealed roles of VEGF in pathological neovascularization (Aiello et al., 1995; Krzystolik et al., 2002; Miller et al., 1994; Pierce et al., 1995). To date, four types of anti-VEGF reagents are clinically available against these neovascular ocular diseases (Gragoudas et al., 2004; Rosenfeld et al., 2006; Takeda et al., 2007). On the other hand, retinal photocoagulation has been established as an effective therapy to suppress the progress of these diseases since decades ago. The therapeutic mechanism is thought to be eliminations of neurons producing angiogenic factor including VEGF. In this matter, the role of VEGF in neovascularization and vascular maintenance is very important in the ophthalmological field (Aiello et al., 1994; Kurihara et al., 2012, 2013). As a machinery to recognize ischemic conditions in retinal cells and to produce angiogenic factors, pathologic hypoxia response can be speculated. We generated astrocyte or retinal pigment epithelium (RPE) cell-specific *Hif-2α* knockout mice, and found a significant suppression of pathological retinal and choroidal angiogenesis in these mice as same as conditional *Vegf*-deficient mice (Kurihara et al., 2012; Weidemann et al., 2010) (Fig. 5). Surprisingly, the conditional RPE-specific *Vegf* knockout mice showed loss of choriocapillaris and cone photoreceptors rapidly after this gene deletion, whereas *Hif-1α*- and/or *Hif-2α*-deficient mice had no phenotype (Kurihara et al., 2012). This strongly suggests that HIF-αs are not needed for the retinal homeostasis because these molecules do not exist in normoxia

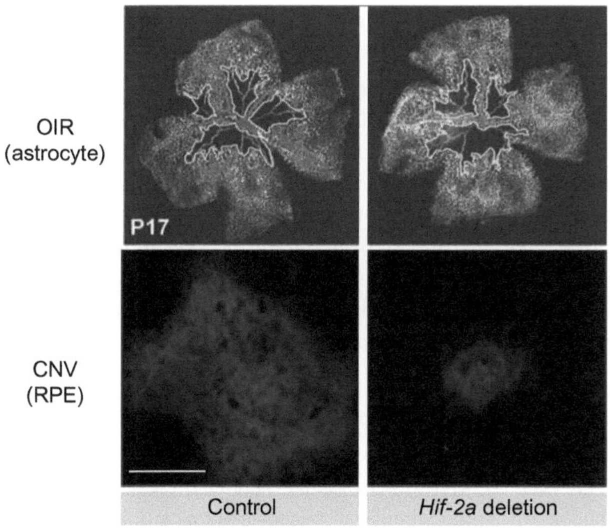

FIG. 5

Pathological angiogenesis suppressed by *Hif-2α* deletion. (*Upper panel*) Pathological angiogenesis (*red* (light gray in the print version)) in oxygen-induced retinopathy model is reduced in the astrocyte-specific *Hif-2α* knockout mice. (*Lower panel*) The volume of the laser-induced choroidal neovascularization (CNV) is decreased in retinal pigment epithelium (RPE)-specific *Hif-2α* knockout mice. Scale bar: 100 μm.

Reprinted with permission from Weidemann, A., Krohne, T.U., Aguilar, E., Kurihara, T., Takeda, N., Dorrell, M.I., Simon, M.C., Haase, V.H., Friedlander, M., Johnson, R.S., 2010. Astrocyte hypoxic response is essential for pathological but not developmental angiogenesis of the retina. Glia 58, 1177–1185; Kurihara, T., Westenskow, P.D., Bravo, S., Aguilar, E., Friedlander, M., 2012. Targeted deletion of Vegfa in adult mice induces vision loss. J. Clin. Invest. 122, 4213–4217.

after the tissue development although the expression of VEGF in normal states is required for not only vascular maintenance but also neuronal activity. In fact, atrophies of photoreceptors and RPE have been observed during long-term anti-VEGF treatment in multiple large clinical study cohorts (Martin et al., 2012; Rosenfeld et al., 2011). To regulate the pathologic hypoxia response considering the pathophysiology, HIF would be an ideal therapeutic target for these neovascular ocular diseases.

5 CONCLUSIONS

The system of hypoxia response has been established to utilize oxygen for our biological activity. During the tissue development, appropriate vascular network is constructed using this hypoxia response. In the process, HIF in neurons, astrocytes, or RPE cells largely affect the phenotypes of vascular development playing important roles in forming and maintenance of neurovascular unit. Most of animals from nematodes to mammals rapidly respond against hypoxia as a critical situation via HIF at cellular and tissue levels. Lately, HIF is activated not only by hypoxia but also in response to various stresses such as inflammation and starvation. It is called "pseudo-hypoxia" because of hypoxia-like responses despite oxygen availability. For example, it has been reported that inactivation of sirtuins, which are related to the longevity, induces activation of HIF leading to pseudo-hypoxia in some age-related diseases (Finley et al., 2011; Gomes et al., 2013; Sebastian et al., 2012; Zhong et al., 2010). Diabetic retinopathy and age-related macular degeneration among ocular diseases and malignant tumors are neovascular diseases which can be induced by a corruption of hypoxia response associated with aging. Therefore, regulation of hypoxia response will be a potentially important therapeutic target for ocular diseases and others in the near future.

ACKNOWLEDGMENTS

I thank Drs. Kazuo Tsubota, Martin Friedlander, Hideyuki Okano, Susumu Ishida, Yoshiaki Kubota, and Yoko Ozawa for their mentorship in the original articles cited in this chapter. T.K. is supported by MEXT/JSPS KAKENHI Grant Number 15K10881, Charitable Trust Fund for Ophthalmic Research in Commemoration of Santen Pharmaceutical's Founder, The Uehara Memorial Foundation, and The Takeda Science Foundation.

REFERENCES

Aiello, L.P., Avery, R.L., Arrigg, P.G., Keyt, B.A., Jampel, H.D., Shah, S.T., Pasquale, L.R., Thieme, H., Iwamoto, M.A., Park, J.E., et al., 1994. Vascular endothelial growth factor in ocular fluid of patients with diabetic retinopathy and other retinal disorders. N. Engl. J. Med. 331, 1480–1487.

Aiello, L.P., Pierce, E.A., Foley, E.D., Takagi, H., Chen, H., Riddle, L., Ferrara, N., King, G.L., Smith, L.E., 1995. Suppression of retinal neovascularization in vivo by inhibition of vascular endothelial growth factor (VEGF) using soluble VEGF-receptor chimeric proteins. Proc. Natl. Acad. Sci. U.S.A. 92, 10457–10461.

Davison, C., Brock, S., Dyke, C.G., 1936. Retinal and central nervous hemangioblastomatosis with visceral changes (von Hippel-Lindau's disease). Bull. Neurol. Inst. N.Y. 5, 72–93.

Finley, L.W., Carracedo, A., Lee, J., Souza, A., Egia, A., Zhang, J., Teruya-Feldstein, J., Moreira, P.I., Cardoso, S.M., Clish, C.B., et al., 2011. SIRT3 opposes reprogramming of cancer cell metabolism through HIF1alpha destabilization. Cancer Cell 19, 416–428.

Forsythe, J.A., Jiang, B.H., Iyer, N.V., Agani, F., Leung, S.W., Koos, R.D., Semenza, G.L., 1996. Activation of vascular endothelial growth factor gene transcription by hypoxia-inducible factor 1. Mol. Cell. Biol. 16, 4604–4613.

Gomes, A.P., Price, N.L., Ling, A.J., Moslehi, J.J., Montgomery, M.K., Rajman, L., White, J.P., Teodoro, J.S., Wrann, C.D., Hubbard, B.P., et al., 2013. Declining NAD(+) induces a pseudohypoxic state disrupting nuclear-mitochondrial communication during aging. Cell 155, 1624–1638.

Gragoudas, E.S., Adamis, A.P., Cunningham Jr., E.T., Feinsod, M., Guyer, D.R., 2004. Pegaptanib for neovascular age-related macular degeneration. N. Engl. J. Med. 351, 2805–2816.

Krzystolik, M.G., Afshari, M.A., Adamis, A.P., Gaudreault, J., Gragoudas, E.S., Michaud, N.A., Li, W., Connolly, E., O'Neill, C.A., Miller, J.W., 2002. Prevention of experimental choroidal neovascularization with intravitreal anti-vascular endothelial growth factor antibody fragment. Arch. Ophthalmol. 120, 338–346.

Kubota, Y., Hirashima, M., Kishi, K., Stewart, C.L., Suda, T., 2008. Leukemia inhibitory factor regulates microvessel density by modulating oxygen-dependent VEGF expression in mice. J. Clin. Invest. 118, 2393–2403.

Kurihara, T., Ozawa, Y., Nagai, N., Shinoda, K., Noda, K., Imamura, Y., Tsubota, K., Okano, H., Oike, Y., Ishida, S., 2008. Angiotensin II type 1 receptor signaling contributes to synaptophysin degradation and neuronal dysfunction in the diabetic retina. Diabetes 57, 2191–2198.

Kurihara, T., Kubota, Y., Ozawa, Y., Takubo, K., Noda, K., Simon, M.C., Johnson, R.S., Suematsu, M., Tsubota, K., Ishida, S., et al., 2010. von Hippel-Lindau protein regulates transition from the fetal to the adult circulatory system in retina. Development 137, 1563–1571.

Kurihara, T., Westenskow, P.D., Krohne, T.U., Aguilar, E., Johnson, R.S., Friedlander, M., 2011. Astrocyte pVHL and HIF-alpha isoforms are required for embryonic-to-adult vascular transition in the eye. J. Cell Biol. 195, 689–701.

Kurihara, T., Westenskow, P.D., Bravo, S., Aguilar, E., Friedlander, M., 2012. Targeted deletion of Vegfa in adult mice induces vision loss. J. Clin. Invest. 122, 4213–4217.

Kurihara, T., Westenskow, P.D., Aguilar, E., Friedlander, M., 2013. VEGF antagonism and age-related macular degeneration: too much of a good thing? Expert Rev. Ophthalmol. 8, 103–105.

Kurihara, T., Westenskow, P.D., Friedlander, M., 2014. Hypoxia-inducible factor (HIF)/vascular endothelial growth factor (VEGF) signaling in the retina. Adv. Exp. Med. Biol. 801, 275–281.

Lang, R., Lustig, M., Francois, F., Sellinger, M., Plesken, H., 1994. Apoptosis during macrophage-dependent ocular tissue remodelling. Development 120, 3395–3403.

Latif, F., Tory, K., Gnarra, J., Yao, M., Duh, F.M., Orcutt, M.L., Stackhouse, T., Kuzmin, I., Modi, W., Geil, L., et al., 1993. Identification of the von Hippel-Lindau disease tumor suppressor gene. Science 260, 1317–1320.

Lindau, A., 1927. Zur Frage der Angiomatosis Retinae und Ihrer Hirncomplikation. Acta Ophthalmol. 4, 193–226.

Martin, D.F., Maguire, M.G., Fine, S.L., Ying, G.S., Jaffe, G.J., Grunwald, J.E., Toth, C., Redford, M., Ferris 3rd., F.L., 2012. Ranibizumab and bevacizumab for treatment of neovascular age-related macular degeneration: two-year results. Ophthalmology 119, 1388–1398.

Maxwell, P.H., Wiesener, M.S., Chang, G.W., Clifford, S.C., Vaux, E.C., Cockman, M.E., Wykoff, C.C., Pugh, C.W., Maher, E.R., Ratcliffe, P.J., 1999. The tumour suppressor protein VHL targets hypoxia-inducible factors for oxygen-dependent proteolysis. Nature 399, 271–275.

Miller, J.W., Adamis, A.P., Shima, D.T., D'Amore, P.A., Moulton, R.S., O'Reilly, M.S., Folkman, J., Dvorak, H.F., Brown, L.F., Berse, B., et al., 1994. Vascular endothelial growth factor/vascular permeability factor is temporally and spatially correlated with ocular angiogenesis in a primate model. Am. J. Pathol. 145, 574–584.

Nakamura-Ishizu, A., Kurihara, T., Okuno, Y., Ozawa, Y., Kishi, K., Goda, N., Tsubota, K., Okano, H., Suda, T., Kubota, Y., 2012. The formation of an angiogenic astrocyte template is regulated by the neuroretina in a HIF-1-dependent manner. Dev. Biol. 363, 106–114.

Pierce, E.A., Avery, R.L., Foley, E.D., Aiello, L.P., Smith, L.E., 1995. Vascular endothelial growth factor/vascular permeability factor expression in a mouse model of retinal neovascularization. Proc. Natl. Acad. Sci. U.S.A. 92, 905–909.

Rosenfeld, P.J., Brown, D.M., Heier, J.S., Boyer, D.S., Kaiser, P.K., Chung, C.Y., Kim, R.Y., 2006. Ranibizumab for neovascular age-related macular degeneration. N. Engl. J. Med. 355, 1419–1431.

Rosenfeld, P.J., Shapiro, H., Tuomi, L., Webster, M., Elledge, J., Blodi, B., 2011. Characteristics of patients losing vision after 2 years of monthly dosing in the phase III ranibizumab clinical trials. Ophthalmology 118, 523–530.

Schodel, J., Oikonomopoulos, S., Ragoussis, J., Pugh, C.W., Ratcliffe, P.J., Mole, D.R., 2011. High-resolution genome-wide mapping of HIF-binding sites by ChIP-seq. Blood 117, e207–e217.

Sebastian, C., Zwaans, B.M., Silberman, D.M., Gymrek, M., Goren, A., Zhong, L., Ram, O., Truelove, J., Guimaraes, A.R., Toiber, D., et al., 2012. The histone deacetylase SIRT6 is a tumor suppressor that controls cancer metabolism. Cell 151, 1185–1199.

Semenza, G.L., Wang, G.L., 1992. A nuclear factor induced by hypoxia via de novo protein synthesis binds to the human erythropoietin gene enhancer at a site required for transcriptional activation. Mol. Cell. Biol. 12, 5447–5454.

Takeda, A.L., Colquitt, J., Clegg, A.J., Jones, J., 2007. Pegaptanib and ranibizumab for neovascular age-related macular degeneration: a systematic review. Br. J. Ophthalmol. 91, 1177–1182.

Usui, Y., Westenskow, P.D., Kurihara, T., Aguilar, E., Sakimoto, S., Paris, L.P., Wittgrove, C., Feitelberg, D., Friedlander, M.S., Moreno, S.K., et al., 2015. Neurovascular crosstalk between interneurons and capillaries is required for vision. J. Clin. Invest. 125, 2335–2346.

Vesalius, A., 1543. De humani corporis fabrica (Basel).

von Hippel, E., 1904. Ueber eine sehr seltene Erkrankung der Netzhaut. Albrecht Von Graefes Arch. Ophthalmol. 59, 83–106.

Wang, G.L., Semenza, G.L., 1995. Purification and characterization of hypoxia-inducible factor 1. J. Biol. Chem. 270, 1230–1237.

Weidemann, A., Krohne, T.U., Aguilar, E., Kurihara, T., Takeda, N., Dorrell, M.I., Simon, M.C., Haase, V.H., Friedlander, M., Johnson, R.S., 2010. Astrocyte hypoxic response is essential for pathological but not developmental angiogenesis of the retina. Glia 58, 1177–1185.

Zhong, L., D'Urso, A., Toiber, D., Sebastian, C., Henry, R.E., Vadysirisack, D.D., Guimaraes, A., Marinelli, B., Wikstrom, J.D., Nir, T., et al., 2010. The histone deacetylase Sirt6 regulates glucose homeostasis via Hif1alpha. Cell 140, 280–293.

CHAPTER

Neurovascular coupling and energy metabolism in the developing brain

10

M. Kozberg[*,1], **E. Hillman**[*,†,‡,1]

*Columbia University, New York, NY, United States

†Kavli Institute for Brain Science, Columbia University, New York, NY, United States

‡Mortimer B. Zuckerman Institute for Mind Brain and Behavior, Columbia University, New York, NY, United States

[1]Corresponding authors: Tel.: +1-212-8542788; Fax: +1-212-8548725, e-mail address: mgk2120@cumc.columbia.edu; eh2245@columbia.edu

Abstract

In the adult brain, increases in local neural activity are almost always accompanied by increases in local blood flow. However, many functional imaging studies of the newborn and developing human brain have observed patterns of hemodynamic responses that differ from adult responses. Among the proposed mechanisms for the observed variations is that neurovascular coupling itself is still developing in the perinatal brain. Many of the components thought to be involved in actuating and propagating this hemodynamic response are known to still be developing postnatally, including perivascular cells such as astrocytes and pericytes. Both neural and vascular networks expand and are then selectively pruned over the first year of human life. Additionally, the metabolic demands of the newborn brain are still evolving. These changes are highly likely to affect early postnatal neurovascular coupling, and thus may affect functional imaging signals in this age group. This chapter will discuss the literature relating to neurovascular development. Potential effects of normal and aberrant development of neurovascular coupling on the newborn brain will also be explored, as well as ways to effectively utilize imaging techniques that rely on hemodynamic modulation such as fMRI and NIRS in younger populations.

Keywords

Neurovascular coupling, fMRI, Optical imaging, Brain development, Brain metabolism

1 INTRODUCTION

In the adult brain, increases in local neural activity result in increases in local blood flow to the activated region (Attwell and Iadecola, 2002; Heeger and Ress, 2002).

Progress in Brain Research, Volume 225, ISSN 0079-6123, http://dx.doi.org/10.1016/bs.pbr.2016.02.002

This influx of oxygenated, arterial blood far exceeds the consumption of oxygen by the tissue, leading to localized increases in hemoglobin oxygenation. The functional magnetic resonance imaging (fMRI) blood-oxygen-level-dependent (BOLD) signal reports these oxygenation changes, representing decreases in local deoxygenated hemoglobin (HbR) concentration as "positive BOLD"—enabling the noninvasive imaging of a proxy of neural activity (see also chapter "Submillimeter-resolution fMRI: Toward understanding local neural processing" by M. Fukuda et al.).

While this process of neurovascular coupling is relatively consistent in the healthy adult brain, key differences have been observed in functional imaging studies of the early postnatal brain in both humans (Arichi et al., 2010; Born et al., 1998; Meek et al., 1998; Yamada et al., 1997) and rodent models (Colonnese et al., 2008; Kozberg et al., 2013; Zehendner et al., 2013). Common observations include a "negative BOLD" response inverted with respect to the adult response (Born et al., 1998; Yamada et al., 1997), a reduced amplitude BOLD response (Arichi et al., 2012), and a biphasic response that gradually becomes entirely positive with progression to adulthood (Kozberg et al., 2013; Zehendner et al., 2013). These variations in response patterns may represent the progressive maturation of neurovascular coupling itself (Arichi et al., 2012; Hillman and Kozberg, 2013), since many components of neurovascular unit are still developing postnatally (Harris et al., 2011). The presence of seemingly conflicting results may correspond to inter-study differences in stimulation and/or imaging paradigms. These accumulated findings have complicated the interpretation of studies of early brain development and limited the utility of functional imaging as a tool to diagnose and assess abnormal brain development.

The postnatal brain is a highly dynamic system undergoing radical changes in neural connectivity and function that extend into childhood and adolescence. The possibility that neurovascular coupling itself may be developing alongside neuronal development in the postnatal brain suggests that in this developmental period the brain may be experiencing differences in energy supply and demand dynamics compared to the adult brain. Understanding these neurovascular and neurometabolic changes could provide new insights into the medical management of infants and children. Furthermore, substantial vascular development takes place during this neural network expansion, and coordination of development between these two systems may be a critical and potentially vulnerable aspect of brain development. Although developmental differences in early neurovascular coupling may confound conventional interpretation of fMRI BOLD-based studies, neurovascular development could provide a biomarker for assessment of normal or abnormal developmental trajectories that could be directly detectable using functional imaging.

This chapter will review postnatal changes in neurovascular responses, the neural and vascular architecture of the brain, the complete neurovascular unit, and how these components relate to one another in the context of brain function, brain development, disease and functional imaging signals.

2 FUNCTIONAL IMAGING OF THE DEVELOPING BRAIN

2.1 IMAGING TECHNIQUES

The two major imaging technologies used to examine postnatal development in the human brain are fMRI BOLD and near-infrared spectroscopy (NIRS; Graham et al., 2014). As mentioned earlier, fMRI BOLD is a measure of the levels of HbR; decreases in HbR lead to an increase in signal which is termed "positive BOLD" and typically interpreted as an increase in neural activity. Thus it is important to note that positive BOLD does not reflect oxygen consumption by the tissue but rather a robust oversupply of oxygenated blood (functional hyperemia). NIRS takes advantage of the different optical absorption spectra of oxygenated hemoglobin (HbO) and HbR to measure relative changes in each across the cortex using different wavelengths of near-infrared light (see also chapter "Hemodynamic signals in fNIRS" by Y. Hoshi). This enables the assessment of both blood oxygenation and volume changes. NIRS is particularly well suited for imaging in the newborn brain because of the thinness of the newborn skull compared to the adult and ease of acquisition in newborns at the bedside (Ferradal et al., 2015; Liao et al., 2010).

2.2 HUMAN STUDIES OF INFANT HEMODYNAMIC RESPONSES

There has been a great deal of variability in the hemodynamic responses reported in the newborn brain (see Table 1). Human studies have broadly demonstrated both negative and positive fMRI BOLD responses to stimulation (Anderson et al., 2001; Arichi et al., 2012; Born et al., 1996; Yamada et al., 1997). Researchers utilizing NIRS have observed adult-like responses (Fig. 1A), as well as two types of "negative BOLD"

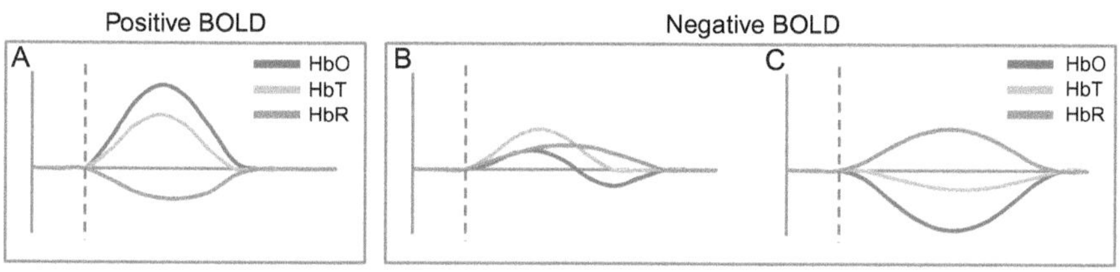

FIG. 1

Diagrams of different types of hemodynamic response to stimulus reported in the developing human brain. (A) Classic adult-like hemodynamic response with an increase in oxygenated (HbO) and total hemoglobin (HbT) and a corresponding decrease in deoxygenated hemoglobin (HbR). This response would appear as "positive BOLD" in fMRI. (B and C) Responses that would both be represented by "negative BOLD" in fMRI. (B) Pattern suggests that functional hyperemia is insufficient to meet oxygen demands of the tissue, resulting in an increase in HbT, HbO, and HbR. (C) A decrease in HbT accompanies an increase in HbR suggesting vasoconstriction amplifying increases in HbR.

Table 1 Hemodynamic Responses Observed in Human Postnatal Development Using fMRI and NIRS

Positive BOLD/Decrease in Deoxygenated Hemoglobin

Stimulation	Study Details	Age	fMRI/ NIRS?	Health Status of Infants	Anesthesia/State	↑ Or ↓ in [HbT]?
Visual	Yamada et al. (1997)	<5 weeks	fMRI	Some preterm, all healthy	Pentobarbital sedation	
	Born et al. (1998, 2000)	<48 weeks	fMRI	Preterm and term, some with visual impairments	Spontaneous sleep or chloral hydrate	
	Karen et al. (2008)	2–10 days	NIRS	All healthy	Spontaneous sleep	Increase
	Liao et al. (2010)	2 days	NIRS	All healthy	Spontaneous sleep	Increase
Auditory	Anderson et al. (2001)	40–50 weeks	fMRI	Some preterm, all healthy	Awake	
Somatosensory	Arichi et al. (2010, 2012)	0–2 weeks	fMRI	Some preterm, all healthy	Awake	

Negative BOLD/Increase in Deoxygenated Hemoglobin

Stimulation	Study Details	Age	fMRI/NIRS?	Health Status of Infants	Anesthesia/State	↑ Or ↓ in [HbT]?
Visual	Yamada et al. (1997)	>5 weeks	fMRI	Some preterm, all healthy	Pentobarbital sedation	
	Born et al. (1998, 2000)	>56 weeks	fMRI	Preterm and term, some with visual impairments	Spontaneous sleep or chloral hydrate	
	Sie et al. (2001)	~18 months	fMRI	Healthy and with periventricular leukomalacia	Promethazine, pethidine, droperidol, diazepam	
	Born et al. (2002)	4–71 months	fMRI BOLD and perfusion	Three healthy, one with cerebral palsy	Chloral hydrate	Decrease (perfusion MRI)
	Meek et al. (1998)	0–14 weeks	NIRS	All healthy	Awake	Increase
	Kusaka et al. (2004)	4–16 weeks	NIRS			Decrease

Auditory	Anderson et al. (2001)	>50 weeks	fMRI	Some preterm, all healthy	Awake	
	Sakatani et al. (1999)	3 days	NIRS	Some preterm, all healthy	Awake	Increase
	Zaramella et al. (2001)	0–7 weeks	NIRS	Admitted for a variety of diagnoses	Awake and spontaneous sleep	Increase
Somatosensory	Erberich et al. (2006)	38–46 weeks	fMRI	Some preterm, all healthy	Isoflurane	
	Heep et al. (2009)	~39 weeks postconceptional age	fMRI	Preterm	Chloral hydrate	

Variable BOLD/Inconsistent Decreases vs Increases in Deoxygenated Hemoglobin

Stimulation	Study Details	Age	fMRI/ NIRS?	Health Status of Infants	Anesthesia/State	Details
Visual	Martin et al. (1999)	1 month to 1 year	fMRI	All healthy	Chloral hydrate, pentobarbital, halothane	<1 month: 30% pos, 30% neg; 1 month to 1 year: 27% pos, 27% neg; 1–6 years: 71% pos, 14% neg
	Hoshi et al. (2000)	4–5 days	NIRS	All healthy	Spontaneous sleep	Increase in HbO and HbT, variable HbR responses

response (informed by simultaneous imaging of HbO): (1) an increase in HbT, HbR, and HbO, suggesting insufficient hyperemia to meet oxygen consumption demands (Fig. 1B) and (2) A decrease in HbT and HbO with an increase in HbR, suggesting local vasoconstriction exacerbating oxygen consumption needs (Fig. 1C).

2.2.1 Observations of negative BOLD responses in the infant brain

Early fMRI studies performed in human newborns identified a negative BOLD response to visual stimulation, hypothesizing that these negative responses were due to oxygen consumption in the presence of minimal functional hyperemia (Born et al., 1996; Joeri et al., 1996). These studies were followed by work from Yamada et al. in 1997, which tracked the progression of fMRI responses to visual stimulation in anesthetized newborns with age (Yamada et al., 1997). They outlined two different responses: in infants under 5 weeks of age, they observed positive BOLD responses, while they observed negative BOLD responses in infants over 5 weeks of age. They attributed this variability to a postnatal increase in cerebral metabolism (see Section 4), hypothesizing that because of low oxygen metabolism rates in younger newborns (Chugani and Phelps, 1986) and a low synaptic density (Huttenlocher et al., 1982), a relatively small increase in blood volume (in comparison to the adult brain) is able to oversupply activated regions in the newborn brain, resulting in a net decrease in the levels of HbR. It was reasoned that older newborns, with a higher cerebral metabolic rate of oxygen ($CMRO_2$) and increased synaptic activity, would use more of the available oxygen, leading to negative BOLD even in the presence of local hyperemia (Muramoto et al., 2002). Subsequently, a number of other groups utilizing fMRI observed negative BOLD responses to visual (Born et al., 1998, 2000; Martin et al., 1999; Sie et al., 2001), auditory (Anderson et al., 2001), and somatosensory stimuli (Erberich et al., 2006) in infants (see Table 1 for details). Studies have also demonstrated negative BOLD responses to somatosensory stimuli in preterm infants (Heep et al., 2009). Since BOLD reports only changes in HbR, negative BOLD responses can be ambiguous and could represent either case shown in Fig. 1B or C. However, a study utilizing perfusion MRI (FAIR) reported concomitant decreases in blood flow in the newborn brain in response to visual stimulation (Born et al., 2002).

Meek et al. published the first hemodynamic study of the awake newborn brain using NIRS (Meek et al., 1998). Their custom NIRS system allowed for infants to sit on examiners' laps and tolerated small amounts of motion, precluding the need for anesthesia or sleeping. Subjects were presented with blocks of visual stimulation. Resulting increases were observed in *both* HbO and HbR (Fig. 1B), indicating an increase in oxygen supply insufficient to meet the oxygen demand of the newborn brain. Their subjects ranged in age from 3 to 14 weeks, aligning relatively well with the Yamada et al.'s negative BOLD age group (5 weeks+), but also suggesting that negative BOLD may also be observed at younger ages. Several groups have supported these findings of an increase in both HbO and HbR in newborns using auditory stimulation (Sakatani et al., 1999; Zaramella et al., 2001). Other groups using NIRS have reported fully inverted responses with increases in HbR and decreases in HbO and HbT in response to visual stimulation (Kusaka et al., 2004) suggesting a blood volume decrease in response to stimulation (Fig. 1C).

2.2.2 Observations of positive BOLD in the newborn infant brain

Some groups using both NIRS and fMRI have identified adult-like responses to visual stimulation in early postnatal populations (Anderson et al., 2001; Arichi et al., 2010, 2012; Born et al., 1998, 2000; Hoshi and Tamura, 1993; Karen et al., 2008; Liao et al., 2010; Taga et al., 2003; Watanabe et al., 2008). Notably, however, groups examining these responses via NIRS who reported increases in HbO accompanied by decreases in HbR noted significantly more variability in their HbR responses than HbO responses—and that these responses were occasionally increases (Hoshi et al., 2000; Liao et al., 2010; Taga et al., 2003).

fMRI studies have reported positive BOLD responses, meaning HbR decreases, in response to visual (Born et al., 1998, 2000), auditory (Anderson et al., 2001), and somatosensory stimulation in the immediate postnatal period (Arichi et al., 2010, 2012) as well as in premature infants (Allievi et al., 2016). As discussed earlier, studies that identified both positive and negative responses generally actually observed positive responses in younger infants (Anderson et al., 2001; Born et al., 1998, 2000; Yamada et al., 1997), potentially corresponding to developmental increases in oxygen metabolism (see Table 1 for age group details).

A recent study detailed changes in the temporal shape of the BOLD response to stimulus as a function of developmental age, reporting a delayed positive BOLD response in preterm infants and a biphasic positive–negative BOLD response in term infants (Fig. 2). The same group has suggested that these forms of the hemodynamic

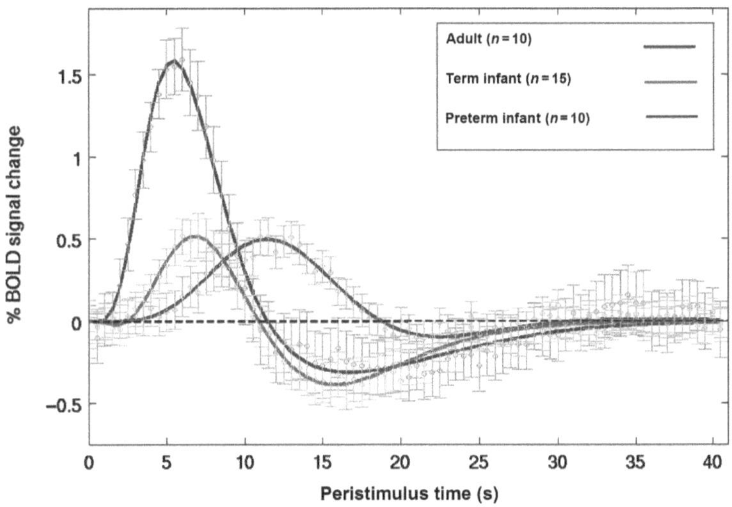

FIG. 2

Human age-dependent hemodynamic response functions to somatosensory stimulation measured using fMRI BOLD.

Figure adapted from Arichi, T., Fagiolo, G., Varela, M., Melendez-calderon, A., Allievi, A., Merchant, N., Tusor, N., Counsell, S.J., Burdet, E., Beckmann, C.F., et al., 2012. Development of BOLD signal hemodynamic responses in the human brain. Neuroimage 63, 663–673.

response function (HRF) should be used for analysis of perinatal fMRI data, rather than the use of a standard adult HRF (Arichi et al., 2012; Cusack et al., 2015).

2.3 RODENT STUDIES OF POSTNATAL NEUROVASCULAR COUPLING

The earlier summaries highlight the wide discrepancies in reports of "normal" hemodynamic responses in the developing human brain. Such studies in human infants pose major challenges, both in terms of image acquisition on neonates that controls for physiological variables, and difficulties in the recruitment of young, healthy subjects for longitudinal studies. Studies thus often include cohorts of premature infants, infants with varied ages or infants with pathologies requiring prolonged postnatal care, making it challenging to establish the underlying cause of conflicting results from different studies.

Studies in rodent models offer the chance to carefully evaluate the underlying cause of changes in responses seen in the human brain. Of course age comparisons between rodents and humans are difficult to validate, particularly in terms of different aspects of brain development, and therefore analysis of developmental trajectory is more meaningful than absolute time points. However, for the work described here, the convention is to assume that the "newborn" human brain is developmentally equivalent to the postnatal day 7 mouse and postnatal day 12 rat, based on the codevelopment of factors such as eye opening and myelination (Quinn, 2005).

In order to reconcile the above findings of positive and negative BOLD in human infants, in recent work from our laboratory, we charted the hemodynamic response to stimulus as a function of postnatal development in both rats and mice using optical imaging of the exposed cortex at high spatiotemporal resolution (Kozberg et al., 2013). We demonstrated that the time courses of hemodynamic responses to neural activity mature postnatally (Kozberg et al., 2013); application of a somatosensory stimulus to postnatal day 12–13 (P12–13) rats was found to result in no or minimal localized increases in blood volume (functional hyperemia) (Fig. 3). This finding was replicated by Zehendner et al. who used whisker stimulation in P7 mice and demonstrated very minimal increases in local blood flow via laser speckle imaging (Zehendner et al., 2013). We found that when functional hyperemia does emerge several days later in development, it is localized to the brain parenchyma (see Fig. 3), meaning either capillary bed and/or diving arterioles, and does not spread to the larger pial arteries until approximately P23 in rats. This result is summarized in Fig. 3.

An important factor in our results was the presence of a delayed, global vasoconstriction in response to stimulus in the developing brain that was present prior to the development of vasodilatory responses (Kozberg et al., 2013; Zehendner et al., 2013). These constrictions were found to remain in place as localized hyperemia develops, resulting in biphasic responses in which there is an increase in total hemoglobin (HbT) followed by a prolonged decrease in HbT as shown in Fig. 3. This global vasoconstriction effect could have a range of causes; a related phenomenon is the "negative surround" commonly interpreted as either "vascular steal" or an indicator of reduced or inhibitory neural activity (Boas et al., 2008; Devor et al., 2008). However, this effect may also be an independent neurovascular coupling

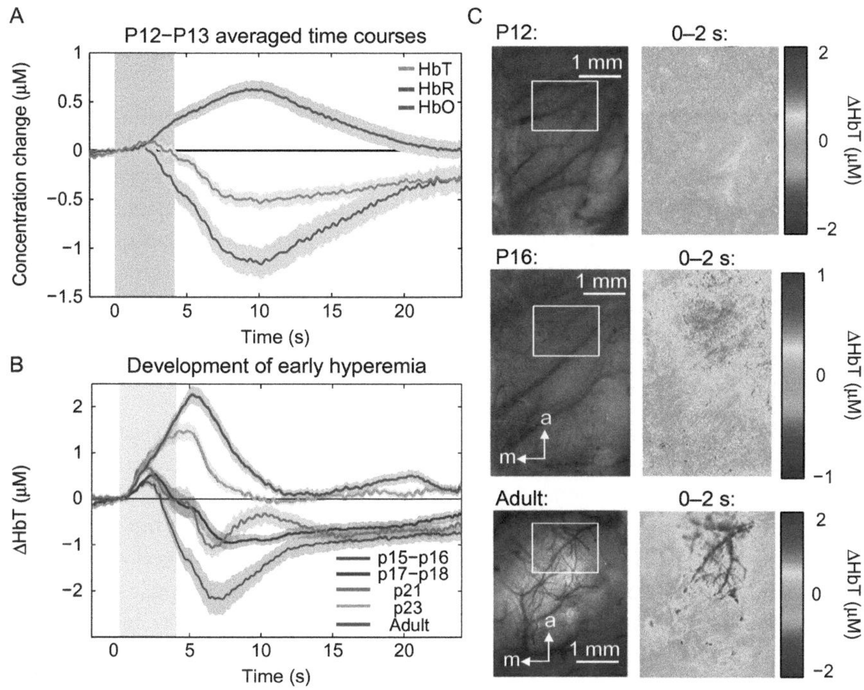

FIG. 3

A functional hyperemia in response to somatosensory stimulation increases significantly with age in early postnatal rats. (A) Observed hemodynamic responses to somatosensory stimulus (*gray bar*) in P12–P13 rats demonstrating decreases in HbT and corresponding increases in HbR and decreases in HbO. (B) Regional blood volume influxes in response to somatosensory stimulus (*gray bar*) increase in amplitude with postnatal age. (C) In younger animals, hemodynamic responses are localized to the capillary bed and/or diving arterioles, while in the adult brain larger pial arterioles dilate as well—generating larger amplitude hemodynamic responses.

Figure adapted from Kozberg, M.G., Chen, B.R., Deleo, S.E., Bouchard, M.B., and Hillman, E.M.C., 2013. Resolving the transition from negative to positive blood oxygen level-dependent responses in the developing brain. Proc. Natl. Acad. Sci. 110, 4380–4385.

mechanism, potentially linked to noradrenergic or cholinergic responses to external stimuli consistent with the hemomodulatory effects of these systems reported in the adult rodent brain (Bekar et al., 2012; Takata et al., 2013). It has been proposed that locus coeruleus-triggered release of norepinephrine (a potent vasoconstrictor) serves to provide counteracting vasoconstriction during stimulation to direct and optimize blood flow to activated regions (Bekar et al., 2012). While this effect requires substantially more investigation, if this vasoconstriction serves to limit the extent of hyperemic responses, its establishment in development prior to hyperemic mechanism maturation could serve to protect fragile perinatal vasculature, which is particularly prone to hemorrhage in premature infants, from uncontrolled vasodilation and systemic blood pressure effects (Volpe, 1997).

These global vasoconstriction effects may be dependent on anesthesia, stimulus type, age, or species; however, as mentioned earlier, some human studies have identified decreases in blood flow (Born et al., 2002) and blood volume (Kusaka et al., 2004) in response to stimulation. A marked arterial vasoconstriction would cause an increase in local HbR (negative BOLD), but observed negative BOLD responses may also be related to underlying oxygen consumption. In the absence of large vasoconstrictions, HbR dynamics would be expected to be a balance of effects relating to oxygen consumption (increasing HbR) and the gradual development of functional hyperemia (decreasing HbR) (eg, the continuum between Fig. 1B and C). This interaction of factors may lead to the increased variability in HbR responses (or BOLD), described earlier.

In our own data, mathematically subtracting this global vasoconstriction effect from acquired data revealed clear evidence of localized oxygen consumption in responding regions of the somatosensory cortex. We observed gradually increasing HbT responses (increasing functional hyperemia as described earlier) with corresponding HbO and HbR dynamics indicating local stimulus-evoked relative hypoxias in early development. As the animals matured, these responses resolved to increases in blood volume balancing consumption effects, until the influx of HbO was sufficient to exceed the oxygen demands of the tissue (Kozberg et al., under review). These results have features consistent with some of the earliest reports of negative BOLD responses in fMRI studies of the human brain (Born et al., 1996).

A study by Colonnese et al. utilized fMRI in a rodent model to explore the postnatal evolution of functional hemodynamics (Colonnese et al., 2008). Although this work reported positive BOLD responses in the postnatal rat brain after P13, responses were prolonged and lower in amplitude compared to the adult brain, and averaged plots do show a marked period of negative BOLD during stimulation and neural activity at P13–15. In rats younger than P11, it is reported that no fMRI BOLD responses (negative or positive) were detected despite the presence of localized neural activity. This condition could indicate that experimental conditions or analysis methods in this study were less susceptible to evoked global vasoconstrictions, such that pre-P11 results confirm an absence of localized functional hyperemia, and minimal (or balanced) effects of oxygen consumption.

2.4 AUTOREGULATION AS A POTENTIAL CONFOUND

In the adult brain, changes in systemic pressure are not directly relayed to the brain due to a process termed autoregulation in which the smooth muscle cells of the brain constrict or relax to counteract increases or decreases in flow, respectively (Iadecola, 2004). However, autoregulation is not fully mature in the postnatal brain (Greisen, 2009). Fluctuations in systemic blood pressure will therefore be reflected in the hemodynamic of the newborn brain (Kozberg et al., 2013) with the potential to profoundly affect the spatiotemporal dynamics and direction of the BOLD signal. Through monitoring systemic blood pressure, we determined that high-amplitude stimuli resulted in large, evoked increases in systemic blood pressure that were closely mirrored by increases in HbT and HbO (and decreases in HbR) in the brain resembling an adult-like "positive BOLD" response. Tuning down stimulus amplitude until no increases in systemic blood pressure were observed yielded a "negative BOLD"-type cortical response that charted the developmental progression shown in Fig. 3. Based on these results, we proposed that a strong startle reflex and immature autoregulation could make the early postnatal brain particularly prone to cortical hyperemia resulting from stimulus-evoked increases in systemic blood pressure. Given the difficulty of monitoring systemic blood pressure in human infants during functional imaging, and difficulties with normalizing stimulus strength in nonverbal subjects who may be anesthetized or asleep, we propose a possible contribution from this blood pressure effect to fMRI and NIRS studies of the immature brain that have reported adult-like fMRI "positive BOLD" responses.

2.5 A POSSIBLE ROLE FOR NEUROVASCULAR DEVELOPMENT

Overall, the studies summarized earlier provide a look at the development of functional hemodynamics in the perinatal brain. Rodent studies have confirmed that the newborn brain exhibits hemodynamic responses that both spatially and temporally differ from those seen in the adult brain. In both Colonnese et al. and recent work from our lab, the neural activity occurring in parallel to these unusual hemodynamic responses has been measured. While the observed neural activity is generally shorter in duration than adult-like responses to the same duration stimuli, this does not fully account for the differences in evoked hemodynamics compared to the adult brain (Colonnese et al., 2008; Kozberg et al., under review). We thus hypothesize that the developmental trajectory of functional hemodynamics in the postnatal brain reflects, at least in part, the gradual development of neurovascular coupling itself. This hypothesis raises the following questions that will be addressed in the remainder of the chapter:

(1) Does early neurovascular immaturity correspond to the physical absence of the mechanisms that link neural activity to blood flow changes? Does neurovascular development represent the establishment of this neurovascular system? Is the primary change made up of structural and functional changes at the level of neurons (or interneurons), vasculature, and/or intermediate cells such as astrocytes and pericytes?

(2) A lack of functional hyperemia during early brain development appears to lead to stimulus-evoked hypoxias local to neural responses to external stimuli. This observation introduces the possibility that the newborn brain exists in a very different metabolic environment to the adult brain, despite being in a period of prolific neural development. What role does this metabolic environment play in normal and abnormal brain development and the effects of therapeutic interventions?

(3) If neurovascular development is a critical component of perinatal brain development, occurring in parallel with neural development, could aberrant neurovascular development be a component in disorders of brain development?

Could this developmental process provide new biomarkers or therapeutic targets? Could fMRI provide a unique tool to track and quantify this developmental trajectory?

(4) Immaturity of neurovascular coupling suggests that fMRI data acquired in premature infants, newborns, and possibly children need to be analyzed differently from adult data. Which factors should be considered in the analysis of these data? How do changes in neurovascular coupling affect resting state fMRI methods?

Many changes occur in the newborn brain during early development. It is thus unsurprising that neurovascular coupling might also be assembling and developing in parallel. The following section details what is currently known about the components that might contribute to the establishment of neurovascular coupling.

3 POSTNATAL DEVELOPMENT OF NEUROVASCULAR CELLS AND STRUCTURES

This section explores the known features of development of cellular components that could represent the building blocks of neurovascular development (see also chapter "Mechanisms of cellular plasticity in cerebral perivascular region" by N. Egawa et al.). Although there is considerable debate about the mechanisms of neurovascular signaling itself, we have included discussion of neurons, vasculature, astrocytes, and pericytes, all of which have been proposed to play a role in neurovascular coupling in the adult brain (see chapters "Neurogenic control of parenchymal arterioles in the cerebral cortex" by H. Hotta, "Involvement of astrocytes in neurovascular communication" by M. Nuriya and H. Hirase, and "Vascular potassium channels in NVC" by K. Yamada).

3.1 POSTNATAL DEVELOPMENT OF NEURAL STRUCTURE AND FUNCTION

Neural networks are far from being fully established at birth, both in the human and rodent brain. The elaboration and refinement of these circuits require inputs from the external environment as well as waves of spontaneous activity within the earlier developed circuitry. Neurons initially undergo a robust period of axonal outgrowth and synaptogenesis postnatally (Cahalane et al., 2011; Lewis et al., 2013; Uylings et al., 1990; Wang et al., 2007), followed by synaptic pruning and circuit refinement with synapse elimination (Huttenlocher et al., 1982). These periods of elaboration and refinement vary regionally, but generally continue in the human brain well into adolescence (Harris et al., 2011).

The cortex is formed through multiple rounds of neurogenesis with almost all cortical neurons originating from radial glial progenitor cells. In the rodent brain at birth, layers 2–4 of the cortex exist in a gradient within the cortical plate; however, layers 5 and 6 are already defined, but still expanding. Over the next few days, the laminar organization becomes more defined, and by postnatal day 4 in the rodent brain, there has been clear structural differentiation into each of the layers (McCandlish et al., 1993). Neurons formed in later rounds of neurogenesis migrate to the superficial layers of cortex, past earlier rounds of neurons (Gao et al., 2014), Generally, deeper neurons (early born) are more connected with subcortical structures while superficial neurons (late born) are projection neurons responsible for intracortical connectivity; postnatal elaboration of neural circuitry primarily involves these more superficial cortical layers. Recent studies in the mouse brain have specifically demonstrated that transcortical projections within layer 2/3 are established throughout the first postnatal week (Hand and Polleux, 2011; Wang et al., 2007). While understanding, the establishment of these long-range connections and network firing patterns is a primary goal of many research programs utilizing functional imaging as a proxy for neural activity in development (Graham et al., 2014), there is a possibility that the connectivity of neurons to the vasculature also varies with the elaboration of neural networks.

Postnatal changes to the neural circuitry are in part activity dependent; the proper development of this circuitry requires external inputs for the process of effectively wiring "like" cells together (Katz and Shatz, 1996). Alterations in experiences can have large effects on neural synapse formation particularly in critical periods of neural development. These critical periods vary in timing regionally (Hensch, 2004). One of the paradigms most widely utilized to examine early cortical plasticity is that of monocular deprivation. If a subject is deprived of input from one eye during a "critical period" of visual cortex development, the region of the cortex receiving inputs from the deprived eye will be significantly smaller than that of the contralateral eye which will be proportionally larger, an effect first demonstrated in kittens by Hubel and Wiesel and subsequently in many other model systems (Hensch, 2004; Hubel and Wiesel, 1963). Recent work has also demonstrated that spontaneous activity may also be essential to circuit development (Feller, 1999). For example, prior to exposure of the visual system to external inputs, ie, prior to eye opening, spontaneous waves of retinal activity occur which are transmitted back to the visual cortex, generating neural activity which may aid in early circuit elaboration and refinement (Ackman et al., 2012).

Refinement of neural circuitry is dependent on the development of inhibitory pathways (Crook et al., 1998). GABAergic interneurons develop at different times postnatally corresponding to region-dependent critical periods (Le Magueresse and Monyer, 2013). Notably, the potassium chloride cotransporter KCC2 is upregulated with age, resulting in a developmental switch of the neurotransmitter GABA from depolarizing to hyperpolarizing at approximately postnatal day 7 in mice (Ben-Ari, 2002), changing the excitatory to inhibitory balance of the brain. These changes have the potential to affect the spatiotemporal profile brain hemodynamics, as some researchers hypothesize that interneurons play an essential role in neurovascular coupling (Cauli et al., 2004).

Lastly, an important facet of brain development is the physical expansion of the cortex. In the rodent brain, the cortical surface area more than doubles within the first 20 days of life (Wang et al., 1996). These differences in size of the cortex themselves may affect the energy demands and distribution in the newborn brain as oxygen diffusion limitations emerge.

3.2 POSTNATAL DEVELOPMENT OF THE BRAIN VASCULATURE

The brain's vasculature itself changes significantly morphologically and functionally during the early postnatal period. The early vascular plexus of the brain is formed embryonically, a disordered plexus of vessels with no true differentiation between arteries, veins, and capillaries. This plexus is used as a scaffold for vascular development with gradual refinement of the vasculature into the different vascular compartments (Herken et al., 1989; Wang et al., 1996); arteries and veins notably only begin to take on their unique characteristics postnatally (Rowan and Maxwell, 1981). Remnants of the early vascular plexus of the brain remain particularly prevalent across the surface of the early postnatal brain, forming a dense meshwork of interconnected vasculature, which is largely pruned postnatally, leaving the larger vessels that become the pial arteries and veins (Wang et al., 1996). While these vessels are pruned, the deeper, intracortical network of vasculature elaborates.

3.2.1 Capillaries

Vascular network expansion via angiogenesis occurs primarily in the bed postnatally (Risser et al., 2009) and the pattern of changes in vascular density in many ways echoes the previously noted observations of neural development. For the first few months of human postnatal development and first postnatal month in mice, the capillary bed expands (Harb et al., 2013; Keep and Jones, 1990; Norman, 1986). The process of vascular remodeling occurs through sprout formation followed by either maintenance or pruning of these sprouts (Harb et al., 2013; Rowan and Maxwell, 1981).

Human studies have examined the cerebral blood flow of the cortex, a metric which is deeply entwined with vascular density, demonstrating that cerebral blood flow is significantly lower at early postnatal ages than in the adult brain, increases until the age of 7, and then decreases to adult levels in the late teen years (Chiron et al., 1992; Takahashi et al., 1999), a pattern notably similar to that of neural synapse formation and elimination. Additionally, rodent studies have demonstrated that postnatal brain vascularization occurs in an inside-to-outside pattern in which deeper cortical layers develop their capillary beds earlier (Norman, 1986; Rowan and Maxwell, 1981), a trend similar to the radial development of the neural architecture of the brain described earlier. Ways in which the parallel development of the neural and vascular systems may influence one another will be discussed later.

Microvascular density has a crucial impact on the oxygenation of the brain. Capillaries are known to be the main locations of oxygen exchange; only a small amount of oxygen exchange occurs along the arterioles (Vovenko, 1999). Efficient nutrient delivery to the adult brain depends in part on the density of the capillary bed; recent studies have demonstrated that in the adult brain islands of relative hypoxia exist in regions furthest from the capillaries, particularly in activated tissue (Sakadžić et al., 2014). Some groups hypothesize that the "oversupply" of oxygenated blood following stimulation may account for this variability in oxygenation based on capillary distribution (Devor et al., 2011). Therefore, secondary to relatively low capillary density, the newborn brain may actually have lower baseline levels of oxygenation than the adult brain.

Capillaries may also be essential for the actuation and execution of neurovascular responses. Recent studies have suggested that functional hyperemia is initiated at the capillary bed level (Hall et al., 2014; Hamilton et al., 2010) before spreading up the diving arterioles (Tian et al., 2010) and reaching the larger pial arteries on the surface of the brain (Chen et al., 2011, 2014). Every neuron is within 8–23 μm of a capillary (Lovick et al., 1999), making them a close signaling contact point; recent work has in fact demonstrated their ability to dilate themselves in response to stimulation, an action accomplished through the relaxation of pericytes and/or pericapillary smooth muscle cells (Hall et al., 2014; Hamilton et al., 2010; Hill et al., 2015). These data suggest that the capacity to perform neurovascular coupling may be diminished in an environment with a low density capillary bed, as in the newborn brain.

3.2.2 Arteries

The arterial distribution of major vessels across the surface of the brain is well established at birth (Wang et al., 1996); however, there are crucial differences in their oxygenation and function from those of the adult. Firstly, arteriovenous shunts which do not exist in the adult brain are present throughout the pial vasculature (Wang et al., 1996), and therefore differences in oxygenation between the pial arteries and veins are diminished as compared to the adult brain. Secondly, as mentioned earlier, arteries only begin to take on their unique characteristics postnatally, meaning that their vascular reactivity may differ (Rowan and Maxwell, 1981).

One of the essential mediators of vasodilation is nitric oxide. A recent study in rats demonstrated that while neural nitric oxide synthase decreases in expression with age, endothelial NOS (eNOS) expression is 85% lower in the 1-month-old brain than in the 14-month-old brain (Bustamante et al., 2008). Furthermore, studies of the piglet brain have demonstrated differences in the arterial response to acetylcholine, known in the adult brain to increase NO levels via eNOS (Andresen et al., 2005). Zuckerman et al. found that acetylcholine in the piglet brain in fact generated a NO-independent, prostanoid-mediated arterial constriction when applied to the newborn brain (Zuckerman et al., 1996).

Recent work from our laboratory demonstrated that the vascular endothelium itself is an essential component of neurovascular signaling and the spatiotemporal shape of the hemodynamic response (Chen et al., 2014). As discussed earlier, hemodynamic responses in the adult brain begin at the capillary bed or diving arteriolar level, and this dilation is back propagated along the arterial tree, eventually reaching the larger arteries on the pial surface. We demonstrated that this signaling is dependent on an intact vascular endothelium, hypothesizing the involvement of both gap junction coupling between endothelial cells enabling a long-range hyperpolarization wave (Ujiie et al., 2003) as well as more local nitric oxide-dependent and prostanoid-dependent signaling (Hillman, 2014). Therefore, changes in endothelial properties postnatally, including gap junction and receptor expression, may have large effects on the spatiotemporal profile of observed neurovascular responses.

3.2.3 Resting vascular tone

The aminergic systems which release norepinephrine, dopamine, and serotonin from the locus coeruleus, ventral tegmental area, and raphe nucleus, respectively, also expand postnatally (Goldman-Rakic and Brown, 1982). These transmitters are all vasoconstrictors and are important for the maintenance of resting vascular tone (Iadecola, 2004) and may even be additionally modulated during a stimulus (Bekar et al., 2012). Developmental changes in these networks could lead to relative "predilation" of the early postnatal arterial system and therefore a diminished capacity for active dilation.

3.3 ASTROCYTES

Astrocytes, ideally positioned surrounding both synapses and blood vessels throughout the brain, have been proposed to play an essential role in functional neurovascular coupling (Attwell et al., 2010). Local increases in extracellular glutamate have been demonstrated to drive increases in astrocytic intracellular calcium (Lind et al., 2013), in turn resulting in vascular dilations in the stimulated regions through the release of vasoactive substances stimulating smooth muscle cell relaxation such as prostaglandins and epoxyeicosatrienoic acids (Takano et al., 2006).

Astrocytes are known to be immature in the postnatal rat brain until postnatal day 21 and do not reach adult density levels until postnatal day 50, with changes in individual astrocyte size, branching and connectivity, as well as in number (Seregi et al., 1987; Stichel et al., 1991). Astrocytes gap junction coupling is established only at postnatal day 11 in the rat visual cortex, which could also affect the ability of astrocytes to signal to one another throughout the cortex (Binmöller and Müller, 1992).

3.4 PERICYTES

As mentioned earlier, any given neuron is within close proximity to the capillary bed, making capillaries a widely accessible signaling target for neurovascular coupling. Pericytes are the mural cells of capillaries (as well as other small vessels of the brain). They are similar to smooth muscle cells in many ways including their expression of contractile proteins, enabling them to change the diameter of capillaries. While smooth muscle cells are separated from the endothelium by the internal elastic lamina, pericytes share a basement membrane with the endothelium, potentially enabling enhanced signaling (Armulik et al., 2005). Recent studies have begun to focus on pericytes as a potential mediator of neurovascular coupling, demonstrating that capillaries dilate prior to precapillary arterioles, and that this dilation is particularly seen in regions with pericyte cell bodies, findings that suggest active dilation (Hall et al., 2014).

Pericytes are deeply involved in the structural patterning of the vasculature of the brain, particularly in the stabilization of vessels during angiogenesis. Throughout the process of sprout formation, pericytes are positioned at endothelial sprout tips. These developmental changes present the possibility that pericytes are unable to actively modulate capillary diameter throughout this dynamic period. Additionally, the investment of the pericytes into the endothelial basement membrane may occur postnatally (Fujimoto, 1995), which may affect communication between the cell types.

4 THE NEWBORN BRAIN'S UNIQUE METABOLIC ENVIRONMENT

Early positron emission tomography studies of the developing brain indicated key differences in postnatal brain metabolism compared to the adult brain (Altman et al., 1993; Chugani and Phelps, 1986; Takahashi et al., 1999). Specifically, the newborn brain was observed to have a lower baseline $CMRO_2$, which increases to adult rates by age 2 and then continues to rise to above adult levels until age 9. These $CMRO_2$ values then decline to adult values by the late teenage years (Chugani and Phelps, 1986). Additional parameters are depicted in Fig. 4.

These findings suggest that the newborn brain may have less oxygen demand than the adult brain, a concept supported by the high tolerance of the neonatal brain to hypoxia (Singer, 1999) as well as prior studies of enzyme expression which have suggested that the early postnatal brain performs primarily nonoxidative glucose metabolism (Bilger and Nehlig, 1991; Bonavita et al., 1962; Booth et al., 1980). A recent study from our laboratory utilizing flavin adenine dinucleotide fluorescence to image rates of oxidative metabolism in the in vivo mouse brain demonstrated that the newborn brain (postnatal day 7–8) does increase its local rate of oxidative metabolism in response to stimulation, but at seemingly lower levels than in the adult (even in the presence of relatively large neural responses) (Kozberg et al., under review). In addition to the potential of the newborn brain to perform nonoxidative metabolism, demand may also be diminished secondary to the lower levels of spontaneous neural activity in the newborn brain (Kozberg et al., under review). Conversely, the level of oxygen availability in the cortex for oxidative metabolism could actually limit the levels of activity in the brain.

Note that while neurovascular coupling is hypothesized to exist in order to bring oxygen and other metabolites to the metabolically active brain (Heeger and Ress, 2002), the execution of hemodynamic responses to neural activity is *not* dependent on metabolic demand by the tissue, at least in terms of both oxygen and glucose availability (Lindauer et al., 2010; Wolf et al., 1997). Additionally, as mentioned earlier, the supply of oxygen to the tissue far exceeds the demand by the tissue (Fox et al., 1988). Therefore, while neurovascular coupling appears to be necessary for normal adult brain function (Iadecola, 2004), the relative lack of neurovascular coupling in the newborn brain does not necessarily imply decreased levels of demand or a completely resource limited state.

4.1 FETAL HEMOGLOBIN

Fetal hemoglobin binds to oxygen more strongly than adult hemoglobin, enabling the transfer of oxygen from mother to fetus prenatally. Oxygen exchange within the tissue is thus affected by the strength of the binding between hemoglobin and oxygen.

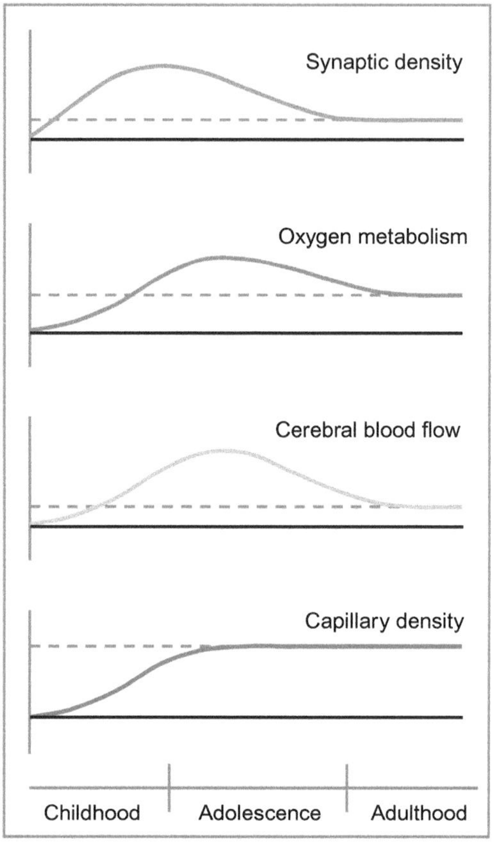

FIG. 4

Developmental changes in neural and vascular density, oxygen metabolism, and cerebral blood flow. Newborn levels are depicted with *solid lines* and adult levels with *dotted lines.*

Based on Chiron, C., Raynaud, C., Mazière, B., Zilbovicius, M., Laflamme, L., Masure, M.C., Dulac, O., Bourguignon, M., and Syrota, A., 1992. Changes in regional cerebral blood flow during brain maturation in children and adolescents. J. Nucl. Med. 33, 696–703; Harb, R., Whiteus, C., Freitas, C., and Grutzendler, J., 2013. in vivo imaging of cerebral microvascular plasticity from birth to death. J. Cereb. Blood Flow Metab. 33, 146–156; Huttenlocher, P., de Courten, C., Garey, L., and van der Loos, H., 1982. Synaptogenesis in human visual cortex—evidence for synapse elimination during normal development. Neurosci. Lett. 33, 247–252; Norman, M.G., 1986. The growth and development of microvasculature in human cerebral cortex. J. Neuropathol. Exp. Neurol. 45, 222–232; Risser, L., Plouraboué, F., Cloetens, P., and Fonta, C., 2009. A 3D-investigation shows that angiogenesis in primate cerebral cortex mainly occurs at capillary level. Int. J. Dev. Neurosci. 27, 185–196; and Takahashi, T., Shirane, R., Sato, S., and Yoshimoto, T., 1999. Developmental changes of cerebral blood flow and oxygen metabolism in children. AJNR Am. J. Neuroradiol. 20, 917–922.

Human infants continue to generate significant amounts of fetal hemoglobin for up to 6 months postnatally (Thein et al., 2009), which impacts (diminishes) oxygen exchange between the vasculature and tissue throughout the body, including in the brain.[1] Of note, the absorption spectra of oxygenated and deoxygenated fetal hemoglobin differ only minimally from those of the adult, and therefore changes in expression of fetal hemoglobin are not expected to significantly affect NIRS analysis (Zijlstra et al., 1991).

4.2 NEUROVASCULAR CO-DEVELOPMENT?

Embryonically, neurons and vasculature may share guidance cues as each is developing (Gelfand et al., 2009). As discussed earlier, neural and vascular networks continue to expand in parallel postnatally and vascular growth appears to be optimized to regions of higher neural activity. Recent studies have focused on determining how modulations in neural activity affect postnatal angiogenesis. Lacoste et al. manipulated inputs into the whisker barrel through both stimulation and selective ablation paradigms and studied the vascularization of the corresponding whisker barrels, finding that regions with fewer neural inputs and activity have decreased vascular density while regions with enhanced activity experience increased levels of postnatal angiogenesis (Lacoste et al., 2014; see Fig. 5). This corresponds well to previous work

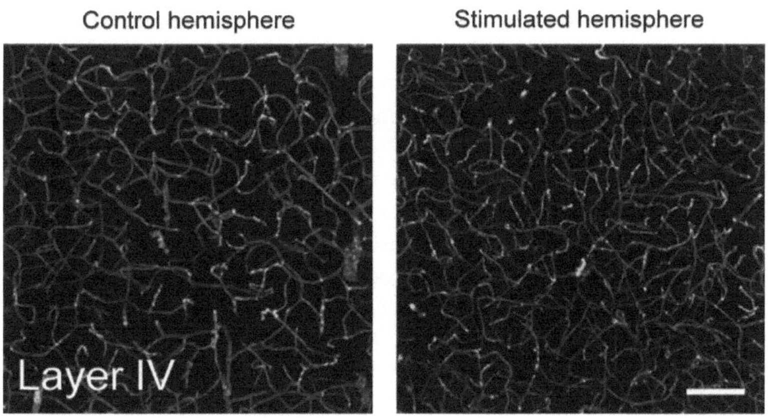

FIG. 5

Whisker stimulation in early postnatal life leading to increased capillary density and branching.

Figure adapted from Lacoste, B., Comin, C.H., Ben-Zvi, A., Kaeser, P.S., Xu, X., Costa, L. da F., and Gu, C., 2014. Sensory-related neural activity regulates the structure of vascular networks in the cerebral cortex. Neuron 83, 1117–1130.

[1]In the rodent, embryonic hemoglobin is largely replaced with adult hemoglobin prior to birth (without an intermediate fetal hemoglobin step as in humans; Iwahara et al., 1996).

from Argandoña and Lafuenta who demonstrated decreased vascularization in the visual cortices of dark-reared rats (Argandoña and Lafuente, 1996) as well as work from Black et al. demonstrating increased vascular density in rats raised in complex environments (Black et al., 1987).

However, a recent study from Whiteus et al. demonstrated that exposing newborn mice to chronic auditory, whisker, or motor stimulation resulted in near *arrest* of normal postnatal angiogenesis in those respective cortices (Whiteus et al., 2014). Unsurprisingly, this decrease in vascular density made the mice more susceptible to hypoxias in regions furthest from the nearest capillary. Of note, contrary to the results of Lacoste et al. stated earlier, they found that smaller fluctuations in activity driven by whisker trimming or moderate whisker stimulation did not result in alterations in the level of angiogenesis. These data suggest that the type and intensity of stimulation may significantly affect its impact on postnatal vascular growth. There may be states of either "too much" or "too little" activity, each of which could lead to decreased angiogenesis.

Increased regional metabolic demand has also been associated with increased vascular density (Riddle et al., 1993; Weber et al., 2008). Detection of this metabolic demand may be in part through hypoxia signaling, which is well-known to drive vascular elaboration, particularly in the developing brain. Harb et al. demonstrated that young adult mice (<3 months old) placed in hypoxia form significantly more vessels, and that these vessels are not eliminated once the animal is brought back to normoxic conditions (Harb et al., 2013). Hypoxia induces increases in vessel growth through the activation of hypoxia-inducible transcription factors (known as HIFs; see chapter "Development and pathological changes of neurovascular unit regulated by hypoxia response in the retina" by T. Kurihara for details), which in turn upregulate expression of signaling molecules involved in angiogenesis such as VEGF (Lacoste and Gu, 2015; Rey and Semenza, 2010).

If hypoxia is a component of directed/activity-dependent vascular patterning, increased oxygenated blood delivery by neurovascular coupling could actually be detrimental to establishing a complete vascular network responsive to the demands of the tissue. The relative hypoxias that occur in response to neural activity in the absence of high-amplitude neurovascular coupling described earlier could be necessary to drive postnatal angiogenesis (Lacoste and Gu, 2015).

5 CONCLUSIONS

Functional imaging studies using fMRI and NIRS in newborns have identified distinctly different response characteristics in newborn infants, and even children exceeding 36 months, compared to stereotypical adult responses. Mounting evidence suggests that these differences in functional responses correspond to the immaturity of neurovascular coupling, the cellular mechanisms that link neural activity to local increases in blood flow in the brain. The substantial cellular and vascular development known to occur during the postnatal period therefore includes both the expansion of neural and vascular networks and the gradual postnatal development of neurovascular coupling mechanisms.

The resultant lack of coupling in the immature brain (<P8 in mice) raises important questions about developing brain metabolism. The newborn brain supports massive growth, neurogenesis, and learning in a state that appears to be incompletely tuned to the brain's metabolic demands. Does the newborn brain encounter an environment of low energy availability that limits the ability of neurons to fire? Moreover, is there a key purpose to the development of the brain's neurovascular system in parallel with the development of neural circuits? From an optimization standpoint, the codevelopment of neural networks and supporting vasculature is logical, as opposed to independent development or initial hardwiring of the vasculature prior to neural expansion and pruning. Some recent results have suggested that relative hypoxic events occur in response to neural activity (in the absence of coupled functional hyperemia), a possible mechanism by which vasculature directs its growth to serve regions of increased neural activity.

5.1 A ROLE IN NORMAL AND ABNORMAL DEVELOPMENT?

The complex (and relatively unexplored) process of neurovascular development continues to occur postnatally in normal development, prompting consideration of the possible role of this developing system in normal and abnormal trajectories of brain development. Since the early postnatal brain appears to operate with a delicate balance of oxygenation, the establishment of functional hyperemia at the right time point in development is likely a key stage of development. Disruptions or delays in the onset of coupled hyperemia could feasibly lead to abnormalities in both neural and vascular development.

The unique metabolic environment of the newborn brain may also have important implications for the care of premature and newborn infants. Several studies have indicated that resuscitation of infants with room air rather than 100% oxygen offers better clinical outcomes and less brain damage, while premature infants may require slight oxygen supplementation (Saugstad, 2010). This oxygen supplementation must be carefully titrated: on the one hand, excess and abnormal vascular development occurs in the eyes of premature infants exposed to high levels of oxygen (retinopathy of prematurity) (Wheatley et al., 2002), while on the other hand ischemia in premature infants is associated with poor neurological outcomes primarily due to cerebral white matter injury (Khwaja and Volpe, 2008). Further studies have raised concerns over the effects of general anesthesia in newborns, indicating that they may have a profound impact on developing neural circuitry (Culley et al., 2012). There is thus an urgent need to understand the optimal metabolic state of the perinatal brain.

5.2 IMPLICATIONS FOR FUTURE FUNCTIONAL IMAGING STUDIES

Both fMRI and NIRS, the primary noninvasive functional imaging modalities used to study human brain development, rely on measurements of hemodynamics as a proxy for neural activity. If indeed, there are periods of early (possibly prenatal) development in which neurovascular coupling is different than in adults, standard analysis and interpretation of fMRI and NIRS data could be misleading. Conversely, with improved knowledge of how to interpret functional imaging signals in these age groups, functional imaging methods could prove invaluable for charting the development of neurovascular coupling.

5.2.1 Functional imaging during stimulation

While observed hemodynamic responses in the newborn brain are varied, there are a few unifying factors. Firstly, most agree that early postnatal hemodynamics differ from the adult in terms of their spatiotemporal dynamics, variations that are likely a result of immature signaling and/or immature vasculature. This necessitates the use of age-specific HRFs in the analysis of functional imaging data (Arichi et al., 2012; Cusack et al., 2015). The question of the age at which human children can be considered to have full "adult-like" hemodynamic responses is still open, with recent studies suggesting that neurovascular responses continue to mature throughout adolescence (Schmithorst et al., 2015) although some studies demonstrate BOLD % signal change equal to the adult in early childhood (Church et al., 2010).

Secondly, unlike in the adult brain in which a large influx in arterial blood in response to stimulation far overcompensates for local oxygen consumption, the balance between local consumption and arterial supply is much more delicate in the developing brain. Because the directionality of the HbR curve (ie, positive vs negative BOLD) is affected by both oxygen metabolism and blood volume changes, interstudy, intersubject, and even intrasubject regional variability may be attributable to slight tips in the balance between these two components.

For example, hyperemia is primarily localized to the capillary bed or diving arterioles of the cortex in the immature brain (Kozberg et al., 2013). This capillary bed is actively expanding in postnatal development following an inside-to-outside maturation pattern suggesting that the deeper layers of the cortex may be best prepared to mount a hemodynamic response. Therefore, in deeper regions of the brain, the balance could be tipped to "positive BOLD" in the early postnatal brain. Notably, this would differentially affect functional imaging modalities as signals from deeper layers of the cortex are less detectable by NIRS, optical intrinsic signal imaging, and speckle than by fMRI. Another important consideration is that the maturation of neurons and vasculature is region dependent, making it likely that there will be stimulus-dependent variability.

Lastly, hemodynamics measurements are often used as a proxy for neural activity, which by nature is an imperfect representation. In the developing brain, it is crucial to remember that while both negative and positive BOLD responses are approximate proxies for neural activity, they are actually reporting two different phenemona, which will have different spatiotemporal distributions.

5.2.2 Resting state functional connectivity mapping

A focus of many recent studies of both the developing and adult brain has been on functional connectivity mapping. This form of imaging takes advantage of spontaneous network activity, neural activity that occurs in the absence of a task. Such activity patterns reveal the underlying architecture of the neural circuitry of the brain with striking accuracy in the healthy adult brain (Fox and Raichle, 2007). These imaging paradigms are much easier to perform in infants than task-based studies because of the speed of imaging and decreased need for subject attention and compliance (Doria et al., 2010; Fransson et al., 2011; Smyser et al., 2010) and have been performed to examine the fetal brain (Ferrazzi et al., 2014). These studies can also be performed using NIRS (Ferradal et al., 2015). However, resting state fMRI data can be more challenging to interpret than stimulus-evoked responses; it remains unknown what precisely drives these slow fluctuations in neural and hemodynamic activity in structurally and functionally connected regions even in the adult brain, and no direct measurement is made to confirm the presence of positive, negative, or absent neurovascular coupling.

Recent work from our laboratory demonstrated diminished amplitude functional hyperemias with accompanying relative local hypoxias in response to spontaneous neural activity in the early postnatal mouse brain, similar to observed hemodynamic responses to stimulus-evoked neural activity (Kozberg et al., under review). This finding indicates that oxygen consumption may also be an important source of signal in functional connectivity mapping of newborns, and potential spatial differences between the localization of oxygen consumption and vascular responses should be considered. Additionally, we identified large, global fluctuations in hemodynamics that were spatiotemporally uncoupled to neural activity. These fluctuations may confound studies reliant on hemodynamic measures of resting state neural activity.

5.3 A LOOK TO THE FUTURE

Neurovascular coupling is an essential and largely underexplored component of early postnatal development that affects both the interpretation of functional imaging studies and the metabolic environment of the developing brain. A more comprehensive understanding of the normal stages of neurovascular coupling maturation could provide important new insights into the etiology of developmental disorders and disorders of prematurity and early childhood and may provide new biomarkers for diagnosis and monitoring and unexplored therapeutic targets for currently intractable conditions.

ACKNOWLEDGMENTS

The authors would like to acknowledge the support from National Institutes of Health grants: 1F31NS084538, 1RO1NS063226, and 1RO1NS076628; and National Science Foundation grant: CBET-0954796.

REFERENCES

Ackman, J.B., Burbridge, T.J., Crair, M.C., 2012. Retinal waves coordinate patterned activity throughout the developing visual system. Nature 490, 219–225.

Allievi, A.G., Arichi, T., Tusor, N., Kimpton, J., Arulkumaran, S., Counsell, S.J., Edwards, A.D., Burdet, E., 2016. Maturation of sensori-motor functional responses in the preterm brain. Cereb. Cortex 26 (1), 402–413.

Altman, D.I., Perlman, J.M., Volpe, J.J., Powers, W.J., 1993. Cerebral oxygen metabolism in newborns. Pediatrics 92, 99–104.

Anderson, A.W., Marois, R., Colson, E.R., Peterson, B.S., Duncan, C.C., Ehrenkranz, R.A., Schneider, K.C., Gore, J.C., Ment, L.R., et al., 2001. Neonatal auditory activation detected by functional magnetic resonance imaging. Magn. Reson. Imaging 19, 1–5.

Andresen, J., Shafi, N.I., Bryan, R.M., 2005. Endothelial influences on cerebrovascular tone. J. Appl. Physiol. 100, 318–327.

Argandoña, E.G., Lafuente, J.V., 1996. Effects of dark-rearing on the vascularization of the developmental rat visual cortex. Brain Res. 732, 43–51.

Arichi, T., Moraux, A., Melendez, A., Doria, V., Groppo, M., Merchant, N., Combs, S., Burdet, E., Larkman, D.J., Counsell, S.J., et al., 2010. Somatosensory cortical activation identified by functional MRI in preterm and term infants. Neuroimage 49, 2063–2071.

Arichi, T., Fagiolo, G., Varela, M., Melendez-calderon, A., Allievi, A., Merchant, N., Tusor, N., Counsell, S.J., Burdet, E., Beckmann, C.F., et al., 2012. Development of BOLD signal hemodynamic responses in the human brain. Neuroimage 63, 663–673.

Armulik, A., Abramsson, A., Betsholtz, C., 2005. Endothelial/pericyte interactions. Circ. Res. 97, 512–523.

Attwell, D., Iadecola, C., 2002. The neural basis of functional brain imaging signals. Trends Neurosci. 25, 621–625.

Attwell, D., Buchan, A.M., Charpak, S., Lauritzen, M., Macvicar, B.A., Newman, E.A., 2010. Glial and neuronal control of brain blood flow. Nature 468, 232–243.

Bekar, L.K., Wei, H.S., Nedergaard, M., 2012. The locus coeruleus-norepinephrine network optimizes coupling of cerebral blood volume with oxygen demand. J. Cereb. Blood Flow Metab. 32, 2135–2145.

Ben-Ari, Y., 2002. Excitatory actions of gaba during development: the nature of the nurture. Nat. Rev. Neurosci. 3, 728–739.

Bilger, A., Nehlig, A., 1991. Quantitative histochemical changes in enzymes involved in energy metabolism in the rat brain during postnatal development. I. Cytochrome oxidase and lactate dehydrogenase. Int. J. Dev. Neurosci. 9, 545–553.

Binmöller, F., Müller, C., 1992. Postnatal development of dye-coupling among astrocytes in the visual cortex. Glia 6, 127–137.

Black, J., Sirevaag, A., Greenough, W., 1987. Complex experience promotes capillary formation in young rat visual cortex. Neurosci. Lett. 83, 351–355.

Boas, D., Jones, S., Devor, A., Huppart, T., Dale, A., 2008. A vascular anatomical network model of the spatio-temporal response to brain activation. Neuroimage 40, 1116–1129.

Bonavita, Vi, Ponte, F., Amore, G., 1962. Lactate dehydrogenase isozymes in the developing rat brain. Nature 196, 576–577.

Booth, R.F., Patel, T.B., Clark, J.B., 1980. The development of enzymes of energy metabolism in the brain of a precocial (guinea pig) and non-precocial (rat) species. J. Neurochem. 34, 17–25.

Born, P., Rostrup, E., Leth, H., Peitersen, B., Lou, H., 1996. Change of visually induced cortical activation patterns during development. Lancet 347, 543.

Born, P., Leth, H., Miranda, M.J., Rostrup, E., Stensgaard, A., Peitersen, B., Larsson, H.B.W., Lou, H.C., 1998. Visual activation in infants and young children studied by functional magnetic resonance imaging. Pediatr. Res. 44, 578–583.

Born, A.P., Miranda, M.J., Rostrup, E., Toft, P.B., Peitersen, B., Larsson, H.B.W., Lou, H.C., 2000. Functional magnetic resonance imaging of the normal and abnormal visual system in early life. Neuropediatrics 31, 24–32.

Born, A.P., Rostrup, E., Miranda, M.J., Larsson, H.B.W., Lou, H.C., 2002. Visual cortex reactivity in sedated children examined with perfusion MRI (FAIR). Magn. Reson. Imaging 20, 199–205.

Bustamante, J., Czerniczyniec, A., Cymeryng, C., Lores-Arnaiz, S., 2008. Age related changes from youth to adulthood in rat brain cortex: nitric oxide synthase and mitochondrial respiratory function. Neurochem. Res. 33, 1216–1223.

Cahalane, D.J., Clancy, B., Kingsbury, M.A., Graf, E., Sporns, O., Finlay, B.L., 2011. Network structure implied by initial axon outgrowth in rodent cortex: empirical measurement and models. PLoS One 6, e16113.

Cauli, B., Tong, X., Rancillac, A., Serluca, N., Lambolez, B., Rossier, J., Hamel, E., 2004. Cortical GABA interneurons in neurovascular coupling: relays for subcortical vasoactive pathways. J. Neurosci. 24, 8940–8949.

Chen, B.R., Bouchard, M.B., Mccaslin, A.F.H., Burgess, S.A., Hillman, E.M.C., 2011. High-speed vascular dynamics of the hemodynamic response. Neuroimage 54, 1021–1030.

Chen, B.R., Kozberg, M.G., Bouchard, M.B., Shaik, M.A., Hillman, E.M.C., 2014. A critical role for the vascular endothelium in functional neurovascular coupling in the brain. J. Am. Heart Assoc. 3, e000787.

Chiron, C., Raynaud, C., Mazière, B., Zilbovicius, M., Laflamme, L., Masure, M.C., Dulac, O., Bourguignon, M., Syrota, A., 1992. Changes in regional cerebral blood flow during brain maturation in children and adolescents. J. Nucl. Med. 33, 696–703.

Chugani, H.T., Phelps, M.E., 1986. Maturational changes in cerebral function in infants determined by 18FDG positron emission tomography. Science 231, 840–843.

Church, J., Peterson, S., Schlagger, B., 2010. The "Task B problem" and other considerations in developmental functional neuroimaging. Hum. Brain Mapp. 31, 852–862.

Colonnese, M.T., Phillips, M.A., Constantine-paton, M., Kaila, K., Jasanoff, A., 2008. Development of hemodynamic responses and functional connectivity in rat somatosensory cortex. Nat. Neurosci. 11, 72–79.

Crook, J.M., Kisvárday, Z.F., Eysel, U.T., 1998. Evidence for a contribution of lateral inhibition to orientation tuning and direction selectivity in cat visual cortex: reversible inactivation of functionally characterized sites combined with neuroanatomical tracing techniques. Eur. J. Neurosci. 10, 2056–2075.

Culley, D.J., Maze, M., Crosby, G., 2012. Reprogramming of the infant brain by surgery with general anesthesia. Mayo Clin. Proc. 87, 110–113.

Cusack, R., Wild, C., Linke, A.C., Arichi, T., Lee, D.S.C., Han, V.K., 2015. Optimizing stimulation and analysis protocols for neonatal fMRI optimizing stimulation and analysis protocols for neonatal fMRI. PLoS One 10, e0120202.

Devor, A., Hillman, E.M.C., Tian, P., Waeber, C., Teng, I.C., Ruvinskaya, L., Shalinsky, M.H., Zhu, H., Haslinger, R.H., Narayanan, S.N., et al., 2008. Stimulus-induced changes in blood

flow and 2-deoxyglucose uptake dissociate in ipsilateral somatosensory cortex. J. Neurosci. 28, 14347–14357.
Devor, A., Sakadz, S., Saisan, P.A., Yaseen, M.A., Roussakis, E., Srinivasan, V.J., Vinogradov, S.A., Rosen, B.R., Buxton, R.B., Dale, A.M., et al., 2011. "Overshoot" of O_2 is required to maintain baseline tissue oxygenation at locations distal to blood vessels. J. Neurosci. 31, 13676–13681.
Doria, V., Beckmann, C.F., Arichi, T., Merchant, N., Groppo, M., Turkheimer, F.E., Counsell, S.J., Murgasova, M., Aljabar, P., Nunes, R.G., et al., 2010. Emergence of resting state networks in the preterm human brain. Proc. Natl. Acad. Sci. U. S. A. 107, 20015–20020.
Erberich, S.G., Panigrahy, A., Friedlich, P., Seri, I., Nelson, M.D., Gilles, F., 2006. Somatosensory lateralization in the newborn brain. Neuroimage 29, 155–161.
Feller, M.B., 1999. Spontaneous correlated activity in developing neural circuits. Neuron 22, 653–656.
Ferradal, S.L., Liao, S.M., Eggebrecht, A.T., Shimony, J.S., Inder, T.E., Culver, J.P., Smyser, C.D., 2015. Functional imaging of the developing brain at the bedside using diffuse optical tomography. Cereb. Cortex, 1–11. Epub ahead of print.
Ferrazzi, G., Kuklisova Murgasova, M., Arichi, T., Malamateniou, C., Fox, M.J., Makropoulos, A., Allsop, J., Rutherford, M., Malik, S., Aljabar, P., et al., 2014. Resting state fMRI in the moving fetus: a robust framework for motion, bias field and spin history correction. Neuroimage 101, 555–568.
Fox, M.D., Raichle, M.E., 2007. Spontaneous fluctuations in brain activity observed with functional magnetic resonance imaging. Nat. Rev. Neurosci. 8, 700–711.
Fox, P.T., Raichle, M.E., Mintun, M.A., Dence, C., Focal, D., 1988. Nonoxidative glucose consumption during focal physiologic neural activity. Science 241, 462–464.
Fransson, P., Aden, U., Blennow, M., Lagercrantz, H., 2011. The functional architecture of the infant brain as revealed by resting-state fMRI. Cereb. Cortex 21, 145–154.
Fujimoto, K., 1995. Pericyte-endothelial gap junctions in developing rat cerebral capillaries: a fine structural study. Anat. Rec. 242, 562–565.
Gao, P., Postiglione, M.P., Krieger, T.G., Hernandez, L., Wang, C., Han, Z., Streicher, C., Papusheva, E., Insolera, R., Chugh, K., et al., 2014. Article deterministic progenitor behavior and unitary production of neurons in the neocortex. Cell 159, 775–788.
Gelfand, M., Hong, S., Gu, C., 2009. Guidance from above: common cues direct distinct signaling outcomes in vascular and neural patterning. Trends Cell Biol. 19, 1199–1216.
Goldman-Rakic, P., Brown, R., 1982. Postnatal development of monoamine content and synthesis in the cerebral cortex of rhesus monkeys. Brain Res. 256, 339–349.
Graham, A.M., Pfeifer, J.H., Fisher, P.A., Lin, W., Gao, W., Fair, D.A., 2014. The potential of infant fMRI research and the study of early life stress as a promising exemplar. Dev. Cogn. Neurosci. 12, 12–39.
Greisen, G., 2009. To autoregulate or not to autoregulate—that is no longer the question. Semin. Pediatr. Neurol. 16, 207–215.
Hall, C.N., Reynell, C., Gesslein, B., Hamilton, N.B., Mishra, A., Sutherland, B.A., O'Farrell, F.M., Buchan, A.M., Lauritzen, M., Attwell, D., 2014. Capillary pericytes regulate cerebral blood flow in health and disease. Nature 508, 55–60.
Hamilton, N.B., Attwell, D., Hall, C.N., 2010. Pericyte-mediated regulation of capillary diameter: a component of neurovascular coupling in health and disease. Front. Neuroenergetics 2, 1–14.

Hand, R., Polleux, F., 2011. Neurogenin2 regulates the initial axon guidance of cortical pyramidal neurons projecting medially to the corpus callosum. Neural Dev. 6, 30.

Harb, R., Whiteus, C., Freitas, C., Grutzendler, J., 2013. In vivo imaging of cerebral microvascular plasticity from birth to death. J. Cereb. Blood Flow Metab. 33, 146–156.

Harris, J.J., Reynell, C., Attwell, D., 2011. The physiology of developmental changes in BOLD functional imaging signals. Dev. Cogn. Neurosci. 1, 199–216.

Heeger, D.J., Ress, D., 2002. What does fMRI tell us about neuronal activity? Nat. Rev. Neurosci. 3, 142–151.

Heep, A., Scheef, L., Jankowski, J., Born, M., Zimmermann, N., Sival, D., Bos, A., Gieseke, J., Bartmann, P., Schild, H., et al., 2009. Functional magnetic resonance imaging of the sensorimotor system in preterm infants. Pediatrics 123, 294–300.

Hensch, T.K., 2004. Critical period regulation. Annu. Rev. Neurosci. 27, 549–579.

Herken, R., Gotz, W., Wattjes, K.-H., 1989. Initial development of capillaries in the neuroepithelium of the mouse. J. Anat. 164, 85–92.

Hill, R.A., Tong, L., Yuan, P., Murikinati, S., Gupta, S., Grutzendler, J., 2015. Regional blood flow in the normal and ischemic brain is controlled by arteriolar smooth muscle cell contractility and not by capillary pericytes. Neuron 87, 95–110.

Hillman, E.M.C., 2014. Coupling mechanism and significance of the BOLD signal: a status report. Annu. Rev. Neurosci. 37, 161–181.

Hillman, E.M.C., Kozberg, M.G., 2013. What secrets can functional MRI reveal about the developing infant brain? Imaging Med. 5, 203–206.

Hoshi, Y., Tamura, M., 1993. Dynamic multichannel near-infrared optical imaging of human brain activity. J. Appl. Physiol. 75, 1842–1846.

Hoshi, Y., Kohri, S., Matsumoto, Y., Cho, K., Matsuda, T., Okajima, S., Fujimoto, S., 2000. Hemodynamic responses to photic stimulation in neonates. Pediatr. Neurol. 23, 323–327.

Hubel, D.H., Wiesel, T.N., 1963. Receptive fields of cells in striate cortex of very young, visually inexperienced kittens. J. Neurophysiol. 26, 994–1002.

Huttenlocher, P., de Courten, C., Garey, L., van der Loos, H., 1982. Synaptogenesis in human visual cortex—evidence for synapse elimination during normal development. Neurosci. Lett. 33, 247–252.

Iadecola, C., 2004. Neurovascular regulation in the normal brain and in Alzheimer's disease. Nat. Rev. Neurosci. 5, 347–360.

Iwahara, S.-I., Abe, Y., Okazaki, T., 1996. Identification of five embryonic hemoglobins of rat and ontogeny of their constituent globins during fetal development. J. Biochem. 119, 360–366.

Joeri, P., Huisman, T., Ekatodramis, D., Loenneker, T., Rumpel, H., Martin, E., 1996. Functional magnetic resonance imaging (fMRI) of the visual cortex in children. Pediatr. Res. 40, 535.

Karen, T., Morren, G., Haensse, D., Bauschatz, A.S., Bucher, H.U., Wolf, M., 2008. Hemodynamic response to visual stimulation in newborn infants using functional near-infrared spectroscopy. Hum. Brain Mapp. 29, 453–460.

Katz, L.C., Shatz, C.J., 1996. Synaptic activity and the construction of cortical circuits. Science 274, 1133–1138.

Keep, R., Jones, H., 1990. Cortical microvessels during brain development: a morphometric study in the rat. Microvasc. Res. 40, 412–426.

Khwaja, O., Volpe, J.J., 2008. Pathogenesis of cerebral white matter injury of prematurity. Arch. Dis. Child. Fetal Neonatal Ed. 93, F153–F161.

Kozberg, M.G., Chen, B.R., Deleo, S.E., Bouchard, M.B., Hillman, E.M.C., 2013. Resolving the transition from negative to positive blood oxygen level-dependent responses in the developing brain. Proc. Natl. Acad. Sci. 110, 4380–4385.

Kozberg, M.G., Ma, Y., Kim, S.H., and Hillman, E.M.C. (under review). Rapid postnatal expansion of neural networks occurs in an environment of altered neurovascular and neurometabolic coupling.

Kusaka, T., Kawada, K., Okubo, K., Nagano, K., Namba, M., Okada, H., Imai, T., Isobe, K., Itoh, S., 2004. Noninvasive optical imaging in the visual cortex in young infants. Hum. Brain Mapp. 22 (2), 122–132.

Lacoste, B., Gu, C., 2015. Control of cerebrovascular patterning by neural activity during postnatal development. Mech. Dev. 138, 43–49.

Lacoste, B., Comin, C.H., Ben-Zvi, A., Kaeser, P.S., Xu, X., Costa Lda, F., Gu, C., 2014. Sensory-related neural activity regulates the structure of vascular networks in the cerebral cortex. Neuron 83, 1117–1130.

Le Magueresse, C., Monyer, H., 2013. GABAergic interneurons shape the functional maturation of the cortex. Neuron 77, 388–405.

Lewis, T.L., Courchet, J., Polleux, F., 2013. Cell biology in neuroscience: cellular and molecular mechanisms underlying axon formation, growth, and branching. J. Cell Biol. 202, 837–848.

Liao, S.M., Gregg, N.M., White, B.R., Zeff, B.W., Bjerkaas, K.A., Inder, T.E., Culver, J.P., 2010. Neonatal hemodynamic response to visual cortex activity: high-density near-infrared spectroscopy study. J. Biomed. Opt. 15, 1–9.

Lind, B.L., Brazhe, A.R., Jessen, S.J., Tan, F.C.C., Lauritzen, M.J., 2013. Rapid stimulus-evoked astrocyte $Ca2+$ elevations and hemodynamic responses in mouse somatosensory cortex in vivo. Proc. Natl. Acad. Sci. 110, E4678–E4687.

Lindauer, U., Leithner, C., Kaasch, H., Rohrer, B., Foddis, M., Füchtemeier, M., Offenhauser, N., Steinbrink, J., Royl, G., Kohl-Bareis, M., et al., 2010. Neurovascular coupling in rat brain operates independent of hemoglobin deoxygenation. J. Cereb. Blood Flow Metab. 30, 757–768.

Lovick, T.A., Brown, L.A., Key, B.J., 1999. Neurovascular relationships in hippocampal slices: physiological and anatomical studies of mechanisms underlying flow-metabolism coupling in intraparenchymal microvessels. Neuroscience 92, 47–60.

Martin, E., Joeri, P., Loenneker, T., Ekatodramis, D., Vitacco, D., Hennig, J., Marcar, V.L., 1999. Visual processing in infants and children studied using functional MRI. Pediatr. Res. 46, 135–140.

McCandlish, C.A., Li, C.X., Waters, R.S., 1993. Early development of the SI cortical barrel field representation in neonatal rats follows a lateral-to-medial gradient: an electrophysiological study. Exp. Brain Res. 92, 369–374.

Meek, J.H., Firbank, M., Elwell, C.E., Atkinson, J., Braddick, O., Wyatt, J.S., 1998. Regional hemodynamic responses to visual stimulation in awake infants. Pediatr. Res. 43, 840–843.

Muramoto, S., Yamada, H., Sadato, N., Kimura, H., Konishi, Y., Kimura, K., Tanaka, M., Kochiyama, T., Yonekura, Y., Ito, H., 2002. Age-dependent change in metabolic response to photic stimulation of the primary visual cortex in infants: functional magnetic resonance imaging study. J. Comput. Assist. Tomogr. 26, 894–901.

Norman, M.G., 1986. The growth and development of microvasculature in human cerebral cortex. J. Neuropathol. Exp. Neurol. 45, 222–232.

Quinn, R., 2005. Comparing rat's to human's age: how old is my rat in people years? Nutrition 21, 775–777.

Rey, S., Semenza, G.L., 2010. Hypoxia-inducible factor-1-dependent mechanisms of vascularization and vascular remodelling. Cardiovasc. Res. 86, 236–242.

Riddle, D.R., Gutierrez, G., Zheng, D., White, L.E., Richards, A., Purves, D., 1993. Differential metabolic and electrical activity in the somatic sensory cortex of juvenile and adult rats. J. Neurosci. 13, 4193–4213.

Risser, L., Plouraboué, F., Cloetens, P., Fonta, C., 2009. A 3D-investigation shows that angiogenesis in primate cerebral cortex mainly occurs at capillary level. Int. J. Dev. Neurosci. 27, 185–196.

Rowan, R.A., Maxwell, D.S., 1981. Patterns of vascular sprouting in the postnatal development of the cerebral cortex of the rat. Am. J. Anat. 160, 247–255.

Sakadžić, S., Mandeville, E.T., Gagnon, L., Musacchia, J.J., Yaseen, M.A., Yucel, M.A., Lefebvre, J., Lesage, F., Anders, M., Eikermann-haerter, K., et al., 2014. Large arteriolar component of oxygen delivery implies a safe margin of oxygen supply to cerebral tissue. Nat. Commun. 5, 5734.

Sakatani, K., Chen, S., Lichty, W., Zuo, H., Wang, Y.P., 1999. Cerebral blood oxygenation changes induced by auditory stimulation in newborn infants measured by near infrared spectroscopy. Early Hum. Dev. 55, 229–236.

Saugstad, O.D., 2010. Resuscitation of newborn infants: from oxygen to room air. Lancet 376, 1970–1971.

Schmithorst, V.J., Vannest, J., Lee, G., Hernandez-Garcia, L., Plante, E., Rajagopal, A., Holland, S.K., 2015. Evidence that neurovascular coupling underlying the BOLD effect increases with age during childhood. Hum. Brain Mapp. 36, 1–15.

Seregi, A., Keller, M., Hertting, G., 1987. Are cerebral prostanoids of astroglial origin? Studies on the prostanoid forming system in developing rat brain and primary cultures of rat astrocytes. Brain Res. 404, 113–120.

Sie, L., Rombouts, S., Valk, J., Hart, A., Scheltens, P., van der Knaap, M., 2001. Functional MRI of visual cortex in infants with or without periventricular leukomalacia. Dev. Med. Child Neurol. 43, 486–490.

Singer, D., 1999. Neonatal tolerance to hypoxia: a comparative-physiological approach. Comp. Biochem. Physiol. A Mol. Integr. Physiol. 123, 221–234.

Smyser, C.D., Inder, T.E., Shimony, J.S., Hill, J.E., Degnan, A.J., Snyder, A.Z., Neil, J.J., 2010. Longitudinal analysis of neural network development in preterm infants. Cereb. Cortex 20, 2852–2862.

Stichel, C.C., Müller, C.M., Zilles, K., 1991. Distribution of glial fibrillary acidic protein and vimentin immunoreactivity during rat visual cortex development. J. Neurocytol. 20, 97–108.

Taga, G., Asakawa, K., Maki, A., Konishi, Y., Koizumi, H., 2003. Brain imaging in awake infants by near-infrared optical topography. Proc. Natl. Acad. Sci. 100, 107222–110727.

Takahashi, T., Shirane, R., Sato, S., Yoshimoto, T., 1999. Developmental changes of cerebral blood flow and oxygen metabolism in children. AJNR Am. J. Neuroradiol. 20, 917–922.

Takano, T., Tian, G.-F., Peng, W., Lou, N., Libionka, W., Han, X., Nedergaard, M., 2006. Astrocyte-mediated control of cerebral blood flow. Nat. Neurosci. 9, 260–267.

Takata, N., Nagai, T., Ozawa, K., Oe, Y., Mikoshiba, K., Hirase, H., 2013. Cerebral blood flow modulation by basal forebrain or whisker stimulation can occur independently of large cytosolic Ca2+ signaling in astrocytes. PLoS One 8, e66525.

Thein, S.L., Menzel, S., Lathrop, M., Garner, C., 2009. Control of fetal hemoglobin: new insights emerging from genomics and clinical implications. Hum. Mol. Genet. 18, R216–R223.

Tian, P., Teng, I.C., May, L.D., Kurz, R., Lu, K., Scadeng, M., Hillman, E.M.C., Crespigny, A.J. De, Arceuil, H.E.D., Mandeville, J.B., et al., 2010. Cortical depth-specific microvascular dilation underlies laminar differences in blood oxygenation level-dependent functional MRI signal. Proc. Natl. Acad. Sci. U. S. A. 107, 15246–15251.

Ujiie, H., Chaytor, A.T., Bakker, L.M., Griffith, T.M., 2003. Essential role of Gap junctions in NO- and prostanoid-independent relaxations evoked by acetylcholine in rabbit intracerebral arteries. Stroke 34, 544–550.

Uylings, H.B., Van Eden, C.G., Parnavelas, J.G., Kalsbeek, A., 1990. The prenatal and postnatal development of rat cerebral cortex. In: Kolb, B., Tees, R.C. (Eds.), The Cerebral Cortex of the Rat. MIT Press, Cambridge, MA, pp. 36–76.

Volpe, J.J., 1997. Brain injury in the premature infant—from pathogenesis to prevention. Brain Dev. 19, 519–534.

Vovenko, E., 1999. Distribution of oxygen tension on the surface of arterioles, capillaries and venules of brain cortex and in tissue in normoxia: an experimental study on rats. Pflugers Arch. 437, 617–623.

Wang, D.-B.B., Blocher, N.C., Spence, M.E., Rovainen, C.M., Woolsey, T.A., 1996. Development and remodeling of cerebral blood vessels and their flow in postnatal mice observed with in vivo videomicroscopy. J. Cereb. Blood Flow Metab. 12, 935–946.

Wang, C.-L., Zhang, L., Zhou, Y., Zhou, J., Yang, X.-J., Duan, S., Xiong, Z.-Q., Ding, Y.-Q., 2007. Activity-dependent development of callosal projections in the somatosensory cortex. J. Neurosci. 27, 11334–11342.

Watanabe, H., Homae, F., Nakano, T., Taga, G., 2008. Functional activation in diverse regions of the developing brain of human infants. Neuroimage 43, 346–357.

Weber, B., Keller, A.L., Reichold, J., Logothetis, N.K., 2008. The microvascular system of the striate and extrastriate visual cortex of the macaque. Cereb. Cortex 18, 2318–2330.

Wheatley, C.M., Dickinson, J.L., Mackey, D.A., Craig, J.E., Sale, M.M., 2002. Retinopathy of prematurity: recent advances in our understanding. Arch. Dis. Child. Fetal Neonatal Ed. 87, F78–F82.

Whiteus, C., Freitas, C., Grutzendler, J., 2014. Perturbed neural activity disrupts cerebral angiogenesis during a postnatal critical period. Nature 505, 407–411.

Wolf, T., Lindauer, U., Villringer, A., Dirnagl, U., 1997. Excessive oxygen or glucose supply does not alter the blood flow response to somatosensory stimulation or spreading depression in rats. Brain Res. 761, 290–299.

Yamada, H., Sadato, N., Konishi, Y., Kimura, K., Tanaka, M., Yonekura, Y., Ishii, Y., 1997. A rapid brain metabolic change in infants detected by fMRI. Neuroreport 8, 3775–3778.

Zaramella, P., Freato, F., Amigoni, A., Salvadori, S., Marangoni, P., Suppjei, A., Schiavo, B., Chiandetti, L., 2001. Brain auditory activation measured by near-infrared spectroscopy (NIRS) in neonates. Pediatr. Res. 49, 213–219.

Zehendner, C.M., Tsohataridis, S., Luhmann, H.J., Yang, J.-W., 2013. Developmental switch in neurovascular coupling in the immature rodent barrel cortex. PLoS One 8, e80749.

Zijlstra, W.G., Buursma, A., Meeuwsen-van der Roest, W.P., 1991. Absorption spectra of human fetal and adult oxyhemoglobin, de-oxyhemoglobin, carboxyhemoglobin, and methemoglobin. Clin. Chem. 37, 1633–1638.

Zuckerman, S.L., Armstead, W., Hsu, P., Shibata, M., Leffler, C., 1996. Age dependence of cerebrovascular response mechanisms in domestic pigs. Am. J. Physiol. Heart Circ. Physiol. 271, H535–H540.

CHAPTER

Exercise and cerebrovascular plasticity

11

T. Nishijima*[,1], I. Torres-Aleman[†], H. Soya[‡]

**Tokyo Metropolitan University, Tokyo, Japan*

[†]Cajal Institute, Madrid, Spain

[‡]University of Tsukuba, Ibaraki, Japan

[1]Corresponding author: Tel.: +81-42-677-2972; Fax: +81-42-677-2961

e-mail address: t-nishijima@tmu.ac.jp

Abstract

Aging impairs cerebrovascular plasticity and subsequently leads cerebral hypoperfusion, which synergistically accelerates aging-associated cognitive dysfunction and neurodegenerative diseases associated with impaired neuronal plasticity. On the other hand, over two decades of researches have successfully demonstrated that exercise, or higher level of physical activity, is a powerful and nonpharmacological approach to improve brain function. Most of the studies have focused on the neuronal aspects and found that exercise triggers improvements in neuronal plasticity, such as neurogenesis; however, exercise can improve cerebrovascular plasticity as well. In this chapter, to understand these beneficial effects of exercise on the cerebral vasculature, we first discuss the issue of changes in cerebral blood flow and its regulation during acute bouts of exercise. Then, how regular exercise improves cerebrovascular plasticity will be discussed. In addition, to shed light on the importance of understanding interactions between the neuron and cerebral vasculature, we describe neuronal activity-driven uptake of circulating IGF-I into the brain.

Keywords

Exercise, Neurogenesis, Angiogenesis, Neurovascular coupling, Neurotrophic coupling, IGF-I

1 INTRODUCTION

With the expansion of the aging population worldwide, aging-associated impairments of brain function (including cognitive decline, dementia, and neurodegenerative diseases) have become serious health concerns. These impairments are considered to be, in part, due to age-associated dysregulation of cerebral microcirculation and subsequent cerebral hypoperfusion (Gorelick et al., 2011). Whereas

Progress in Brain Research, Volume 225, ISSN 0079-6123, http://dx.doi.org/10.1016/bs.pbr.2016.03.010

numerous studies are demonstrating that exercise or higher levels of physical activity can improve brain health, for example, preventing age-associated cognitive decline (Colcombe et al., 2003; Kramer et al., 1999; Yaffe et al., 2001), reducing the risk of depression (Brown et al., 2005; Ku et al., 2012; Paffenbarger et al., 1994) and Alzheimer's disease (Buchman et al., 2012; Hamer and Chida, 2009; Lautenschlager et al., 2008).

To reveal the underlying mechanisms of how exercise improves brain function, the majority of studies have focused on neuronal plasticity, such as neurogenesis and trophic factor expression. Cerebrovascular plasticity is another important target that can be improved by exercise (Barnes, 2015; Lange-Asschenfeldt and Kojda, 2008; Tarumi and Zhang, 2014), although this has received relatively less attention (Bolduc et al., 2013). So far, regular exercise has been demonstrated to increase angiogenesis (Black et al., 1990; Clark et al., 2009; Ding et al., 2006; Van der Borght et al., 2009), ameliorate an age-related decline in cerebral blood flow (CBF) (Ainslie et al., 2008; Viboolvorakul and Patumraj, 2014), increase cerebrovascular reactivity (Barnes et al., 2013; Murrell et al., 2013), and improve tolerance to ischemic stroke (Arrick et al., 2012; Endres et al., 2003). In this chapter, to understand these beneficial effects of exercise on cerebrovascular plasticity, we start from the issue of changes in CBF and its regulation during acute bouts of exercise.

2 CHANGES IN CBF DURING EXERCISE

During exercise, to meet the substantial increase of oxygen demand of skeletal muscles, blood flow to contracting muscles increases in an intensity-dependent manner and reaches 15–20 times greater at maximal exercise, compared to the resting value (Powers and Howley, 2012). The increase in muscle blood flow is achieved by both increased cardiac output (CA) and redistribution of CA from less-active visceral organs to contracting skeletal muscles. At rest, skeletal muscles receive approximately 15–20% of total CA, which increases up to 80–85% during maximal exercise (Powers and Howley, 2012). Even in this situation when drastic whole-body hemodynamic changes are triggered by exercise, consistent blood flow to the brain is necessary to supply oxygen and energy substrates for maintaining brain function. Therefore, changes in CBF during exercise have garnered great interest in the field of exercise physiology.

Traditionally, global and regional CBF has been believed to be relatively stable during exercise in both experimental animals (Foreman et al., 1976; Globus et al., 1983; Gross et al., 1980; Hohimer et al., 1983) and humans (Scheinberg et al., 1954; Zobl et al., 1965). For example, an initial study by Foreman et al. (1976) examined changes in total and regional CBF during mild (75% of maximal heart rate) and exhaustive treadmill exercise in miniature swine by a radioactive microsphere method. They found no significant changes in both total and regional CBF in the cortex, thalamus, hippocampus, medulla oblongata, and pons, but regional CBF increases only in the cerebellum gray matter (Foreman et al., 1976). However, this classical concept of the stable CBF during exercise has been now revised (Ogoh and Ainslie, 2009; Querido and Sheel, 2007; Secher et al., 2008). In particular, Delp et al. (2001) examined in detail CBF changes in 23 brain regions during treadmill running at mild and maximum speeds (at 70% and 100% of $\dot{V}O_2$ max, respectively) in miniature swine by using radioactive microsphere, and found significant increases in both total and regional CBF, especially in the subcortical regions involved in motor and cardiorespiratory control, maintenance of equilibrium, and vision (Delp et al., 2001). These findings are further supported by the study demonstrating that slow-speed walking on the rotating rod (3.1 cm/s) increased regional CBF in the neural circuits involving motor and sensory function (primary motor cortex, primary somatosensory cortex, dorsolateral striatum, ventrolateral thalamus, cerebellum, and visual cortex) in rats when using [^{14}C]-iodoantipyrine method (Holschneider et al., 2003).

The radioactive tracers methods, used in the above-mentioned studies, have an advantage when examining the distribution of regional CBF across the brain, but cannot continuously monitor changes in CBF during exercise because they require postmortem analysis. Alternatively, laser-Doppler flowmetry (LDF) enables continuous and real-time monitoring of regional CBF, even in conscious and exercising rodents. With using LDF, walking/running on the treadmill has been demonstrated to increase regional CBF in the somatosensory cortex (Gu et al., 2003), hippocampus (Nakajima et al., 2003; Nishijima and Soya, 2006; Nishijima et al., 2012; Osborne, 1997), and striatum (Osborne, 1997), but not in the olfactory bulb (Nishijima et al., 2012) in rats (Fig. 1). LDF studies also demonstrated that hippocampal CBF increased immediately after the initiation of walking, and returned to basal levels after cessation of short-term walking (2.4 m/min, for 30 s) (Nakajima et al., 2003), whereas, increased hippocampal CBF with long-term running (10 m/min, for 100 min) remained elevated for 10–20 min after its cessation (Nishijima and Soya, 2006). Importantly, based on the concepts of neurovascular coupling, these animal studies also contribute to identify the regions those neuronal activities are activated by exercise (Soya et al., 2011).

Exercise-induced increases in total and regional CBF have been demonstrated in humans as well, which are well reviewed elsewhere (Ogoh and Ainslie, 2009; Querido and Sheel, 2007; Secher et al., 2008). Because fMRI and PET require physical restriction, most studies use transcranial Doppler ultrasound (TCD) and examine changes in blood flow velocity in the basal cerebral arteries, including the middle (MCA), anterior (ACA), and posterior cerebral artery (PCA) as surrogates for CBF. It is reported that unilateral handgrip exercise increases contralateral MCA velocity (MCAv) (Jorgensen et al., 1993; Linkis et al., 1995). This suggests that the exercise-induced changes in MCAv reflect neuronal activation within the cortical region those CBF is supplied by MCA. However, this assumption should be carefully applied to dynamic whole body exercise, especially if the exercise intensity is high. Observed changes in MCAv do not simply reflect the changes in neuronal activation.

Overwhelming evidence indicates that, during incremental exercise up to ~60% of $\dot{V}O_2$ max, MCAv linearly increases by 10–20%, then decreases toward baseline at heavy intensity (Ide and Secher, 2000; Ogoh and Ainslie, 2009; Querido and Sheel,

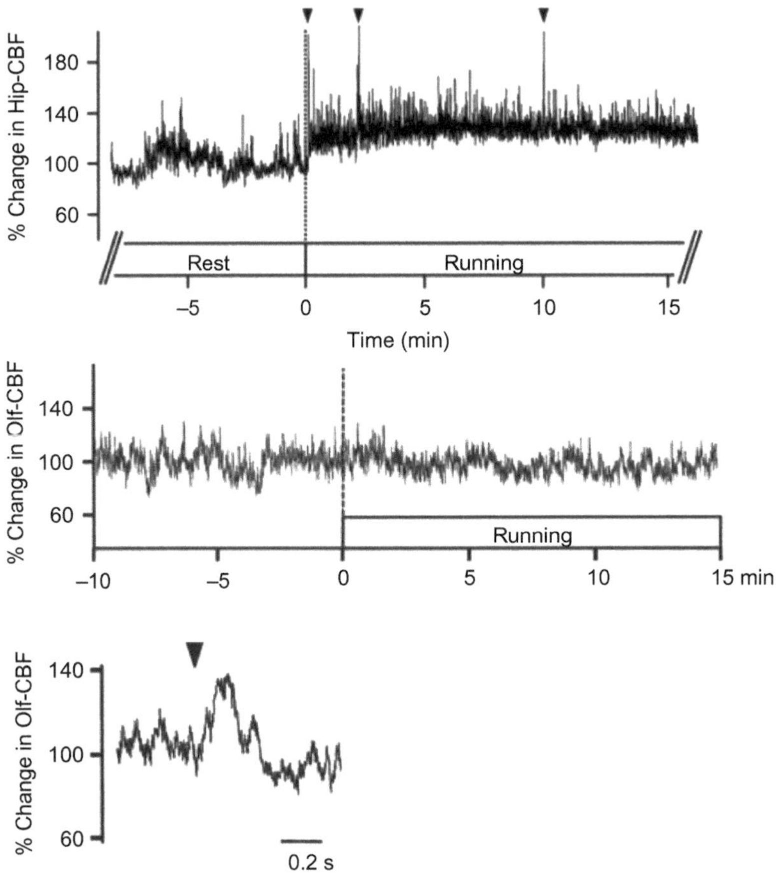

FIG. 1

Typical row data of laser-Doppler recording of hippocampal CBF (Hip-CBF) and olfactory CBF (Olf-CBF) during treadmill running at mild speed (10 m/min). Hip-CBF increased from onset of the treadmill running and maintained during exercise. Movement-induced artifacts indicated by arrowheads were occasionally observed, but easily distinguishable and omitted from data analysis (top). Olf-CBF did not change with treadmill running (middle), but instantaneously increased with odor stimuli (bottom). An arrowhead indicates a point that a rat sniffed ethanol immersed cotton.

Top panel adopted from Nishijima, T., Soya, H., 2006. Evidence of functional hyperemia in the rat hippocampus during mild treadmill running. Neurosci. Res. 54, 186–191. Middle and bottom panels adopted from Nishijima, T., Okamoto, M., Matsui, T., Kita, I., Soya, H., 2012. Hippocampal functional hyperemia mediated by NMDA receptor/NO signaling in rats during mild exercise. J. Appl. Physiol. 112, 197–203.

2007; Sato et al., 2011) (Fig. 2). Importantly, exercise alters physiological factors influencing CBF in an intensity-dependent manner (Fig. 3). Mean arterial pressure (MAP) and CA increases linearly to exercise intensity (ie, $\dot{V}O_2$ max) and partial pressure of arterial oxygen (PaO_2) slightly decreases, which would contribute to the

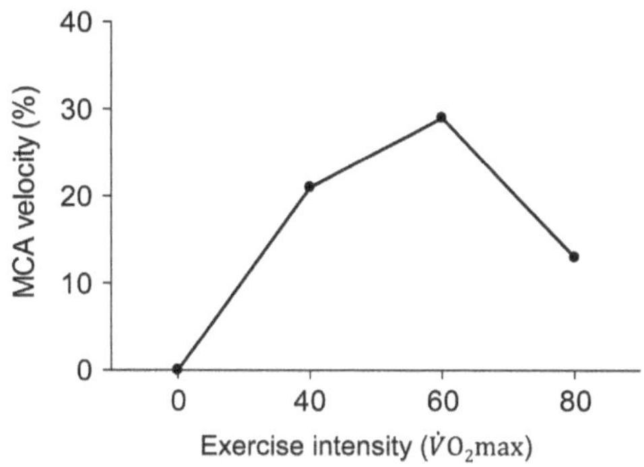

FIG. 2

Exercise intensity-dependent changes in MCA velocity during exercise. Healthy nonathletic subjects performed cycling exercise at 40%, 60%, and 80% $\dot{V}O_2$ max, and changes in MCA velocity were monitored by transcranial Doppler ultrasound. MCA velocity increases up to ~60% of $\dot{V}O_2$ max, then decreases toward baseline at heavy intensity.

Adapted from Sato, K., Ogoh, S., Hirasawa, A., Oue, A., Sadamoto, T., 2011. The distribution of blood flow in the carotid and vertebral arteries during dynamic exercise in humans. J. Physiol. 589, 2847–2856.

increase in MCAv. Importantly, partial pressure of arterial carbon dioxide ($PaCO_2$), a potent vasodilative factor, increases at low-to-moderate intensity, but subsequently decreases even lower than resting value during heavy-intensity exercise because of hyperventilation. Therefore, the drop of MCAv at high-intensity exercise is considered to be due to the hyperventilation-induced hypocapnia, despite further increase in neuronal activation (Ogoh and Ainslie, 2009; Querido and Sheel, 2007). The decrease in CBF at heavy-and-exhaustive exercise might become a limiting factor for exercise execution, so called central fatigue (Nybo and Nielsen, 2001); however, this issue still remains controversial (Ide and Secher, 2000; Secher et al., 2008).

Because studies using TCD cannot provide direct evidence about changes in regional CBF in the brain microcirculation during exercise, several PET studies have been conducted (Christensen et al., 2000; Hiura et al., 2014; Williamson et al., 1999). In the recent study by Hiura et al. (2014), male subjects performed cycling exercise (50–60% of estimated heart rate max for 15 min) on the scanner bed in supine position—in order to restrict a head movement—and changes in rCBF were quantitatively investigated. As expected, exercise increased rCBF in motor-related regions (primary motor and somatosensory cortex for the bilateral legs, supplementary motor area, cerebellar vermis and cortex, and left insula cortex). Importantly, this study performed PET scans at initial and latter phase of exercise (3 and 13 min after the onset of exercise, respectively), and found that the increases in rCBF were significantly lower at the latter phase of exercise than the initial phase. Furthermore, at the latter phase, the significant increases from resting value were only found in the primary motor and somatosensory cortex and cerebellar vermis. Because CA and $PaCO_2$ were stable during exercise, the transient rCBF increases suggest a specific mechanism regulating rCBF during the onset of exercise. This study clearly indicates the importance of examining time-course changes in rCBF during exercise, which might

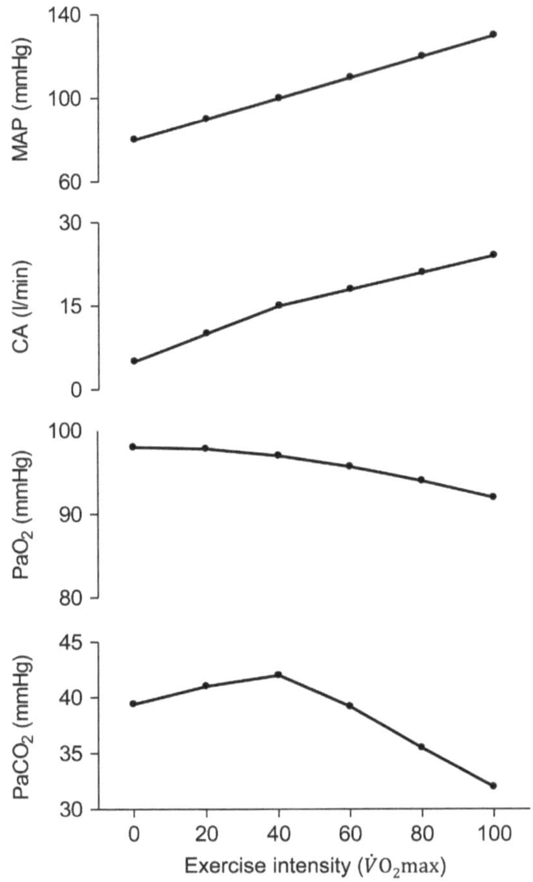

FIG. 3

Changes in arterial blood parameters regulating CBF during acute bouts of dynamic exercise in intensity-dependent manners. Note that the absolute value and amount of responses of these parameters change with age and fitness level of subjects. *MAP*, mean arterial pressure; *CA*, cardiac output.

reveal a higher cortical center (central command network) involved in autonomic adaptation during the onset of exercise (ie, before cardiovascular response reaches a state of equilibrium) (Hiura et al., 2014).

3 REGULATION OF CBF DURING EXERCISE

In addition to the above-mentioned physiological factors (MAP, CA, PaO_2, and $PaCO_2$), many other factors (including pH, glucose, lactate, sympathetic and parasympathetic nerve activity, muscle mechanoreceptor, and body temperature) are involved in CBF regulation, which are excellently reviewed elsewhere

(Ogoh and Ainslie, 2009; Querido and Sheel, 2007). Therefore, regulation of CBF during exercise is extremely complex, not only because of the multifactorial regulation, but also because these physiological factors change depending on the intensity and duration of exercise. Details of the CBF regulation by these systemic factors are out of focus and do not describe in this chapter. More importantly, in addition to these systemic factors, neuronal activity must be critically involved in regulating cerebral vascular tone during exercise; however, definitive evidence is surprisingly lacking.

Nakajima et al. (2003) demonstrated with rats that increases in hippocampal CBF (Hip-CBF) during treadmill walking was partially inhibited by both nicotinic acetylcholine receptor antagonist (mecamylamine) and NO synthase inhibitor (L-NAME) (Nakajima et al., 2003). These results suggest that Hip-CBF during walking is, in part, regulated by cholinergic vasodilation and NO production (see also chapter "Neurogenic control of parenchymal arterioles in the cerebral cortex" by Hotta). However, because these inhibitors were injected intravenously before exercise, these systemic drug treatments altered resting MAP and suppressed exercise-induced increase in MAP, which made it difficult to interpret the results.

In order to examine neuronal regulation of CBF during exercise, we have developed a unique rat model that combines LDF with microdialysis (Fig. 4) (Nishijima and Soya, 2006; Nishijima et al., 2012). Hip-CBF was continuously monitored during mild treadmill running (10 m/min) for over 100 min, and local drug administration was performed by microdialysis from 20 to 30 min after the start of running. Hip-CBF increased from the onset of running and maintained its higher level (about 27% on average compared to resting state) throughout the running bouts. We successfully found that the increase in Hip-CBF was almost completely suppressed by the infusion of tetrodotoxin (TTX), NMDA receptor antagonist (MK-801), and NO synthase inhibitor (L-NAME). Importantly, because the running speed was very slow, rats easily performed the long-term running and physiological factors (MAP, PaO_2, $PaCO_2$, blood glucose, and lactate) were kept stable throughout the running. Furthermore, the local drug treatments into the hippocampus did not affect these physiological factors (Kaiser and During, 1995; Nishijima et al., 2012). Thus, these results clearly demonstrate that neuronal activity and neurogenic mediators are functionally involved in the regulation of Hip-CBF during mild exercise. Although neuronal regulation during high-intensity exercise was not examined (because rats cannot perform treadmill running for a long time, and vigorous head movements produce mechanical artifacts in LDF recording), our results suggest that, at least in submaximal exercise, neuronal activity is a dominant factor in regulating regional CBF during exercise.

4 NVC DURING EXERCISE

During exercise, especially in competitive ball games such as football and basketball, perception and integration of various external stimuli are necessary for conducting appropriate motor action, or better sports performance. Therefore, one of the challenging issues in sports science is how central nervous system (CNS) function

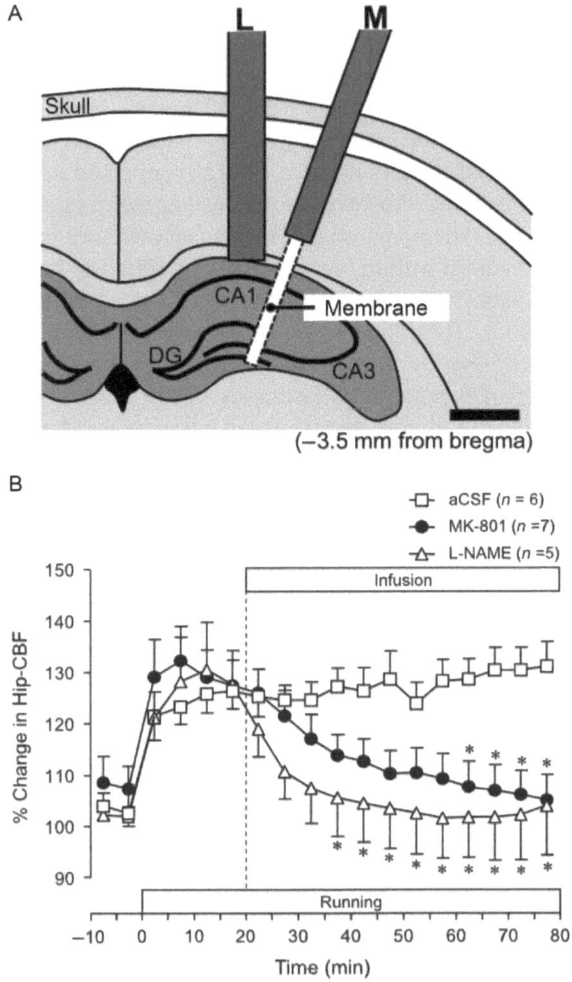

FIG. 4

Neuronal regulation of hippocampal CBF (Hip-CBF) during mild treadmill running in rats. (A) Schematic illustration showing the implantation of paired guide cannulae into the dorsal hippocampus of rat. Because the measuring volume of LDF is approximately 1 mm^3 from the tip of the probe, the close proximity is necessary to deliver exogenous drug into the area where CBF is monitored. *L*, guide cannula for laser-Doppler flowmetry (LDF); *M*, microdialysis probe. Scale bar represents 1 mm. (B) Percent changes in hippocampal CBF (Hip-CBF) vs baseline during mild treadmill running. Intrahippocampal infusion of MK-801 (1 mM) and L-NAME (2 mM) performed 20 min after the stat of running significantly suppressed the increase in Hip-CBF during exercise.

Adopted from Nishijima, T., Okamoto, M., Matsui, T., Kita, I., Soya, H., 2012. Hippocampal functional hyperemia mediated by NMDA receptor/NO signaling in rats during mild exercise. J. Appl. Physiol. 112, 197–203.

is altered during exercise. As mentioned above, acute bouts of exercise itself activate neuronal activity and increases both regional and global CBF. Accordingly, Willie et al. (2011) examined an intriguing hypothesis that "CBF reserve might be attenuated leading to a reduction in neurovascular coupling during exercise." To test the hypothesis, healthy subjects performed a visual stimulation task (reading a scientific journal article, 40 s for each trial) both at rest and during cycling exercise at moderate intensity (~70% of estimated heart rate max, 152.2 ± 19.1 bpm) and blood flow velocity in the PCA—that supplies blood to the occipital visual cortex—was measured using TCD. As expected, exercise increased mean PCA velocity by about 26%. However, contrary to their hypothesis, visual stimulation similarly increased PCA velocity both at rest (20% from prestimulation value) and during exercise (19%), and time-course profiles of PCA velocity were also identical in both conditions in spite of the increased baseline (Fig. 5). These results indicate that NVC, in response to visual stimulation, is maintained during submaximal exercise (Willie et al., 2011).

This finding was further supported by a series of studies examining different modes and intensities of exercise (Yamaguchi et al., 2014, 2015a,b). They found that NVC, ie, an increase in PCA velocity evoked by visual stimulation (a reversed checker board), was also maintained during static handgrip exercise (at 30% of maximal voluntary contraction, for 2 min) (Yamaguchi et al., 2014) and low-to-moderate intensity cycling exercise (~70% of estimated heart rate max) (Yamaguchi et al., 2015b). However, the visual stimulation-induced NVC was no longer maintained during high intensity (about 80% of estimated heart rate max) and exhaustive cycling

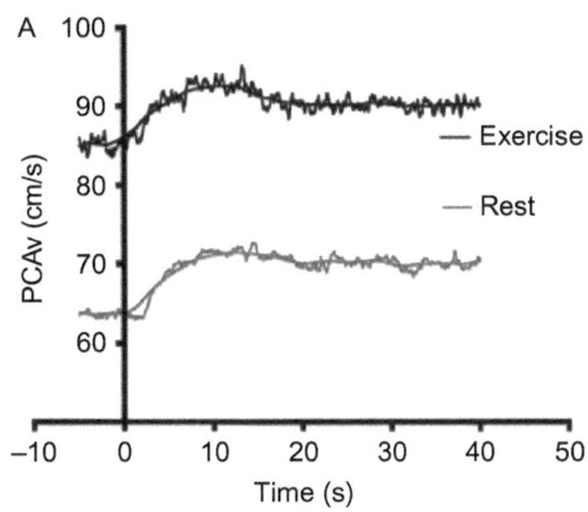

FIG. 5

Changes in posterior common artery (PCA) velocity in response to visual stimulation was not different between at rest and during exercise in spite of the increased baseline with exercise. Because subjects closed their eyes before visual stimulation, PCA velocity at the late phase of visual stimulation remained higher than baseline.

Adopted from Willie, C.K., Cowan, E.C., Ainslie, P.N., Taylor, C.E., Smith, K.J., Sin, P.Y., Tzeng, Y.C., 2011. Neurovascular coupling and distribution of cerebral blood flow during exercise. J. Neurosci. Methods 198, 270–273.

exercise (Yamaguchi et al., 2015a,b). During low-to-moderate intensity exercise, visual stimulation also increased MAP (ie, pressure response), which contributes, in part, to the visual stimulation-induced NVC. In contrast, during high-intensity exercise, the pressure response to visual stimulation was almost completely abolished, which is considered to be a condition of the suppressed NVC responding to visual stimulation.

Although studies examining NVC in other brain regions in response to different stimuli/context are still lacking, the impaired NVC during high-intensity exercise led us to speculate that CNS function would be impaired during high-intensity exercise. Indeed, Ando et al. (2012) have demonstrated that premotor reaction time (time to onset of the electromyographic activity) in response to visual stimulation, which represents the time that CNS perceives and processes a visual stimulus and conducts a motor command, became significantly longer during cycling exercise at high-intensity (75% $\dot{V}O_2$ max), but not at low-intensity (40% $\dot{V}O_2$ max). Although it is uncertain whether the impaired NVC is a causal factor of CNS dysfunction, or whether neuronal reactivity itself is blunted during high-intensity exercise, further studies will be needed to understand how acute exercise, especially at high-intensity, affects NVC and CNS function.

5 EFFECTS OF REGULAR EXERCISE ON THE BRAIN FUNCTION

5.1 WHY IT DOES MATTER?

As Marcus Tullius Cicero (ancient Roman philosopher, BC106–43) said that "it is exercise alone that supports the spirits and keeps the mind in vigor", it has been empirically believed that exercise, or higher levels of physical activity, can improve cognitive function and mental health. Since the historical discovery that exercise increases brain-derived neurotrophic factor (BDNF) gene expression in the rat hippocampus (Neeper et al., 1995), extensive researches have been conducted on both human and experimental rodents, demonstrating incontrovertible evidence that regular exercise improves not only muscle, but also brain function (Dishman et al., 2006; Erickson et al., 2014; Hillman et al., 2008; Intlekofer and Cotman, 2013; Loprinzi et al., 2013; Mattson, 2012).

On the contrary, populations worldwide are increasingly becoming physically inactive both in the developed and developing country (Dumith et al., 2011; Guthold et al., 2008; Kohl et al., 2012). Indeed, the global prevalence of physical inactivity among adults over 15 years old is estimated around 31% (Hallal et al., 2012). Physical inactivity is, without doubt, established as the definitive risk factor for somatic health problems, including cardiovascular diseases, type 2 diabetes, and some types of cancers (Lee et al., 2012), and is currently rated as the fourth leading risk factor for global mortality (WHO, 2009). Importantly, recent epidemiological studies have been revealing that physical inactivity also relates to psychological health problems, especially depression (Brown et al., 2005; Camacho et al., 1991; Ku et al., 2012;

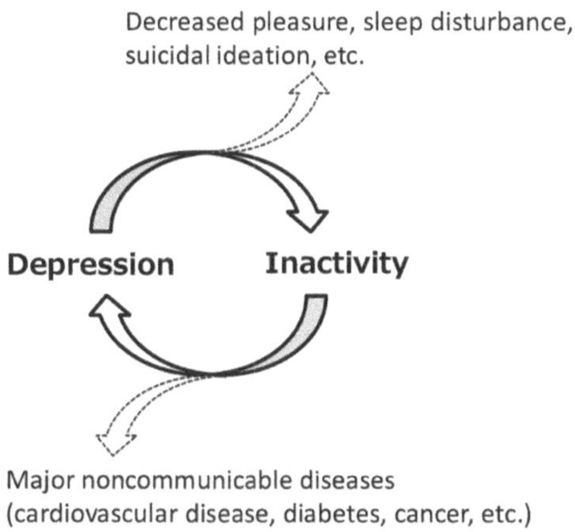

FIG. 6

Epidemiological studies have demonstrated that physical inactivity and depression are bidirectional, which strongly suggests that these two form a negative chain reaction.
Adopted from Nishijima, T., Kita, I., 2015. Deleterious effects of physical inactivity on the hippocampus: new insight into the increasing prevalence of stress-related depression. J. Phys. Fitness Sports Med. 4, 253–258.

Lucas et al., 2011; Motl et al., 2004; Paffenbarger et al., 1994; Teychenne et al., 2008). Because associations between physical inactivity and depression are bi-directional (Fig. 6), brain health promotion through exercise will become more and more important to combat against today's inactive lifestyle (Nishijima and Kita, 2015).

So far, we have discussed CBF responses, and its regulation during acute bouts of exercise. In the next section, we briefly outline how regular exercise affects neuronal plasticity. Following this, effects of regular exercise on cerebrovascular plasticity will be discussed in more detail.

5.2 EFFECTS OF REGULAR EXERCISE ON NEURONAL PLASTICITY

A recent meta-analysis showed that regular exercise can decrease the incidence of neurodegenerative diseases, including dementia and Alzheimer's disease (Hamer and Chida, 2009). Furthermore, Sofi et al. (2011) has performed a meta-analysis of prospective studies that examined association between physical activity and cognitive decline in nondemented subjects. They found that the risk of cognitive decline was significantly lower (hazard ratio (HR) = 0.62, 95% confidence interval (CI) = 0.54–0.70) in subjects performing high levels of physical exercise compared to sedentary subjects (Sofi et al., 2011). Importantly, the protective effect was also evident in subjects performing low-to-moderate levels of physical activity (HR = 0.65, 95% CI = 0.57–0.75). It should be noted that the latter meta-analysis revealed that regular exercise is protective even against the nonclinical cognitive decline, which plays a fundamental role in maintaining quality of life. These cognitive enhancements through regular exercise are found in young subjects as well. For example, Hillman et al. (2014) reported that 9 months of after-school exercise programs improved executive function in children (8- to 9-years old) (Hillman et al., 2014).

As potential mechanisms underlying these beneficial effects of exercise on cognitive function, regular exercise is known to improve long-term potentiation (LTP), the basis of synaptic plasticity for learning and memory (Farmer et al., 2004; van Praag et al., 1999a). Furthermore, the most well-characterized plastic change induced by regular exercise is hippocampal neurogenesis, which plays a significant role in hippocampus dependent learning, memory, and antidepressant action (Sahay and Hen, 2007). In 1999, van Praag et al. published the first experiments describing that voluntary wheel running increased cell proliferation, survival, and neuronal differentiation in the hippocampal dentate gyrus in mice (van Praag et al., 1999b). This milestone discovery has fascinated both neuroscientist and sports scientists alike, and the exercise-induced neurogenesis has been replicated with abundant studies and confirmed throughout their lifespan, from young (Helfer et al., 2009; Li et al., 2013) to old (Kronenberg et al., 2006; van Praag et al., 2005), including Alzheimer's model animals as well (Mirochnic et al., 2009; Rodriguez et al., 2011). Importantly, exercise-induced neurogenesis seems to contribute, in part, to the improved hippocampus-dependent learning and memory (Clark et al., 2008; van Praag et al., 1999a), spatial pattern separation (Creer et al., 2010), and antidepressant effects (Ernst et al., 2006). Although the exercise-induced neurogenesis has not yet been confirmed in humans, several studies demonstrated that regular exercise can increase the volume and perfusion of the human hippocampus (Erickson et al., 2011; Maass et al., 2015; Pereira et al., 2007; ten Brinke et al., 2015). In addition, regular exercise can enhance morphology of the hippocampal neuron, such as synaptogenesis and dendritic arborization (Dietrich et al., 2008; Redila and Christie, 2006).

Although the molecular mechanisms for exercise-induced neurogenesis and cognitive improvement are still fully uncertain, there is abundant evidence indicating significant roles of neurotrophins and growth factors (Intlekofer and Cotman, 2013). As mentioned before, BDNF is a molecule that is consistently shown to increase with regular exercise. BDNF itself has neurogenic effects (Scharfman et al., 2005), and blocking the BDNF action suppresses the exercise-induced neurogenesis and hippocampal dependent functions (Li et al., 2008; Vaynman et al., 2004). Insulin-like growth factor I (IGF-I), a potent neuroprotective hormone produced mostly in the liver, is another pivotal molecule mediating the effects of exercise on the brain (Fernandez and Torres-Aleman, 2012). Series of studies by Torres-Aleman et al. have demonstrated that blocking an action of serum IGF-I by systemic injection of IGF-1 antiserum, or reducing serum level of IGF-I, inhibits the effects of exercise on the hippocampus, including neurogenesis (Carro et al., 2000, 2001; Lopez-Lopez et al., 2004; Trejo et al., 2001). Importantly, these findings suggest an active transport mechanism of serum IGF-I into the brain parenchyma through the blood–brain barrier (BBB), which will be discussed later. A number of other factors, such as inflammatory cytokines (Nichol et al., 2008), neurotransmitters (Garcia et al., 2003), redox signaling (Radak et al., 2013), sex hormones (Okamoto et al., 2012), and nutrients (van Praag, 2009), are known to modulate the effects of exercise on the brain, although details are not described here. Taken together, regular exercise induces profound molecular and structural changes associated with neuronal plasticity, which contributes to gain functional improvement and resilience to cognitive and mental deficits associated with aging and stress (Stranahan and Mattson, 2012).

5.3 EFFECTS OF REGULAR EXERCISE ON CEREBROVASCULAR PLASTICITY

Neurons and the cerebrovascular bed are inseparable. Like a chicken-and-egg dilemma, loss of neurons causes cerebral hypoperfusion, whereas an insufficient blood supply causes neuronal dysfunction, which subsequently increases the risk for cognitive impairments and neurodegenerative diseases (Gorelick et al., 2011). Conversely, it is plausible to say that improved neuronal plasticity should be accompanied with improved cerebrovascular plasticity. Indeed, recent researches are revealing the remarkable impacts of regular exercise on the cerebrovascular plasticity (Barnes, 2015; Lange-Asschenfeldt and Kojda, 2008; Tarumi and Zhang, 2014).

Basal levels of CBF gradually decline with aging, but regular exercise can prevent the reduction. Ainslie et al. (2008) measured basal MCAv of sedentary and endurance-trained men aged 19–79 years old. MCAv had progressively declined with age in both groups (~1% per year), but the endurance-trained subjects consistently demonstrated higher levels of MCAv (Fig. 7) (Ainslie et al., 2008).

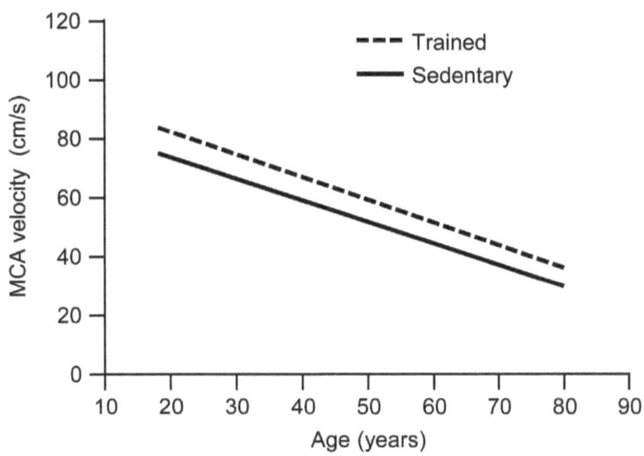

FIG. 7

Decline in MCA velocity with aging in sedentary and endurance-trained men.

Adapted from Ainslie, P.N., Cotter, J.D., George, K.P., Lucas, S., Murrell, C., Shave, R., Thomas, K.N., Williams, M.J., Atkinson, G., 2008. Elevation in cerebral blood flow velocity with aerobic fitness throughout healthy human ageing. J. Physiol. 586, 4005–4010.

The protective effect of exercise against age-related decline in CBF was also confirmed in the cortex of rats (Viboolvorakul and Patumraj, 2014). Furthermore, Pereira et al. (2007) have demonstrated with MRIs that aerobic exercise training for 3 months increases the cerebral blood volume (CBV) in the hippocampal dentate gyrus in healthy adults. They also have found that 6 weeks of voluntary wheel running increased the CBV in the hippocampal dentate gyrus in mice, which was positively correlated with the exercise-induced neurogenesis (Pereira et al., 2007).

As a potential mechanism of the increase in cerebral perfusion, evidence indicates that regular exercise induces angiogenesis in several brain regions, including the hippocampus (Clark et al., 2009; Van der Borght et al., 2009), striatum (Clark et al., 2009), cerebellum (Black et al., 1990; Lopez-Lopez et al., 2004), and cortex (Swain et al., 2003; Viboolvorakul and Patumraj, 2014). The angiogenic effects of exercise seem to be maintained in aged animals (Ding et al., 2006). Exercise is also known to increase the expression of vascular endothelial growth factor, which plays a significant role in both angiogenesis and neurogenesis (Fabel et al., 2003; Udo et al., 2008).

Regular exercise also improves cerebrovascular reactivity in response to CO_2, which is a well-established index for evaluating cerebrovascular function, and known to be impaired with aging and neurodegenerative Alzheimer's disease (den Abeelen et al., 2014; Glodzik et al., 2013). For example, an increase in MCAv to CO_2 inhalation, ie, hypercapnia, was impaired in aged-sedentary subjects, but regular exercise can improve the blunted cerebrovascular reactivity (Bailey et al., 2013). The cerebrovascular reactivity to CO_2 can be improved with several months of aerobic training both in young and aged humans (Murrell et al., 2013). Ivey et al. (2011) have also demonstrated that 6 months of aerobic training intervention improved the cerebrovascular reactivity even in the stroke survivor (Ivey et al., 2011). The improvement in cerebrovascular reactivity through regular exercise is considered to be physiologically linked with improved cognitive function (Bailey et al., 2013).

One of the most well-known benefits of regular exercise on the neurovascular unit would be a protective effect against stroke, which is the third leading causes of death and markedly decreases quality of life due to permanent disabilities (Middleton et al., 2013; Schmidt et al., 2014). Epidemiological studies have demonstrated that moderate-to-high levels of physical activity were associated with a lower risk of ischemic stroke both in men (Willey et al., 2009) and women (Sattelmair et al., 2010). Abundant animal studies have also demonstrated that exercise preconditioning contributes to reduce infarction volume, edema, BBB permeability, and improves neurological deficits followed by transient focal ischemia, such as MCA occlusion (Arrick et al., 2012, 2014; Ding et al., 2004; Endres et al., 2003; Gertz et al., 2006; Guo et al., 2008; He et al., 2014). Exercise performed following stroke is also demonstrated to be protective and to accelerate recovery from the stroke (Austin et al., 2014).

As for the molecular mechanisms of the improved cerebrovascular reactivity, and the stroke protection induced by regular exercise, endothelial NO synthase (eNOS) seems to be critically involved. Importantly, Endres et al. (2003) have demonstrated that 3 weeks of exercise reduces cerebral infract volume and improved neurological sensory–motor deficits following 1-h MCA occlusion and reperfusion in control 129AV wild-type mice; however, the positive effects were abolished in eNOS knockout mice. They also found that exercise increased eNOS expression and activity in the aorta and improved endothelium-dependent reactivity of isolated femoral arteries in wild-type mice (Endres et al., 2003). Furthermore, exercise has been demonstrated to increase production and circulation levels of endothelial progenitor cells, an important mediator for inducing vascularization and improving endothelial function (Gertz et al., 2006; Laufs et al., 2004). Although evidence is limited, exercise might increase eNOS expression in the rat hippocampus (Chen and Xiao, 2014). Taken together, these findings suggest that regular exercise confers protective effects against ischemic stroke through eNOS-dependent mechanisms (Bolduc et al., 2013).

5.4 OPPOSING EFFECTS OF AGING AND REGULAR EXERCISE ON THE NEURONAL AND CEREBROVASCULAR PLASTICITY

We have summarized potential mechanisms of how aging causes cognitive dysfunction and increases the risks of neurodegenerative diseases, whereas regular exercise can counteract the deleterious processes (Fig. 8). As mentioned above, aging not only impairs neuronal plasticity, but cerebrovascular plasticity as well. Importantly, impaired cerebrovascular plasticity (including vascular density, endothelial reactivity, and eNOS-dependent function) subsequently leads to cerebral hypoperfusion, which becomes an alternative risk for neuronal plasticity (Barnes, 2015). Thus, aging-associated cognitive dysfunction and neurodegenerative diseases due to impaired neuronal plasticity would be synergistically accelerated by impaired

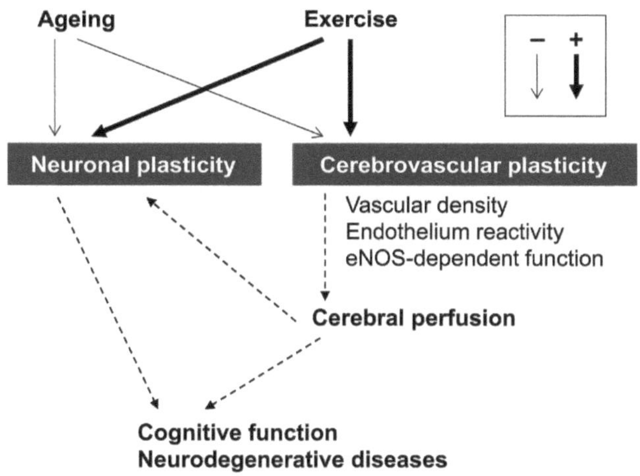

FIG. 8

Opposing effects of aging and regular exercise on the neuronal and cerebrovascular plasticity. Details are in the text. The *dashed arrows* indicate actions depending on changes in their upstream factors.

cerebrovascular plasticity (Gorelick et al., 2011). Whereas, over two decades of research have solidified our convictions that regular exercise improves neuronal plasticity. In addition, recent studies are unveiling that regular exercise can improve cerebrovascular plasticity, although researches are still insufficient. Contrary to the aging processes, the exercise-induced improvement in cerebrovascular plasticity increases cerebral perfusion, which would contribute to keep/improve cognitive function and gain neuroprotection. Presumably, these synergistic actions would be a reason why regular exercise exerts multi-beneficial effects on the brain.

6 NEUROTROPHIC COUPLING, A POSSIBLE MODULATOR OF EXERCISE EFFECTS ON THE BRAIN

So far, scientists tend to focus only on one aspect of the brain, ie, neuronal or cerebrovascular plasticity, in their studies. Because these two are interdependent, interactive studies that simultaneously examine the response/adaptation of neuron and cerebral vasculature to regular exercise will become more important for deeper understanding of the effects of exercise on the brain. As a unique example, we have described here that neuronal activity triggers uptake of circulating IGF-I into the brain. Understanding the significance of neurovascular interaction will open a new avenue in the research of exercise neuroscience.

IGF-I has multipotent neuroprotective effects and has been demonstrated as a potent mediator of the multi-beneficial effects of exercise on the brain (Fernandez and Torres-Aleman, 2012). However, it should be noted here that IGF-I mRNA expression is extremely low but the protein level remain more stable in the adult brain, indicating a mismatch between production and protein levels of IGF-I. In addition, as mentioned before, blocking an action of circulating IGF-I abolished the positive effects of exercise (Carro et al., 2000, 2001; Lopez-Lopez et al., 2004; Trejo et al., 2001). Therefore, and here is a fundamental question, in order to act in the brain parenchyma, how does serum IGF-I pass through the BBB, which generally limits free passage of blood proteins into the brain (Dziegielewska and Saunders, 2002)? Previous studies have shown that serum IGF-I can cross the BBB (Pardridge, 1993; Yu et al., 2006); however, its detailed regulatory mechanisms remained uncertain. Since neuronal activity is a key impetus promoting neuronal functions, we have hypothesized that neuronal activity would locally regulate entrance of circulating IGF-I through trans-endothelial transport mechanisms. That is, active areas can reap the benefits of IGF-I. To test the hypothesis, we conducted a series of *in vivo* and *in vitro* experiments and found the activity-dependent entry of IGF-I into the brain and its mechanisms, which are summarized below and in Fig. 9 (Nishijima et al., 2010).

1. Neuronal activity stimulates the release of NO from neurons. Concomitantly, ATP, prostaglandin E2 (PGE2), and epoxyeicosatrienoic acid (EET) are also released from astrocyte. These vasodilatory mediators increase regional CBF (neurovascular coupling), which eventually increases the local availability of the circulating IGF-1 at the activated area.

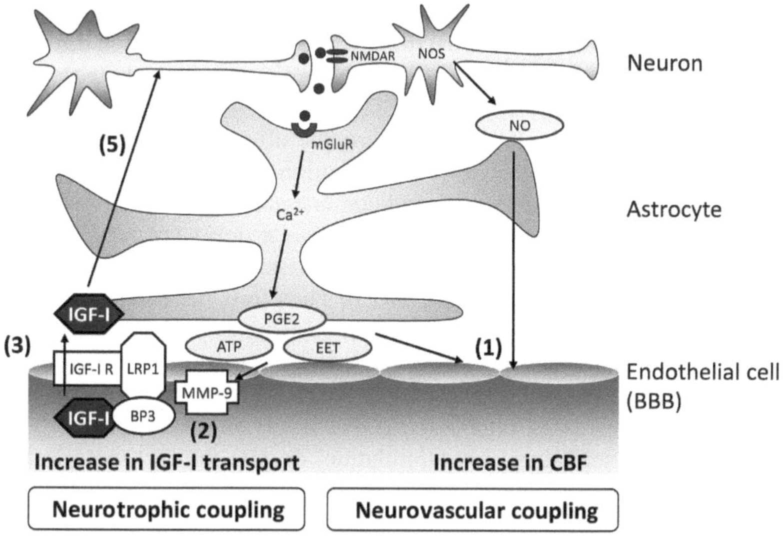

FIG. 9

Schematic illustration of neuronal activity-driven entry of serum IGF-I into the brain (neurotrophic coupling). Details are in the text.

2. The astrocyte-derived ATP, PGE2, and EET concomitantly activate matrix metalloproteinase-9 (MMP-9) activity, which cleavages IGF-I binding protein-3 (BP3). Unless IGF-I is released from BP3, IGF-I cannot cross the BBB presumably because IGF-I bound with BP3 would be unable to interact with IGF-I receptors.
3. Cleaved IGF-I passes the BBB through the low-density lipoprotein receptor-related protein-1 (LRP-1)-dependent transcytosis into the brain parenchyma.
4. Increased IGF-I input exerts the neuroprotective effects of IGF-I on the area where neuronal activity is activated.

Since the concept of neurovascular coupling was first described more than a century ago (Roy and Sherrington, 1890), underlying molecular mechanisms have been extensively studied and, more importantly, neurovascular coupling is applied as a basic principal of neuroimaging (see chapters "Bridging macroscopic and microscopic methods for the measurements of cerebral blood flow: Toward finding the determinants in maintaining the CBF homeostasis" by Kanno and Masamoto; "New horizons in neurometabolic and neurovascular coupling from calibrated fMRI" by Shu et al.; "Submillimeter-resolution fMRI: Toward understanding local neural processing" by Fukuda et al.; and "Hemodynamic signals in fNIRS" by Hoshi). Thus, as a synonym, we termed the activity-dependent IGF-I uptake as neurotrophic coupling (see also chapter "Mechanisms of cellular plasticity in cerebral perivascular region" by Egawa et al.). Further studies are required to expand the concept of neurotrophic coupling through examination of whether neuronal activity similarly triggers uptake of other blood-borne growth factors and cytokines into the brain. These studies will further highlight the importance of brain activation through physically, cognitively, and socially active life, which eventually contributes to increased cognitive reserves (Arenaza-Urquijo et al., 2015).

In order to gain more of IGF-I's neurotrophic effects, a simple idea would be to increase the level of blood IGF-I concentration. However, it should be noted that increases in the blood IGF-I concentration is associated with the risk of various type of cancers because of its potent proliferative action (Arnaldez and Helman, 2012). In addition, higher blood IGF-I might decreases longevity, which is observed in several animal models (Junnila et al., 2013). Therefore, increases in IGF-I uptake by a highly-active lifestyle would be a more realistic approach, rather than exogenous IGF-I treatment. On the other hand, the circulating IGF-I concentration progressively decreases with age (Sonntag et al., 2013). Interestingly, a recent study showed that decreased circulating IGF-I levels attenuated the cortical CBF response evoked by somatosensory stimulation, despite the fact that the response in neuronal activity was identical to that of control mice, suggesting that IGF-I is necessary for proper neurovascular coupling (Toth et al., 2015). Correctively, the combination of preventing the age-dependent decline of blood IGF-I with a highly-active lifestyle would be a key strategy to promote brain function via IGF-I's neurotrophic actions.

7 CONCLUSIONS AND PERSPECTIVE

Acute bouts of exercise triggers neuronal activation and leads subsequent hemodynamic changes in the brain. If the exercise is repeated, functional adaptation happens in the neuron and cerebral vasculature. Therefore, the diverse beneficial effects of regular exercise on the brain would be synergistically mediated by both improved neuronal and cerebrovascular plasticity—although the latter has received relatively less attention. Besides IGF-I, peripheral molecules seem to play an important role in mediating exercise effects. Indeed, a recent study demonstrated that muscle-derived PGC-1α1 mediates the antidepressive effects of exercise through modulating the kynurenine metabolism (Agudelo et al., 2014). Collectively, future studies will be required to take an interactive approach to further reveal underlying mechanisms of neuroprotective effects of exercise, which would help us combat the current sedentary lifestyle.

REFERENCES

Agudelo, L.Z., Femenia, T., Orhan, F., Porsmyr-Palmertz, M., Goiny, M., Martinez-Redondo, V., Correia, J.C., Izadi, M., Bhat, M., Schuppe-Koistinen, I., Pettersson, A.T., Ferreira, D.M., Krook, A., Barres, R., Zierath, J.R., Erhardt, S., Lindskog, M., Ruas, J.L., 2014. Skeletal muscle PGC-1alpha1 modulates kynurenine metabolism and mediates resilience to stress-induced depression. Cell 159, 33–45.

Ainslie, P.N., Cotter, J.D., George, K.P., Lucas, S., Murrell, C., Shave, R., Thomas, K.N., Williams, M.J., Atkinson, G., 2008. Elevation in cerebral blood flow velocity with aerobic fitness throughout healthy human ageing. J. Physiol. 586, 4005–4010.

Ando, S., Kokubo, M., Nakae, S., Kimura, M., Hojo, T., Ebine, N., 2012. Effects of strenuous exercise on visual perception are independent of visual resolution. Physiol. Behav. 106, 117–121.

Arenaza-Urquijo, E.M., Wirth, M., Chetelat, G., 2015. Cognitive reserve and lifestyle: moving towards preclinical Alzheimer's disease. Front. Aging Neurosci. 7, 134.

Arnaldez, F.I., Helman, L.J., 2012. Targeting the insulin growth factor receptor 1. Hematol. Oncol. Clin. North Am. 26, 527–542.

Arrick, D.M., Sun, H., Mayhan, W.G., 2012. Influence of exercise training on ischemic brain injury in type 1 diabetic rats. J. Appl. Physiol. (1985) 113, 1121–1127.

Arrick, D.M., Yang, S., Li, C., Cananzi, S., Mayhan, W.G., 2014. Vigorous exercise training improves reactivity of cerebral arterioles and reduces brain injury following transient focal ischemia. Microcirculation 21, 516–523.

Austin, M.W., Ploughman, M., Glynn, L., Corbett, D., 2014. Aerobic exercise effects on neuroprotection and brain repair following stroke: a systematic review and perspective. Neurosci. Res. 87, 8–15.

Bailey, D.M., Marley, C.J., Brugniaux, J.V., Hodson, D., New, K.J., Ogoh, S., Ainslie, P.N., 2013. Elevated aerobic fitness sustained throughout the adult lifespan is associated with improved cerebral hemodynamics. Stroke 44, 3235–3238.

Barnes, J.N., 2015. Exercise, cognitive function, and aging. Adv. Physiol. Educ. 39, 55–62.

Barnes, J.N., Taylor, J.L., Kluck, B.N., Johnson, C.P., Joyner, M.J., 2013. Cerebrovascular reactivity is associated with maximal aerobic capacity in healthy older adults. J. Appl. Physiol. 114, 1383–1387.

Black, J.E., Isaacs, K.R., Anderson, B.J., Alcantara, A.A., Greenough, W.T., 1990. Learning causes synaptogenesis, whereas motor activity causes angiogenesis, in cerebellar cortex of adult rats. Proc. Natl. Acad. Sci. U.S.A. 87, 5568–5572.

Bolduc, V., Thorin-Trescases, N., Thorin, E., 2013. Endothelium-dependent control of cerebrovascular functions through age: exercise for healthy cerebrovascular aging. Am. J. Physiol. Heart Circ. Physiol. 305, H620–H633.

Brown, W.J., Ford, J.H., Burton, N.W., Marshall, A.L., Dobson, A.J., 2005. Prospective study of physical activity and depressive symptoms in middle-aged women. Am. J. Prev. Med. 29, 265–272.

Buchman, A.S., Boyle, P.A., Yu, L., Shah, R.C., Wilson, R.S., Bennett, D.A., 2012. Total daily physical activity and the risk of AD and cognitive decline in older adults. Neurology 78, 1323–1329.

Camacho, T.C., Roberts, R.E., Lazarus, N.B., Kaplan, G.A., Cohen, R.D., 1991. Physical activity and depression: evidence from the Alameda County Study. Am. J. Epidemiol. 134, 220–231.

Carro, E., Nunez, A., Busiguina, S., Torres-Aleman, I., 2000. Circulating insulin-like growth factor I mediates effects of exercise on the brain. J. Neurosci. 20, 2926–2933.

Carro, E., Trejo, J.L., Busiguina, S., Torres-Aleman, I., 2001. Circulating insulin-like growth factor I mediates the protective effects of physical exercise against brain insults of different etiology and anatomy. J. Neurosci. 21, 5678–5684.

Chen, Q., Xiao, D.S., 2014. Long-term aerobic exercise increases redox-active iron through nitric oxide in rat hippocampus. Nitric Oxide 36, 1–10.

Christensen, L.O., Johannsen, P., Sinkjaer, T., Petersen, N., Pyndt, H.S., Nielsen, J.B., 2000. Cerebral activation during bicycle movements in man. Exp. Brain Res. 135, 66–72.

Clark, P.J., Brzezinska, W.J., Thomas, M.W., Ryzhenko, N.A., Toshkov, S.A., Rhodes, J.S., 2008. Intact neurogenesis is required for benefits of exercise on spatial memory but not motor performance or contextual fear conditioning in C57BL/6J mice. Neuroscience 155, 1048–1058.

Clark, P.J., Brzezinska, W.J., Puchalski, E.K., Krone, D.A., Rhodes, J.S., 2009. Functional analysis of neurovascular adaptations to exercise in the dentate gyrus of young adult mice associated with cognitive gain. Hippocampus 19, 937–950.

Colcombe, S.J., Erickson, K.I., Raz, N., Webb, A.G., Cohen, N.J., McAuley, E., Kramer, A.F., 2003. Aerobic fitness reduces brain tissue loss in aging humans. J. Gerontol. A Biol. Sci. Med. Sci. 58, 176–180.

Creer, D.J., Romberg, C., Saksida, L.M., van Praag, H., Bussey, T.J., 2010. Running enhances spatial pattern separation in mice. Proc. Natl. Acad. Sci. U. S. A. 107, 2367–2372.

Delp, M.D., Armstrong, R.B., Godfrey, D.A., Laughlin, M.H., Ross, C.D., Wilkerson, M.K., 2001. Exercise increases blood flow to locomotor, vestibular, cardiorespiratory and visual regions of the brain in miniature swine. J. Physiol. 533, 849–859.

den Abeelen, A.S., Lagro, J., van Beek, A.H., Claassen, J.A., 2014. Impaired cerebral autoregulation and vasomotor reactivity in sporadic Alzheimer's disease. Curr. Alzheimer Res. 11, 11–17.

Dietrich, M.O., Andrews, Z.B., Horvath, T.L., 2008. Exercise-induced synaptogenesis in the hippocampus is dependent on UCP2-regulated mitochondrial adaptation. J. Neurosci. 28, 10766–10771.

Ding, Y., Li, J., Luan, X., Ding, Y.H., Lai, Q., Rafols, J.A., Phillis, J.W., Clark, J.C., Diaz, F.G., 2004. Exercise pre-conditioning reduces brain damage in ischemic rats that may be associated with regional angiogenesis and cellular overexpression of neurotrophin. Neuroscience 124, 583–591.

Ding, Y.H., Li, J., Zhou, Y., Rafols, J.A., Clark, J.C., Ding, Y., 2006. Cerebral angiogenesis and expression of angiogenic factors in aging rats after exercise. Curr. Neurovasc. Res. 3, 15–23.

Dishman, R.K., Berthoud, H.R., Booth, F.W., Cotman, C.W., Edgerton, V.R., Fleshner, M.R., Gandevia, S.C., Gomez-Pinilla, F., Greenwood, B.N., Hillman, C.H., Kramer, A.F., Levin, B.E., Moran, T.H., Russo-Neustadt, A.A., Salamone, J.D., Van Hoomissen, J.D., Wade, C.E., York, D.A., Zigmond, M.J., 2006. Neurobiology of exercise. Obesity (Silver Spring) 14, 345–356.

Dumith, S.C., Hallal, P.C., Reis, R.S., Kohl 3rd., H.W., 2011. Worldwide prevalence of physical inactivity and its association with human development index in 76 countries. Prev. Med. 53, 24–28.

Dziegielewska, K.M., Saunders, N.R., 2002. The ins and outs of brain-barrier mechanisms. Trends Neurosci. 25, 69–71.

Endres, M., Gertz, K., Lindauer, U., Katchanov, J., Schultze, J., Schrock, H., Nickenig, G., Kuschinsky, W., Dirnagl, U., Laufs, U., 2003. Mechanisms of stroke protection by physical activity. Ann. Neurol. 54, 582–590.

Erickson, K.I., Voss, M.W., Prakash, R.S., Basak, C., Szabo, A., Chaddock, L., Kim, J.S., Heo, S., Alves, H., White, S.M., Wojcicki, T.R., Mailey, E., Vieira, V.J., Martin, S.A., Pence, B.D., Woods, J.A., McAuley, E., Kramer, A.F., 2011. Exercise training increases size of hippocampus and improves memory. Proc. Natl. Acad. Sci. U. S. A. 108, 3017–3022.

Erickson, K.I., Leckie, R.L., Weinstein, A.M., 2014. Physical activity, fitness, and gray matter volume. Neurobiol. Aging 35 (Suppl. 2), S20–S28.

Ernst, C., Olson, A.K., Pinel, J.P., Lam, R.W., Christie, B.R., 2006. Antidepressant effects of exercise: evidence for an adult-neurogenesis hypothesis? J. Psychiatry Neurosci. 31, 84–92.

Fabel, K., Tam, B., Kaufer, D., Baiker, A., Simmons, N., Kuo, C.J., Palmer, T.D., 2003. VEGF is necessary for exercise-induced adult hippocampal neurogenesis. Eur. J. Neurosci. 18, 2803–2812.

Farmer, J., Zhao, X., van Praag, H., Wodtke, K., Gage, F.H., Christie, B.R., 2004. Effects of voluntary exercise on synaptic plasticity and gene expression in the dentate gyrus of adult male Sprague-Dawley rats in vivo. Neuroscience 124, 71–79.

Fernandez, A.M., Torres-Aleman, I., 2012. The many faces of insulin-like peptide signalling in the brain. Nat. Rev. Neurosci. 13, 225–239.

Foreman, D.L., Sanders, M., Bloor, C.M., 1976. Total and regional cerebral blood flow during moderate and severe exercise in miniature swine. J. Appl. Physiol. 40, 191–195.

Garcia, C., Chen, M.J., Garza, A.A., Cotman, C.W., Russo-Neustadt, A., 2003. The influence of specific noradrenergic and serotonergic lesions on the expression of hippocampal brain-derived neurotrophic factor transcripts following voluntary physical activity. Neuroscience 119, 721–732.

Gertz, K., Priller, J., Kronenberg, G., Fink, K.B., Winter, B., Schrock, H., Ji, S., Milosevic, M., Harms, C., Bohm, M., Dirnagl, U., Laufs, U., Endres, M., 2006. Physical activity improves long-term stroke outcome via endothelial nitric oxide synthase-dependent augmentation of neovascularization and cerebral blood flow. Circ. Res. 99, 1132–1140.

Globus, M., Melamed, E., Keren, A., Tzivoni, D., Granot, C., Lavy, S., Stern, S., 1983. Effect of exercise on cerebral circulation. J. Cereb. Blood Flow Metab. 3, 287–290.

Glodzik, L., Randall, C., Rusinek, H., de Leon, M.J., 2013. Cerebrovascular reactivity to carbon dioxide in Alzheimer's disease. J. Alzheimers Dis. 35, 427–440.

Gorelick, P.B., Scuteri, A., Black, S.E., DeCarli, C., Greenberg, S.M., Iadecola, C., Launer, L.J., Laurent, S., Lopez, O.L., Nyenhuis, D., Petersen, R.C., Schneider, J.A., Tzourio, C., Arnett, D.K., Bennett, D.A., Chui, H.C., Higashida, R.T., Lindquist, R., Nilsson, P.M., Roman, G.C., Sellke, F.W., Seshadri, S., Council, A.H.A.S., Prevention, C.E., Nursing, C.C., Inter, C.C.R., Anesthesia, C.C.S., 2011. Vascular contributions to cognitive impairment and dementia a statement for healthcare professionals from the American Heart Association/ American Stroke Association. Stroke 42, 2672–2713.

Gross, P.M., Marcus, M.L., Heistad, D.D., 1980. Regional distribution of cerebral blood flow during exercise in dogs. J. Appl. Physiol. Respir. Environ. Exerc. Physiol. 48, 213–217.

Gu, W., Jiang, W., Wester, P., 2003. Real-time cortical cerebral blood flow follow-up in conscious, freely moving rats by laser Doppler flowmetry. Methods 30, 172–177.

Guo, M., Cox, B., Mahale, S., Davis, W., Carranza, A., Hayes, K., Sprague, S., Jimenez, D., Ding, Y., 2008. Pre-ischemic exercise reduces matrix metalloproteinase-9 expression and ameliorates blood-brain barrier dysfunction in stroke. Neuroscience 151, 340–351.

Guthold, R., Ono, T., Strong, K.L., Chatterji, S., Morabia, A., 2008. Worldwide variability in physical inactivity a 51-country survey. Am. J. Prev. Med. 34, 486–494.

Hallal, P.C., Andersen, L.B., Bull, F.C., Guthold, R., Haskell, W., Ekelund, U., 2012. Global physical activity levels: surveillance progress, pitfalls, and prospects. Lancet 380, 247–257.

Hamer, M., Chida, Y., 2009. Physical activity and risk of neurodegenerative disease: a systematic review of prospective evidence. Psychol. Med. 39, 3–11.

He, Z., Wang, X., Wu, Y., Jia, J., Hu, Y., Yang, X., Li, J., Fan, M., Zhang, L., Guo, J., Leung, M.C., 2014. Treadmill pre-training ameliorates brain edema in ischemic stroke via down-regulation of aquaporin-4: an MRI study in rats. PLoS One 9, e84602.

Helfer, J.L., Goodlett, C.R., Greenough, W.T., Klintsova, A.Y., 2009. The effects of exercise on adolescent hippocampal neurogenesis in a rat model of binge alcohol exposure during the brain growth spurt. Brain Res. 1294, 1–11.

Hillman, C.H., Erickson, K.I., Kramer, A.F., 2008. Be smart, exercise your heart: exercise effects on brain and cognition. Nat. Rev. Neurosci. 9, 58–65.

Hillman, C.H., Pontifex, M.B., Castelli, D.M., Khan, N.A., Raine, L.B., Scudder, M.R., Drollette, E.S., Moore, R.D., Wu, C.T., Kamijo, K., 2014. Effects of the FITKids randomized controlled trial on executive control and brain function. Pediatrics 134, e1063–e1071.

Hiura, M., Nariai, T., Ishii, K., Sakata, M., Oda, K., Toyohara, J., Ishiwata, K., 2014. Changes in cerebral blood flow during steady-state cycling exercise: a study using oxygen-15-labeled water with PET. J. Cereb. Blood Flow Metab. 34, 389–396.

Hohimer, A.R., Hales, J.R., Rowell, L.B., Smith, O.A., 1983. Regional distribution of blood flow during mild dynamic leg exercise in the baboon. J. Appl. Physiol. Respir. Environ. Exerc. Physiol. 55, 1173–1177.

Holschneider, D.P., Maarek, J.M., Yang, J., Harimoto, J., Scremin, O.U., 2003. Functional brain mapping in freely moving rats during treadmill walking. J. Cereb. Blood Flow Metab. 23, 925–932.

Ide, K., Secher, N.H., 2000. Cerebral blood flow and metabolism during exercise. Prog. Neurobiol. 61, 397–414.

Intlekofer, K.A., Cotman, C.W., 2013. Exercise counteracts declining hippocampal function in aging and Alzheimer's disease. Neurobiol. Dis. 57, 47–55.

Ivey, F.M., Ryan, A.S., Hafer-Macko, C.E., Macko, R.F., 2011. Improved cerebral vasomotor reactivity after exercise training in hemiparetic stroke survivors. Stroke 42, 1994–2000.

Jorgensen, L.G., Perko, G., Payne, G., Secher, N.H., 1993. Effect of limb anesthesia on middle cerebral response to handgrip. Am. J. Physiol. 264, H553–H559.

Junnila, R.K., List, E.O., Berryman, D.E., Murrey, J.W., Kopchick, J.J., 2013. The GH/IGF-1 axis in ageing and longevity. Nat. Rev. Endocrinol. 9, 366–376.

Kaiser, M.G., During, M.J., 1995. Combining laser Doppler flowmetry with microdialysis: a novel approach to investigate the coupling of regional cerebral blood flow to neuronal activity. J. Neurosci. Methods 60, 165–173.

Kohl 3rd., H.W., Craig, C.L., Lambert, E.V., Inoue, S., Alkandari, J.R., Leetongin, G., Kahlmeier, S., 2012. The pandemic of physical inactivity: global action for public health. Lancet 380, 294–305.

Kramer, A.F., Hahn, S., Cohen, N.J., Banich, M.T., McAuley, E., Harrison, C.R., Chason, J., Vakil, E., Bardell, L., Boileau, R.A., Colcombe, A., 1999. Ageing, fitness and neurocognitive function. Nature 400, 418–419.

Kronenberg, G., Bick-Sander, A., Bunk, E., Wolf, C., Ehninger, D., Kempermann, G., 2006. Physical exercise prevents age-related decline in precursor cell activity in the mouse dentate gyrus. Neurobiol. Aging 27, 1505–1513.

Ku, P.W., Fox, K.R., Chen, L.J., Chou, P., 2012. Physical activity and depressive symptoms in older adults: 11-year follow-up. Am. J. Prev. Med. 42, 355–362.

Lange-Asschenfeldt, C., Kojda, G., 2008. Alzheimer's disease, cerebrovascular dysfunction and the benefits of exercise: from vessels to neurons. Exp. Gerontol. 43, 499–504.

Laufs, U., Werner, N., Link, A., Endres, M., Wassmann, S., Jurgens, K., Miche, E., Bohm, M., Nickenig, G., 2004. Physical training increases endothelial progenitor cells, inhibits neointima formation, and enhances angiogenesis. Circulation 109, 220–226.

Lautenschlager, N.T., Cox, K.L., Flicker, L., Foster, J.K., van Bockxmeer, F.M., Xiao, J., Greenop, K.R., Almeida, O.P., 2008. Effect of physical activity on cognitive function in older adults at risk for Alzheimer disease: a randomized trial. JAMA 300, 1027–1037.

Lee, I.M., Shiroma, E.J., Lobelo, F., Puska, P., Blair, S.N., Katzmarzyk, P.T., 2012. Effect of physical inactivity on major non-communicable diseases worldwide: an analysis of burden of disease and life expectancy. Lancet 380, 219–229.

Li, Y., Luikart, B.W., Birnbaum, S., Chen, J., Kwon, C.H., Kernie, S.G., Bassel-Duby, R., Parada, L.F., 2008. TrkB regulates hippocampal neurogenesis and governs sensitivity to antidepressive treatment. Neuron 59, 399–412.

Li, H., Liang, A., Guan, F., Fan, R., Chi, L., Yang, B., 2013. Regular treadmill running improves spatial learning and memory performance in young mice through increased hippocampal neurogenesis and decreased stress. Brain Res. 1531, 1–8.

Linkis, P., Jorgensen, L.G., Olesen, H.L., Madsen, P.L., Lassen, N.A., Secher, N.H., 1995. Dynamic exercise enhances regional cerebral artery mean flow velocity. J. Appl. Physiol. (1985) 78, 12–16.

Lopez-Lopez, C., LeRoith, D., Torres-Aleman, I., 2004. Insulin-like growth factor I is required for vessel remodeling in the adult brain. Proc. Natl. Acad. Sci. U. S. A. 101, 9833–9838.

Loprinzi, P.D., Herod, S.M., Cardinal, B.J., Noakes, T.D., 2013. Physical activity and the brain: a review of this dynamic, bi-directional relationship. Brain Res. 1539, 95–104.

Lucas, M., Mekary, R., Pan, A., Mirzaei, F., O'Reilly, E.J., Willett, W.C., Koenen, K., Okereke, O.I., Ascherio, A., 2011. Relation between clinical depression risk and physical activity and time spent watching television in older women: a 10-year prospective follow-up study. Am. J. Epidemiol. 174, 1017–1027.

Maass, A., Duzel, S., Brigadski, T., Goerke, M., Becke, A., Sobieray, U., Neumann, K., Lovden, M., Lindenberger, U., Backman, L., Braun-Dullaeus, R., Ahrens, D., Heinze, H.J., Muller, N.G., Lessmann, V., Sendtner, M., Duzel, E., 2015. Relationships of peripheral IGF-1, VEGF and BDNF levels to exercise-related changes in memory, hippocampal perfusion and volumes in older adults. Neuroimage (in press).

Mattson, M.P., 2012. Energy intake and exercise as determinants of brain health and vulnerability to injury and disease. Cell Metab. 16, 706–722.

Middleton, L.E., Corbett, D., Brooks, D., Sage, M.D., Macintosh, B.J., McIlroy, W.E., Black, S.E., 2013. Physical activity in the prevention of ischemic stroke and improvement of outcomes: a narrative review. Neurosci. Biobehav. Rev. 37, 133–137.

Mirochnic, S., Wolf, S., Staufenbiel, M., Kempermann, G., 2009. Age effects on the regulation of adult hippocampal neurogenesis by physical activity and environmental enrichment in the APP23 mouse model of Alzheimer disease. Hippocampus 19, 1008–1018.

Motl, R.W., Birnbaum, A.S., Kubik, M.Y., Dishman, R.K., 2004. Naturally occurring changes in physical activity are inversely related to depressive symptoms during early adolescence. Psychosom. Med. 66, 336–342.

Murrell, C.J., Cotter, J.D., Thomas, K.N., Lucas, S.J., Williams, M.J., Ainslie, P.N., 2013. Cerebral blood flow and cerebrovascular reactivity at rest and during sub-maximal exercise: effect of age and 12-week exercise training. Age (Dordr.) 35, 905–920.

Nakajima, K., Uchida, S., Suzuki, A., Hotta, H., Aikawa, Y., 2003. The effect of walking on regional blood flow and acetylcholine in the hippocampus in conscious rats. Auton. Neurosci. 103, 83–92.

Neeper, S.A., Gomez-Pinilla, F., Choi, J., Cotman, C., 1995. Exercise and brain neurotrophins. Nature 373, 109.

Nichol, K.E., Poon, W.W., Parachikova, A.I., Cribbs, D.H., Glabe, C.G., Cotman, C.W., 2008. Exercise alters the immune profile in Tg2576 Alzheimer mice toward a response coincident with improved cognitive performance and decreased amyloid. J. Neuroinflammation 5, 13.

Nishijima, T., Kita, I., 2015. Deleterious effects of physical inactivity on the hippocampus: new insight into the increasing prevalence of stress-related depression. J. Phys. Fitness Sports Med. 4, 253–258.

Nishijima, T., Soya, H., 2006. Evidence of functional hyperemia in the rat hippocampus during mild treadmill running. Neurosci. Res. 54, 186–191.

Nishijima, T., Piriz, J., Duflot, S., Fernandez, A.M., Gaitan, G., Gomez-Pinedo, U., Verdugo, J.M., Leroy, F., Soya, H., Nunez, A., Torres-Aleman, I., 2010. Neuronal activity drives localized blood-brain-barrier transport of serum insulin-like growth factor-I into the CNS. Neuron 67, 834–846.

Nishijima, T., Okamoto, M., Matsui, T., Kita, I., Soya, H., 2012. Hippocampal functional hyperemia mediated by NMDA receptor/NO signaling in rats during mild exercise. J. Appl. Physiol. 112, 197–203.

Nybo, L., Nielsen, B., 2001. Middle cerebral artery blood velocity is reduced with hyperthermia during prolonged exercise in humans. J. Physiol. 534, 279–286.

Ogoh, S., Ainslie, P.N., 2009. Regulatory mechanisms of cerebral blood flow during exercise: new concepts. Exerc. Sport Sci. Rev. 37, 123–129.

Okamoto, M., Hojo, Y., Inoue, K., Matsui, T., Kawato, S., McEwen, B.S., Soya, H., 2012. Mild exercise increases dihydrotestosterone in hippocampus providing evidence for androgenic mediation of neurogenesis. Proc. Natl. Acad. Sci. U. S. A. 109, 13100–13105.

Osborne, P.G., 1997. Hippocampal and striatal blood flow during behavior in rats: chronic laser Doppler flowmetry study. Physiol. Behav. 61, 485–492.

Paffenbarger Jr., R.S., Lee, I.M., Leung, R., 1994. Physical activity and personal characteristics associated with depression and suicide in American college men. Acta Psychiatr. Scand. Suppl. 377, 16–22.

Pardridge, W.M., 1993. Transport of insulin-related peptides and glucose across the blood-brain barrier. Ann. N. Y. Acad. Sci. 692, 126–137.

Pereira, A.C., Huddleston, D.E., Brickman, A.M., Sosunov, A.A., Hen, R., McKhann, G.M., Sloan, R., Gage, F.H., Brown, T.R., Small, S.A., 2007. An in vivo correlate of exercise-induced neurogenesis in the adult dentate gyrus. Proc. Natl. Acad. Sci. U. S. A. 104, 5638–5643.

Powers, S.K., Howley, E.T., 2012. Exercise Physiology: Theory and Application to Fitness and Performance, eighth ed. McGraw-Hill, NY.

Querido, J.S., Sheel, A.W., 2007. Regulation of cerebral blood flow during exercise. Sports Med. 37, 765–782.

Radak, Z., Zhao, Z., Koltai, E., Ohno, H., Atalay, M., 2013. Oxygen consumption and usage during physical exercise: the balance between oxidative stress and ROS-dependent adaptive signaling. Antioxid. Redox Signal. 18, 1208–1246.

Redila, V.A., Christie, B.R., 2006. Exercise-induced changes in dendritic structure and complexity in the adult hippocampal dentate gyrus. Neuroscience 137, 1299–1307.

Rodriguez, J.J., Noristani, H.N., Olabarria, M., Fletcher, J., Somerville, T.D., Yeh, C.Y., Verkhratsky, A., 2011. Voluntary running and environmental enrichment restores impaired hippocampal neurogenesis in a triple transgenic mouse model of Alzheimer's disease. Curr. Alzheimer Res. 8, 707–717.

Roy, C.S., Sherrington, C.S., 1890. On the regulation of the blood-supply of the brain. J. Physiol. 11. 85–108.

Sahay, A., Hen, R., 2007. Adult hippocampal neurogenesis in depression. Nat. Neurosci. 10, 1110–1115.

Sato, K., Ogoh, S., Hirasawa, A., Oue, A., Sadamoto, T., 2011. The distribution of blood flow in the carotid and vertebral arteries during dynamic exercise in humans. J. Physiol. 589, 2847–2856.

Sattelmair, J.R., Kurth, T., Buring, J.E., Lee, I.M., 2010. Physical activity and risk of stroke in women. Stroke 41, 1243–1250.

Scharfman, H., Goodman, J., Macleod, A., Phani, S., Antonelli, C., Croll, S., 2005. Increased neurogenesis and the ectopic granule cells after intrahippocampal BDNF infusion in adult rats. Exp. Neurol. 192, 348–356.

Scheinberg, P., Blackburn, L.I., Rich, M., Saslaw, M., 1954. Effects of vigorous physical exercise on cerebral circulation and metabolism. Am. J. Med. 16, 549–554.

Schmidt, A., Wellmann, J., Schilling, M., Strecker, J.K., Sommer, C., Schabitz, W.R., Diederich, K., Minnerup, J., 2014. Meta-analysis of the efficacy of different training strategies in animal models of ischemic stroke. Stroke 45, 239–247.

Secher, N.H., Seifert, T., Van Lieshout, J.J., 2008. Cerebral blood flow and metabolism during exercise: implications for fatigue. J. Appl. Physiol. (1985) 104, 306–314.

Sofi, F., Valecchi, D., Bacci, D., Abbate, R., Gensini, G.F., Casini, A., Macchi, C., 2011. Physical activity and risk of cognitive decline: a meta-analysis of prospective studies. J. Intern. Med. 269, 107–117.

Sonntag, W.E., Deak, F., Ashpole, N., Toth, P., Csiszar, A., Freeman, W., Ungvari, Z., 2013. Insulin-like growth factor-1 in CNS and cerebrovascular aging. Front. Aging Neurosci. 5, 27.

Soya, H., Okamoto, M., Matsui, T., Lee, M., Inoue, K., Soya, S., Nishikawa, S., Fujikawa, T., Chang, H., Nishijima, T., 2011. Brain activation via exercise: exercise conditions leading to neuronal activation & hippocampal neurogenesis. J. Exer. Nutr. Biochem. 15, 1–10.

Stranahan, A.M., Mattson, M.P., 2012. Recruiting adaptive cellular stress responses for successful brain ageing. Nat. Rev. Neurosci. 13, 209–216.

Swain, R.A., Harris, A.B., Wiener, E.C., Dutka, M.V., Morris, H.D., Theien, B.E., Konda, S., Engberg, K., Lauterbur, P.C., Greenough, W.T., 2003. Prolonged exercise induces angiogenesis and increases cerebral blood volume in primary motor cortex of the rat. Neuroscience 117, 1037–1046.

Tarumi, T., Zhang, R., 2014. Cerebral hemodynamics of the aging brain: risk of Alzheimer disease and benefit of aerobic exercise. Front. Physiol. 5, 6.

ten Brinke, L.F., Bolandzadeh, N., Nagamatsu, L.S., Hsu, C.L., Davis, J.C., Miran-Khan, K., Liu-Ambrose, T., 2015. Aerobic exercise increases hippocampal volume in older women with probable mild cognitive impairment: a 6-month randomised controlled trial. Br. J. Sports Med. 49, 248–254.

Teychenne, M., Ball, K., Salmon, J., 2008. Physical activity and likelihood of depression in adults: a review. Prev. Med. 46, 397–411.

Toth, P., Tarantini, S., Ashpole, N.M., Tucsek, Z., Milne, G.L., Valcarcel-Ares, N.M., Menyhart, A., Farkas, E., Sonntag, W.E., Csiszar, A., Ungvari, Z., 2015. IGF-1 deficiency impairs neurovascular coupling in mice: implications for cerebromicrovascular aging. Aging Cell 14, 1034–1044.

Trejo, J.L., Carro, E., Torres-Aleman, I., 2001. Circulating insulin-like growth factor I mediates exercise-induced increases in the number of new neurons in the adult hippocampus. J. Neurosci. 21, 1628–1634.

Udo, H., Yoshida, Y., Kino, T., Ohnuki, K., Mizunoya, W., Mukuda, T., Sugiyama, H., 2008. Enhanced adult neurogenesis and angiogenesis and altered affective behaviors in mice overexpressing vascular endothelial growth factor 120. J. Neurosci. 28, 14522–14536.

Van der Borght, K., Kobor-Nyakas, D.E., Klauke, K., Eggen, B.J., Nyakas, C., Van der Zee, E.A., Meerlo, P., 2009. Physical exercise leads to rapid adaptations in hippocampal vasculature: temporal dynamics and relationship to cell proliferation and neurogenesis. Hippocampus 19, 928–936.

van Praag, H., 2009. Exercise and the brain: something to chew on. Trends Neurosci. 32, 283–290.

van Praag, H., Christie, B.R., Sejnowski, T.J., Gage, F.H., 1999a. Running enhances neurogenesis, learning, and long-term potentiation in mice. Proc. Natl. Acad. Sci. U. S. A. 96, 13427–13431.

van Praag, H., Kempermann, G., Gage, F.H., 1999b. Running increases cell proliferation and neurogenesis in the adult mouse dentate gyrus. Nat. Neurosci. 2, 266–270.

van Praag, H., Shubert, T., Zhao, C., Gage, F.H., 2005. Exercise enhances learning and hippocampal neurogenesis in aged mice. J. Neurosci. 25, 8680–8685.

Vaynman, S., Ying, Z., Gomez-Pinilla, F., 2004. Hippocampal BDNF mediates the efficacy of exercise on synaptic plasticity and cognition. Eur. J. Neurosci. 20, 2580–2590.

Viboolvorakul, S., Patumraj, S., 2014. Exercise training could improve age-related changes in cerebral blood flow and capillary vascularity through the upregulation of VEGF and eNOS. Biomed. Res. Int. 2014, 230791.

WHO, 2009. Global Health Risks: Mortality and Burden of Disease Attributable to Selected Major Risks. World Health Organization, Geneva.

Willey, J.Z., Moon, Y.P., Paik, M.C., Boden-Albala, B., Sacco, R.L., Elkind, M.S., 2009. Physical activity and risk of ischemic stroke in the Northern Manhattan Study. Neurology 73, 1774–1779.

Williamson, J.W., McColl, R., Mathews, D., Ginsburg, M., Mitchell, J.H., 1999. Activation of the insular cortex is affected by the intensity of exercise. J. Appl. Physiol. (1985) 87, 1213–1219.

Willie, C.K., Cowan, E.C., Ainslie, P.N., Taylor, C.E., Smith, K.J., Sin, P.Y., Tzeng, Y.C., 2011. Neurovascular coupling and distribution of cerebral blood flow during exercise. J. Neurosci. Methods 198, 270–273.

Yaffe, K., Barnes, D., Nevitt, M., Lui, L.Y., Covinsky, K., 2001. A prospective study of physical activity and cognitive decline in elderly women: women who walk. Arch. Intern. Med. 161, 1703–1708.

Yamaguchi, Y., Kashima, H., Fukuba, Y., Hayashi, N., 2014. Cerebral blood flow and neurovascular coupling during static exercise. J. Physiol. Sci. 64, 195–201.

Yamaguchi, Y., Ikemura, T., Hayashi, N., 2015a. Exhaustive exercise attenuates the neurovascular coupling by blunting the pressor response to visual stimulation. Biomed. Res. Int. 2015, 671063.

Yamaguchi, Y., Ikemura, T., Kashima, H., Hayashi, N., 2015b. Effects of vasodilatation and pressor response on neurovascular coupling during dynamic exercise. Eur. J. Appl. Physiol. 115, 619–625.

Yu, Y., Kastin, A.J., Pan, W., 2006. Reciprocal interactions of insulin and insulin-like growth factor I in receptor-mediated transport across the blood-brain barrier. Endocrinology 147, 2611–2615.

Zobl, E.G., Talmers, F.N., Christensen, R.C., Baer, L.J., 1965. Effect of exercise on the cerebral circulation and metabolism. J. Appl. Physiol. 20, 1289–1293.

CHAPTER

Neurovascular coupling—What next?

12

K. Masamoto*,[1], H. Hirase†, K. Yamada‡, I. Kanno§

**Brain Science Inspired Life Support Research Center, University of Electro-Communications, Tokyo, Japan*

†RIKEN Brain Science Institute, Wako, Saitama, Japan

‡Hirosaki University Graduate School of Medicine, Hirosaki, Aomori, Japan

§Molecular Imaging Center, National Institute of Radiological Sciences, Chiba, Japan

[1]Corresponding author: Tel./Fax: +81-42-443-5930, e-mail address: masamoto@mce.uec.ac.jp

Keywords

Cerebral microcirculation, Blood flow, fMRI, PET, Optical imaging, Neurovascular unit

The brain is a highly demanding organ as far as energy metabolism is concerned. The cerebral microcirculation is the most fundamental infrastructure that actively communicates with neural activity. The importance of neurovascular coupling (NVC) is well substantiated by the fact that it is conserved across mammalian species. As explored in this volume and elsewhere, the cellular and molecular mechanism for NVC has been a subject of intensive research over the past decades owing to the progress in imaging techniques and molecular biology. We now know that multiple cell types and mechanisms are involved in neurovascular communication, although the lead players in the capillary bed where the neurons and glia are most closely associated have been still a matter of intensive debate (eg, a contribution of pericytes Hall et al., 2014; Hill et al., 2015).

The multiple mechanisms in the NVCs could well act cooperatively. If so, why we need such redundant mechanisms? A particular mechanism might play a key role in certain environmental conditions or in a cell-type-specific manner. For example, neural activity changes local cellular environment, such as oxygen partial pressure, pH, temperature, neurohormones, neuropeptides, and ion balances. Each of those may independently set off a chain reaction of the vasodilation/vasoconstriction or capillary flow in multiple time and spatial ranges. An alternative scenario could be that an activation of a specific neural circuit leads secondarily to multiple cell processing depending on a type of the cells involved. In this scenario, a deficiency of certain types of the neurovascular communications may eventually trigger a breakdown of neural functions, although it may require long time for development

Progress in Brain Research, Volume 225, ISSN 0079-6123, http://dx.doi.org/10.1016/bs.pbr.2016.03.007

of neuronal symptoms following the onset of cerebrovascular disturbance, making difficult to prove the causal relationship.

Given the current progress in the field, what further progress might lead to breakthroughs? For better quantification of cerebral microcirculation, more complete measurements of flow dynamism would be required in three-dimensional microvascular networks. The scan speed and depth penetration of current multiphoton microscopy limits the quantification of blood flow within a volume of 1 mm^3. Noninvasive imaging methods such as fMRI and PET have great depth penetration and area coverage, but still lack sufficient temporal and spatial resolution. Although these techniques complement each other, further technical improvement that allows the quantification of complete network flow structures is awaited. Very recently, a new method was put forward by Errico et al. (2015) which make use of ultrafast scanning of ultrasound that reflects microbubbles in the blood flow. Use of sound information in combination with optical methods might well lead to better visualization and quantification of the dynamism in cerebral blood flow (CBF).

For neuronal and glial activity, an ultimate goal would be simultaneous recordings of all the activity in the brain like being achieved for neuronal Ca^{2+} imaging in the *Caenorhabditis elegans* and zebrafish (Ahrens et al., 2013; Prevedel et al., 2014; Schrödel et al., 2013). However, such optical methods could only be used for small creatures due to a limitation in light penetration. Perhaps a more imminent problem is that imaging so many pixels with a high frame rate requires so much data storage! Once raw data for whole cerebral microcirculation and neural activity are acquired, we will probably need to rely on artificial intelligence to extract the relationship between cerebral microcirculation and neural activity at the whole brain level. Recent developments in *deep learning* might help discovering new rules or unexpected NVC from highly convoluted sets of data, for instance, mood disorder, or early symptom of dementia, neurodegeneration, and cerebral apoplexy. Once such relationship is extracted, opto- or pharmacogenetic techniques would be among choices to proceed to the next step of developing treatments for such conditions.

Finally, plasticity of the brain has been shown to be critically influenced by the blood content; multiple studies have shown that plasma replacement from young animals will improve brain plasticity and learning (Katsimpardi et al., 2014; Villeda et al., 2011, 2014). Historically, glial cells have long been thought as mere supporting elements for neurons until the era of Ca^{2+} imaging and gliotransmission, bidirectional communication between glial cells and neurons is now one of the most active areas of research. What about cerebrovascular systems? Are they mere a supporting structure? The next generation technologies would answer the questions.

1 CONFERENCE HISTORY ON A TOPIC OF NVC

Since Roy and Sherrington (1890), enormous amounts of investigation have been carried out to understand the NVC mechanism. Lassen and Ingvar organized the first workshop that focused on NVC (Brain Work: the coupling of function, metabolism, and blood flow in the brain, Copenhagen, 26–30 May 1974, proceedings of the Alfred Benzon Symposium VIII, edited by David H. Ingvar and Niels A. Lassen, Munksgaard, Distributed by Academic Press, 1975). Sixteen years later, they organized the second workshop on the update of NVC (Brain Work and mental activity: quantitative studies with radioactive tracers, Copenhagen, 12–16 August 1990, proceedings of the Alfred Benzon Symposium 31, edited by Niels A. Lassen et al., 1991). These two workshops gathered most scientists at each time and they presented the state-of-the-art data on NVC. Minoru Tomita organized workshop on NVC imaging with all scientists in this field (Brain Activation and CBF Control: held in Tokyo, 5–8 June 2001, proceedings of the Satellite meeting on Brain Activation and Cerebral Blood Flow Control, edited by Minoru Tomita, Iwao Kanno, and Edith Hamel, Elsevier 2002). The activity followed to the biennial Gordon Research Conference since 2004 (https://www.grc.org/conferences.aspx?id=0000521). Although active discussions are held in biennial Brain Symposium, we may need some renewals on this topic.

REFERENCES

Ahrens, M.B., Orger, M.B., Robson, D.N., Li, J.M., Keller, P.J., 2013. Whole-brain functional imaging at cellular resolution using light-sheet microscopy. Nat. Methods 10 (5), 413–420.

Errico, C., Pierre, J., Pezet, S., Desailly, Y., Lenkei, Z., Couture, O., Tanter, M., 2015. Ultrafast ultrasound localization microscopy for deep super-resolution vascular imaging. Nature 527 (7579), 499–502.

Hall, C.N., Reynell, C., Gesslein, B., Hamilton, N.B., Mishra, A., Sutherland, B.A., O'Farrell, F.M., Buchan, A.M., Lauritzen, M., Attwell, D., 2014. Capillary pericytes regulate cerebral blood flow in health and disease. Nature 508 (7494), 55–60.

Hill, R.A., Tong, L., Yuan, P., Murikinati, S., Gupta, S., Grutzendler, J., 2015. Regional blood flow in the normal and ischemic brain is controlled by arteriolar smooth muscle cell contractility and not by capillary pericytes. Neuron 87 (1), 95–110.

Katsimpardi, L., Litterman, N.K., Schein, P.A., Miller, C.M., Loffredo, F.S., Wojtkiewicz, G.R., Chen, J.W., Lee, R.T., Wagers, A.J., Rubin, L.L., 2014. Vascular and neurogenic rejuvenation of the aging mouse brain by young systemic factors. Science 344 (6184), 630–634.

Prevedel, R., Yoon, Y.G., Hoffmann, M., Pak, N., Wetzstein, G., Kato, S., Schrödel, T., Raskar, R., Zimmer, M., Boyden, E.S., Vaziri, A., 2014. Simultaneous whole-animal 3D imaging of neuronal activity using light-field microscopy. Nat. Methods 11 (7), 727–730.

Roy, C.S., Sherrington, C.S., 1890. On the regulation of the blood-supply of the brain. J. Physiol. 11 (1–2), 85–158.

Schrödel, T., Prevedel, R., Aumayr, K., Zimmer, M., Vaziri, A., 2013. Brain-wide 3D imaging of neuronal activity in Caenorhabditis elegans with sculpted light. Nat. Methods 10 (10), 1013–1020.

Villeda, S.A., Luo, J., Mosher, K.I., Zou, B., Britschgi, M., Bieri, G., Stan, T.M., Fainberg, N., Ding, Z., Eggel, A., Lucin, K.M., Czirr, E., Park, J.S., Couillard-Després, S., Aigner, L., Li, G., Peskind, E.R., Kaye, J.A., Quinn, J.F., Galasko, D.R., Xie, X.S., Rando, T.A.,

Wyss-Coray, T., 2011. The ageing systemic milieu negatively regulates neurogenesis and cognitive function. Nature 477 (7362), 90–94.
Villeda, S.A., Plambeck, K.E., Middeldorp, J., Castellano, J.M., Mosher, K.I., Luo, J., Smith, L.K., Bieri, G., Lin, K., Berdnik, D., Wabl, R., Udeochu, J., Wheatley, E.G., Zou, B., Simmons, D.A., Xie, X.S., Longo, F.M., Wyss-Coray, T., 2014. Young blood reverses age-related impairments in cognitive function and synaptic plasticity in mice. Nat. Med. 20 (6), 659–663.

Index

Note: Page numbers followed by "*f*" indicate figures, and "*t*" indicate tables.

D

E

F

G

H

I

K

L

M

N

O

P

Q

R

Volume 167: Stress Hormones and Post Traumatic Stress Disorder: Basic Studies and Clinical Perspectives, by E.R. de Kloet, M.S. Oitzl and E. Vermetten (Eds.) – 2008, ISBN 978-0-444-53140-7.
Volume 168: Models of Brain and Mind: Physical, Computational and Psychological Approaches, by R. Banerjee and B.K. Chakrabarti (Eds.) – 2008, ISBN 978-0-444-53050-9.
Volume 169: Essence of Memory, by W.S. Sossin, J.-C. Lacaille, V.F. Castellucci and S. Belleville (Eds.) – 2008, ISBN 978-0-444-53164-3.
Volume 170: Advances in Vasopressin and Oxytocin – From Genes to Behaviour to Disease, by I.D. Neumann and R. Landgraf (Eds.) – 2008, ISBN 978-0-444-53201-5.
Volume 171: Using Eye Movements as an Experimental Probe of Brain Function—A Symposium in Honor of Jean Büttner-Ennever, by Christopher Kennard and R. John Leigh (Eds.) – 2008, ISBN 978-0-444-53163-6.
Volume 172: Serotonin–Dopamine Interaction: Experimental Evidence and Therapeutic Relevance, by Giuseppe Di Giovanni, Vincenzo Di Matteo and Ennio Esposito (Eds.) – 2008, ISBN 978-0-444-53235-0.
Volume 173: Glaucoma: An Open Window to Neurodegeneration and Neuroprotection, by Carlo Nucci, Neville N. Osborne, Giacinto Bagetta and Luciano Cerulli (Eds.) – 2008, ISBN 978-0-444-53256-5.
Volume 174: Mind and Motion: The Bidirectional Link Between Thought and Action, by Markus Raab, Joseph G. Johnson and Hauke R. Heekeren (Eds.) – 2009, 978-0-444-53356-2.
Volume 175: Neurotherapy: Progress in Restorative Neuroscience and Neurology — Proceedings of the 25th International Summer School of Brain Research, held at the Royal Netherlands Academy of Arts and Sciences, Amsterdam, The Netherlands, August 25–28, 2008, by J. Verhaagen, E.M. Hol, I. Huitinga, J. Wijnholds, A.A. Bergen, G.J. Boer and D.F. Swaab (Eds.) –2009, ISBN 978-0-12-374511-8.
Volume 176: Attention, by Narayanan Srinivasan (Ed.) – 2009, ISBN 978-0-444-53426-2.
Volume 177: Coma Science: Clinical and Ethical Implications, by Steven Laureys, Nicholas D. Schiff and Adrian M. Owen (Eds.) – 2009, 978-0-444-53432-3.
Volume 178: Cultural Neuroscience: Cultural Influences On Brain Function, by Joan Y. Chiao (Ed.) – 2009, 978-0-444-53361-6.
Volume 179: Genetic models of schizophrenia, by Akira Sawa (Ed.) – 2009, 978-0-444-53430-9.
Volume 180: Nanoneuroscience and Nanoneuropharmacology, by Hari Shanker Sharma (Ed.) – 2009, 978-0-444-53431-6.
Volume 181: Neuroendocrinology: The Normal Neuroendocrine System, by Luciano Martini, George P. Chrousos, Fernand Labrie, Karel Pacak and Donald W. Pfaff (Eds.) – 2010, 978-0-444-53617-4.
Volume 182: Neuroendocrinology: Pathological Situations and Diseases, by Luciano Martini, George P. Chrousos, Fernand Labrie, Karel Pacak and Donald W. Pfaff (Eds.) – 2010, 978-0-444-53616-7.
Volume 183: Recent Advances in Parkinson's Disease: Basic Research, by Anders Björklund and M. Angela Cenci (Eds.) – 2010, 978-0-444-53614-3.
Volume 184: Recent Advances in Parkinson's Disease: Translational and Clinical Research, by Anders Björklund and M. Angela Cenci (Eds.) – 2010, 978-0-444-53750-8.
Volume 185: Human Sleep and Cognition Part I: Basic Research, by Gerard A. Kerkhof and Hans P.A. Van Dongen (Eds.) – 2010, 978-0-444-53702-7.
Volume 186: Sex Differences in the Human Brain, their Underpinnings and Implications, by Ivanka Savic (Ed.) – 2010, 978-0-444-53630-3.
Volume 187: Breathe, Walk and Chew: The Neural Challenge: Part I, by Jean-Pierre Gossard, Réjean Dubuc and Arlette Kolta (Eds.) – 2010, 978-0-444-53613-6.
Volume 188: Breathe, Walk and Chew; The Neural Challenge: Part II, by Jean-Pierre Gossard, Réjean Dubuc and Arlette Kolta (Eds.) – 2011, 978-0-444-53825-3.
Volume 189: Gene Expression to Neurobiology and Behaviour: Human Brain Development and Developmental Disorders, by Oliver Braddick, Janette Atkinson and Giorgio M. Innocenti (Eds.) – 2011, 978-0-444-53884-0.

Volume 190: Human Sleep and Cognition Part II: Clinical and Applied Research, by Hans P.A. Van Dongen and Gerard A. Kerkhof (Eds.) – 2011, 978-0-444-53817-8.
Volume 191: Enhancing Performance for Action and perception: Multisensory Integration, Neuroplasticity and Neuroprosthetics: Part I, by Andrea M. Green, C. Elaine Chapman, John F. Kalaska and Franco Lepore (Eds.) – 2011, 978-0-444-53752-2.
Volume 192: Enhancing Performance for Action and Perception: Multisensory Integration, Neuroplasticity and Neuroprosthetics: Part II, by Andrea M. Green, C. Elaine Chapman, John F. Kalaska and Franco Lepore (Eds.) – 2011, 978-0-444-53355-5.
Volume 193: Slow Brain Oscillations of Sleep, Resting State and Vigilance, by Eus J.W. Van Someren, Ysbrand D. Van Der Werf, Pieter R. Roelfsema, Huibert D. Mansvelder and Fernando H. Lopes da Silva (Eds.) – 2011, 978-0-444-53839-0.
Volume 194: Brain Machine Interfaces: Implications For Science, Clinical Practice And Society, by Jens Schouenborg, Martin Garwicz and Nils Danielsen (Eds.) – 2011, 978-0-444-53815-4.
Volume 195: Evolution of the Primate Brain: From Neuron to Behavior, by Michel A. Hofman and Dean Falk (Eds.) – 2012, 978-0-444-53860-4.
Volume 196: Optogenetics: Tools for Controlling and Monitoring Neuronal Activity, by Thomas Knöpfel and Edward S. Boyden (Eds.) – 2012, 978-0-444-59426-6.
Volume 197: Down Syndrome: From Understanding the Neurobiology to Therapy, by Mara Dierssen and Rafael De La Torre (Eds.) – 2012, 978-0-444-54299-1.
Volume 198: Orexin/Hypocretin System, by Anantha Shekhar (Ed.) – 2012, 978-0-444-59489-1.
Volume 199: The Neurobiology of Circadian Timing, by Andries Kalsbeek, Martha Merrow, Till Roenneberg and Russell G. Foster (Eds.) – 2012, 978-0-444-59427-3.
Volume 200: Functional Neural Transplantation III: Primary and stem cell therapies for brain repair, Part I, by Stephen B. Dunnett and Anders Björklund (Eds.) – 2012, 978-0-444-59575-1.
Volume 201: Functional Neural Transplantation III: Primary and stem cell therapies for brain repair, Part II, by Stephen B. Dunnett and Anders Björklund (Eds.) – 2012, 978-0-444-59544-7.
Volume 202: Decision Making: Neural and Behavioural Approaches, by V.S. Chandrasekhar Pammi and Narayanan Srinivasan (Eds.) – 2013, 978-0-444-62604-2.
Volume 203: The Fine Arts, Neurology, and Neuroscience: Neuro-Historical Dimensions, by Stanley Finger, Dahlia W. Zaidel, François Boller and Julien Bogousslavsky (Eds.) – 2013, 978-0-444-62730-8.
Volume 204: The Fine Arts, Neurology, and Neuroscience: New Discoveries and Changing Landscapes, by Stanley Finger, Dahlia W. Zaidel, François Boller and Julien Bogousslavsky (Eds.) – 2013, 978-0-444-63287-6.
Volume 205: Literature, Neurology, and Neuroscience: Historical and Literary Connections, by Anne Stiles, Stanley Finger and François Boller (Eds.) – 2013, 978-0-444-63273-9.
Volume 206: Literature, Neurology, and Neuroscience: Neurological and Psychiatric Disorders, by Stanley Finger, François Boller and Anne Stiles (Eds.) – 2013, 978-0-444-63364-4.
Volume 207: Changing Brains: Applying Brain Plasticity to Advance and Recover Human Ability, by Michael M. Merzenich, Mor Nahum and Thomas M. Van Vleet (Eds.) – 2013, 978-0-444-63327-9.
Volume 208: Odor Memory and Perception, by Edi Barkai and Donald A. Wilson (Eds.) – 2014, 978-0-444-63350-7.
Volume 209: The Central Nervous System Control of Respiration, by Gert Holstege, Caroline M. Beers and Hari H. Subramanian (Eds.) – 2014, 978-0-444-63274-6.
Volume 210: Cerebellar Learning, Narender Ramnani (Ed.) – 2014, 978-0-444-63356-9.
Volume 211: Dopamine, by Marco Diana, Gaetano Di Chiara and Pierfranco Spano (Eds.) – 2014, 978-0-444-63425-2.
Volume 212: Breathing, Emotion and Evolution, by Gert Holstege, Caroline M. Beers and Hari H. Subramanian (Eds.) – 2014, 978-0-444-63488-7.
Volume 213: Genetics of Epilepsy, by Ortrud K. Steinlein (Ed.) – 2014, 978-0-444-63326-2.
Volume 214: Brain Extracellular Matrix in Health and Disease, by Asla Pitkänen, Alexander Dityatev and Bernhard Wehrle-Haller (Eds.) – 2014, 978-0-444-63486-3.

Volume 215: The History of the Gamma Knife, by Jeremy C. Ganz (Ed.) – 2014, 978-0-444-63520-4.
Volume 216: Music, Neurology, and Neuroscience: Historical Connections and Perspectives, by François Boller, Eckart Altenmüller, and Stanley Finger (Eds.) – 2015, 978-0-444-63399-6.
Volume 217: Music, Neurology, and Neuroscience: Evolution, the Musical Brain, Medical Conditions, and Therapies, by Eckart Altenmüller, Stanley Finger, and François Boller (Eds.) – 2015, 978-0-444-63551-8.
Volume 218: Sensorimotor Rehabilitation: At the Crossroads of Basic and Clinical Sciences, by Numa Dancause, Sylvie Nadeau, and Serge Rossignol (Eds.) – 2015, 978-0-444-63565-5.
Volume 219: The Connected Hippocampus, by Shane O'Mara and Marian Tsanov (Eds.) – 2015, 978-0-444-63549-5.
Volume 220: New Trends in Basic and Clinical Research of Glaucoma: A Neurodegenerative Disease of the Visual System, by Giacinto Bagetta and Carlo Nucci (Eds.) – 2015, 978-0-444-63566-2.
Volume 221: New Trends in Basic and Clinical Research of Glaucoma: A Neurodegenerative Disease of the Visual System, by Giacinto Bagetta and Carlo Nucci (Eds.) – 2015, 978-0-12-804608-1.
Volume 222: Computational Neurostimulation, by Sven Bestmann (Ed.) – 2015, 978-0-444-63546-4.
Volume 223: Neuroscience for Addiction Medicine: From Prevention to Rehabilitation - Constructs and Drugs, by Hamed Ekhtiari and Martin Paulus (Eds.) – 2016, 978-0-444-63545-7.
Volume 224: Neuroscience for Addiction Medicine: From Prevention to Rehabilitation - Methods and Interventions, by Hamed Ekhtiari and Martin P. Paulus (Eds.) – 2016, 978-0-444-63716-1.

Papier | Fördert gute Waldnutzung
FSC® C083411

Kolibri 360 GmbH
Ferdinand-Jühlke-Straße 7
99095 Erfurt, Deutschland
produktsicherheit@kolibri360.de